ADVANCED
DATA-TRANSMISSION
SYSTEMS

ADVANCED DATA-TRANSMISSION SYSTEMS

A. P. CLARK, MA, PhD, DIC, MIERE, CEng
*Department of Electronic and Electrical Engineering,
Loughborough University of Technology*

PENTECH PRESS
London : Plymouth

First published 1977
by Pentech Press Limited
Estover Road, Plymouth PL6 7PZ

ISBN 0 7273 0101 2

Printed and bound in Great Britain by
Billing & Sons Limited,
Guildford, London and Worcester

Preface

This book is the outcome partly of some of the research activities in digital communication systems, in the Department of Electronic and Electrical Engineering of Loughborough University of Technology, and partly also of some lecture courses in digital communications presented over the past five years to postgraduate students in this Department, many of these lecture courses being themselves developed from the research activities of the Department.

The book is written mainly for private study by practising and student engineers who are interested in the design and development of data-transmission systems. It is concerned primarily with basic principles and techniques rather than with practical considerations or details of equipment design. Again, although parts of the book might suggest that it is a treatise on applied mathematics, the emphasis throughout is on engineering or physical principles rather than on the mathematics itself. Care has been taken in the mathematical analysis to include only those derivations that help to illustrate and elucidate the principles involved. Where no simple or helpful derivation is known to the author, the result is presented without proof.

The real understanding of any result (or relationship) involves not just the ability to derive the result mathematically, but also the appreciation of its meaning in physical terms, together with a concept of its importance or magnitude. With this in view, an attempt has been made, wherever possible, to get behind the mathematics through simple intuitive arguments and so to explain the meaning of the result obtained. To assist further in the explanation of the more important principles and techniques, these are illustrated by simple worked examples. Unnecessary mathematical complexity is avoided by considering throughout only *real* signals rather than the more general complex signals. The latter have in fact been extensively studied. Once the principles and techniques have been mastered for real signals, it is a relatively simple step to extend these to complex signals, with no unexpected changes in any of the relationships involved. However, the initial study of complex rather than real signals could for many readers be the last straw that broke the camel's back!

The emphasis of the book is on *future* systems and in particular those designed for the greatest efficiency and effectiveness in operation, consistent with an acceptable degree of equipment complexity. Many of the systems described here have never actually been built and tested as such. All systems have however been studied extensively by computer simulation and theoretical analysis, usually to a much deeper level and under much more varied conditions than those considered in this book. Detailed designs have furthermore been carried out on some of the more important techniques studied, and these have been costed and compared with alternative techniques to give some idea of the relative cost-effectiveness of the corresponding systems. An account of this work is however beyond the scope of the book.

The book contains much novel material that has not been published elsewhere, and a large part of the material that has been published has appeared only quite recently in different papers. The book in fact contains a collection of all the various basic techniques known to the author that are relevant to its central theme, and should therefore be a useful source of ideas for the modem designer. It is *not* however a comprehensive handbook on data transmission, and indeed a conscious effort has been made to avoid unnecessary duplication of material that is adequately covered in other books and published papers. For instance, the important topics of error detection and correction, synchronization techniques, and analogue modulation methods, have been largely ignored. Again, no attempt has been made to cover the more fundamental topics normally found in textbooks on communications, such as Fourier transforms, Laplace transforms, random variables, random processes, signal transmission through linear systems, the sampling theorem, information theory, and so on. What has been done instead has been to concentrate on one particular aspect of data transmission which, in the view of the author, is the most important and far reaching of recent developments. This is *the application of computer-like techniques to a sampled digital signal.* The recent developments of microcomputers and microprocessors mean that it is now often simpler to handle signals as sample values rather than as continuous waveforms. Following the sampling of the received signal, the receiver operates on a *sequence of numbers* and therefore it becomes a special purpose digital computer. Particular techniques must now be used to analyse and characterize the sampled signal, and many different digital techniques become available for implementing the signal detection process. An incidental advantage in dealing with a sampled signal is that most of the more serious of the mathematical difficulties, involved with the analysis of continuous waveforms, entirely disappear! Indeed, it is hoped that this book may convince the reader of the remarkable simplicity and elegance that may be achieved in the analysis of a sampled signal, given only the use of the elementary theory of polynomials, elementary matrix algebra, and the properties of a multi-dimensional Euclidean vector space.

The first part of the book (Chapters 1–3) is an introduction to the important topics of signal design, linear signal-distortion, and the optimum detection and estimation of digital signals. The second part of the book (Chapters 4–6) is concerned with various techniques for detecting distorted digital signals. The treatment here starts with the simplest and in general least effective techniques, and shows how these may be developed in stages towards the more complex and most effective techniques. Many different techniques are studied, with particular emphasis on those techniques that are either particularly simple to implement or else help to clarify the understanding of other techniques. The book concentrates throughout on what could or should be done in order to achieve the most efficient or effective data-transmission system, rather than on what is or has been done in practice, and it should therefore be of particular interest to those concerned with the design and development of the next generation of data-transmission systems.

It is assumed that the reader has a basic understanding of Fourier transforms, probability theory, random variables, random processes and in particular the Gaussian random process, signal transmission through linear systems, the sampling theorem, linear modulation methods, matrix algebra, vector spaces and linear transformations. No very advanced understanding of any of these topics is however required.

Each chapter is reasonably self-contained and suitable for presentation as a separate course of lectures in a first-year postgraduate course on digital communications.

The main purpose of the techniques studied in this book is to enable the maximum transmission rate to be achieved over a linear channel, for an acceptable tolerance to noise and an acceptable degree of equipment complexity. However, it is sometimes said that with the advent of specially designed data-networks or wideband channels, such as possibly fibre-optic links, there will no longer be a need for data-transmission systems that make a really efficient use of bandwidth. There are two serious fallacies in this point of view. Firstly, most transmission channels cost an appreciable amount of money, either in rental or else in their initial cost and subsequent maintenance. It does not therefore make economic sense to use such channels at transmission rates substantially below the maximum available, particularly when an increased transmission rate can be achieved at no significant increase in cost. Secondly, it is a well-known fact that if a facility is provided, this of itself often tends to increase the demand for the facility, sometimes even when no such demand apparently existed before the facility was introduced. Thus, even if the facilities for data transmission were vastly increased, hence initially avoiding the need for any but the very simplest techniques, the demand for data transmission would then very probably increase, simply because the facilities were there. Sooner or later the demand would catch up with the available capacity, and lead once again to a search for the more efficient use of the channels. The general techniques studied in this book are therefore likely to maintain their present importance for some time to come.

Finally, it is of interest to consider briefly the contents of the individual chapters of the book and the prior knowledge required for their proper understanding.

Chapter 1 presents some introductory background material and defines various terms that are used in the book. It also summarizes some of the more important results (theorems and experimental observations) that are used but not derived or studied in the book.

Chapter 2 develops the theory of the discrete Fourier transform (DFT) from first principles, using matrix algebra, and then applies the DFT to the analysis and description of signal distortion in a sampled baseband signal.

Chapter 3 develops the theory of the optimum detection and estimation of a sampled baseband digital signal, from first principles, and applies the theory to particular and important cases where the noise is Gaussian.

Chapter 4 analyses and compares various techniques for equalizing a time invariant or slowly time-varying baseband channel, in a synchronous serial data-transmission system.

Chapter 5 analyses and compares various techniques for the

adaptive detection of slowly time-varying baseband signals in a synchronous serial data-transmission system.

Chapter 6 follows the same lines as Chapter 5 but considers parallel systems using code-division multiplexing, in place of serial systems.

The initial development of some of the basic techniques described in Chapters 5 and 6 was carried out in the Department of Electrical Engineering of the Imperial College of Science and Technology, University of London, from 1965 to 1968, this work being sponsored jointly by the Science Research Council and the Plessey Company. The remainder of the original work reported in the book was carried out from 1970 to 1975 in the Department of Electronic and Electrical Engineering of Loughborough University of Technology. Of this work, that reported in Section 4 of Chapter 5 has been sponsored by the Science Research Council. The author gratefully acknowledges the financial support and facilities provided by these various organizations. The author would also like to acknowledge the valuable contributions of Dr. A. K. Mukherjee, Dr. F. Ghani, Dr. U. S. Tint, Dr. M. N. Serinken, Mr. M. Shum, Mr. L. H. W. Tang, Mr. I. B. Ismail, Dr. A. Clements, Mr. J. D. Harvey, Dr. R. B. Hanes and Dr. J. D. Daley in the study of the different systems by computer simulation, and in the detailed development of these systems. Finally, the author would like to thank Professor J. W. R. Griffiths for his constant and enthusiastic encouragement of the research projects and his help in providing the facilities needed for the successful completion of this work.

Contents

1 Introduction

1.1 BACKGROUND, DEFINITIONS AND ASSUMPTIONS

One of the most striking of recent developments in technology has been the rapid growth of digital communication systems.[7] In such systems the transmitted signal is a *waveform* which may, for instance, be carried by the electric current or voltage in a pair of wires, by electromagnetic radiation in the atmosphere, by light (visible electromagnetic radiation) in a glass fibre, or by high frequency sound waves in water or the atmosphere. The essential feature of a *digital* communication system is that the transmitted waveform is itself composed of separate *signal-elements* (often referred to as symbols, digits, bits or pulses) and these signal-elements carry the data (information) which it is required to transmit. A signal-element is thus a *unit* of the transmitted waveform.

The data carried by an individual signal-element form a *symbol* or digit, which is often one of the numerals 0, 1, 2, ..., and the data carried by the entire transmitted group of signal elements form a *message*. The symbols are themselves usually arranged in separate groups, each group forming a *character* (or word). Most often the transmitted message is 'alpha-numeric' which means that it is composed of a sequence of characters, each of which is a letter or numeral. Some of these characters may additionally be punctuation marks and instructions such as 'space' and 'new line'.

The signal elements are normally transmitted in sequence, one following another, to give a *serial* system, but they may sometimes be sent in separate groups, the signal elements in each group being transmitted simultaneously, to give a *parallel* system. Most often in a serial system the signal elements are transmitted at a steady rate of a given number of elements per second (bauds), the receiver being held in time synchronism with the received signal. Such a system is known as a *synchronous serial system* and, in applications where a relatively high transmission rate is required over a given channel, it is the most commonly used of the different systems.[10]

The *channel* contains the *transmission path*, connecting the transmitter to the receiver, and includes also the filter at the output of the transmitter and the filter at the input to the receiver, both of which shape the transmitted signal. The shape of any given signal-element may be considerably altered as a result of its transmission over the channel and, for certain transmission paths, the resultant shape may also vary with time.[10] Thus the shape of a received signal-element is not necessarily related directly or even in any very simple manner to the symbol represented by that element. It is for this reason that a clear distinction is made here between a *signal-element*, which is the actual waveform transmitted, and the corresponding *symbol*, which is the data represented by that waveform. In order to identify the two, the *value* of a signal-element is taken to be the *symbol* represented by that element. Clearly, the *element value* does not refer to the *shape* of the signal-element. Often, to simplify the terminology used, both the value of a signal-element and the symbol represented by that element are taken to be the actual value of the corresponding portion of the *transmitted* signal. This can be done because the particular possible values allocated to a symbol may be selected arbitrarily, so long as there are the correct number of possible values and these are all different.

If the symbol represented by an individual signal-element has one of two possible values (say 0 and 1), the signal is said to be *binary coded* or *binary*. Each signal-element now has one of two possible *shapes*, and the two possible shapes of a received signal-element may, of course, vary with time. If a symbol has one of m possible values, the signal is said to be m-level. This does not, of course, mean that a transmitted signal-element literally has one of m possible *levels*.

If a received signal-element has one of two possible values which are equally likely, the signal-element contains *one bit of information*. A *bit* is a unit measure of *information*, and is not to be confused with a symbol or signal-element. If the element values in a received signal are statistically independent and equally likely to have any of m possible values, the information content per signal-element is $\log_2 m$ bits. If the signal-element rate is A bauds, the transmission rate (information rate) of the signal is $A \log_2 m$ bits/s. Clearly, the transmission rate can be increased by raising the signal-element rate A, the number of levels m, or both of these together.

The simplest digital data signal is a sequence of binary signal-elements as shown in Figure 1.1. A signal element here has the value $-k$ or k over the whole of its duration, representing the symbol 0 or 1, respectively. Each signal element has the same duration of T seconds and follows immediately after the preceding element, so that the signal-element rate is $1/T$ bauds. This is a binary antipodal baseband signal such as could be used in a synchronous serial data-transmission system. In a binary antipodal signal, one of

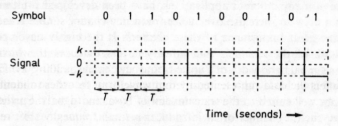

Fig. 1.1 *Rectangular binary baseband signal*

the possible shapes of a signal element is the *negative* of the other. A baseband signal is one whose spectrum usually extends to zero frequency (dc) or to very low frequencies, and in which the symbol carried by a signal element is identified by the *actual value* of that element over a certain time interval or at a certain instant in time. Thus the binary baseband signal in Figure 1.1 could alternatively be shaped as in Figure 1.2. The frequency band (bandwidth) occupied

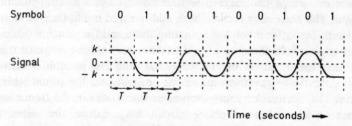

Fig. 1.2 *Rounded binary baseband signal*

by this waveform is much narrower than that occupied by the waveform in Figure 1.1.[10] At the receiver, each signal-element is now sampled at its mid-point in time and each sample has the value $\pm k$. To avoid the conversion from the sample values $-k$ and k to the corresponding symbols 0 and 1, the symbol represented by a signal element can be taken to be its sample value $-k$ or k. Thus the detection of a received signal-element involves just the determination of its value at the appropriate sampling instant, this value uniquely determining the datum carried by that element.

Among the most important of the transmission paths that are used for the transmission of data are voice-frequency channels over the telephone network and HF radio links.[8-10] Most of the data-transmission systems studied in this book, although of potential value in many different applications, have been developed primarily with a view to their use over voice-frequency channels. One reason for the great importance of these channels is the highly developed network of the channels that already covers much of the world. The nominal bandwidth of a voice-frequency channel is 300–3000 Hz, the effective bandwidth being often much less, so that these channels are not well suited to the transmission of baseband signals. Furthermore, such a channel often introduces a small frequency shift into the spectrum of the transmitted signal.[10] This causes serious and additional complications in the detection of the received signal. The latter may differ considerably from the transmitted signal, and the relationship between the two is a function of time. To overcome these problems, it is necessary now to use a *modulated-carrier signal*. Amplitude, frequency or phase modulation (AM, FM or PM) may be used, the carrier frequency being selected to place the transmitted signal spectrum in the centre of the available frequency band.[8-10]

Telephone circuits and HF radio links introduce both noise and distortion which can have a serious effect on a transmitted data signal. The noise may include both additive and multiplicative components, the latter involving both amplitude and frequency modulation effects.[10] In the case of additive noise, the noise waveform is *added* to the signal waveform, whereas, in the case of multiplicative noise, the noise waveform multiplies or *modulates* the signal waveform. The attenuation and group-delay distortions in the frequency characteristics of a telephone circuit may reduce the effective bandwidth of the channel to about half its nominal bandwidth, thus seriously degrading the performance of the data-transmission system. HF radio links are time-varying channels and may from time to time introduce considerably more serious distortion than telephone circuits. Thus a data-transmission system requires a good tolerance to both noise and distortion if it is to operate satisfactorily over telephone circuits and HF radio links.[10]

For relatively low transmission rates of up to 1200 bits/s over telephone circuits, a serial binary FM system has been adopted mainly because of its great versatility and relative simplicity.[6] For the higher transmission rate of 2400 bits/s over the better telephone circuits (private lines), a synchronous serial 4-level PM system has been adopted.[4,5] This may be used only at the given transmission rate and is less flexible than the FM system, which may be used over a range of transmission rates. However, the PM system achieves a more efficient use of bandwidth and has a better tolerance to both noise and distortion.[4-6] 2400 bits/s is probably the highest transmission rate that can be achieved successfully with simple techniques in non-adaptive systems, and then only over the better telephone circuits.[10] On the other hand, purely theoretical considerations suggest that transmission rates at least as high as 20,000 bits/s should be achieved successfully over telephone circuits by means of the appropriate techniques.[2] Clearly, there is scope for very considerable improvement over conventional techniques.

In order to achieve the most efficient use of bandwidth, together with simple and effective detection processes, it is necessary to use a modulation method in which both the modulation and demodulation of the data signal are *linear* operations. There are two alternative linear systems that are generally preferrred. The first of these uses a vestigial sideband suppressed carrier AM signal, and the second uses two double-sideband suppressed carrier AM signals whose carriers are in phase quadrature (at the same frequency and at a phase angle of 90°). Coherent demodulation is used in each case.[10] A vestigial-sideband signal is a modulated-carrier signal in which a large part of one sideband is removed, so that only one sideband and a vestige of the other are transmitted. In an AM signal, the carrier cannot itself be used to assist in the detection of the received data signal, so that for the best tolerance to additive noise, for a given transmitted signal power level, the signal carrier is removed. Special arrangements must however now be made to ensure the carrier-phase synchronization of the coherent demodulator on to the received signal. Usually these involve the addition of a low-level pilot carrier to the transmitted signal.[10]

When a vestigial-sideband suppressed carrier AM signal is transmitted, the transmission path together with the modulator (at the transmitter) and the demodulator (at the receiver) appear as a linear baseband channel.[11] The theoretical analysis of the resultant system therefore reduces to that of a simple linear baseband system, just so long as the correct synchronization of the coherent demodulator on to the received signal is assumed. When two double-sideband suppressed carrier AM signals are transmitted in phase quadrature, the transmission path together with the two modulators at the transmitter and the two coherent demodulators at the receiver, appear as two linear baseband channels, which are coupled (inter-connected) in an appropriate manner. Furthermore, by considering one of these baseband channels as carrying a 'real' signal and the other as carrying an 'imaginary' signal, the two baseband channels can be considered as a single baseband channel carrying a 'complex' signal (a signal having both real and imaginary components). The theoretical analysis of the resultant system therefore again reduces to that of a linear baseband channel. Since the emphasis throughout this book is on data-transmission systems that make an efficient use of bandwidth, it is assumed that whenever the transmission path is a bandpass channel, a suppressed-carrier AM signal is used and, except where otherwise stated, this is assumed to be a vestigial-sideband signal. The resultant channel is therefore in every case a baseband channel, so that only baseband signals are used. Modulated-carrier signals will not therefore be considered further. They are studied in detail elsewhere.[8-10]

The problems of synchronizing the receiver on to the received signal, for both element timing and carrier phase, will not be considered further in this book. These are major problems in their own right whose investigation is by no means simple and is not closely related to the main topics studied in this book. Considerable work has been done in this area.[7,10] It is therefore assumed throughout this book that the receiver is correctly synchronized on to the received signal, for both element timing and carrier phase.

Suppose that a received signal-element has a given shape (basic waveform) $y(t)$ which is nonzero only over the time interval $0-T$, and suppose that the element value (symbol) is carried by a multiplying parameter s, so that the resultant received signal-element is $sy(t)$. This signal is very typical of those that are normally used over linear baseband channels of the type considered in this book. If the signal is received in the presence of additive white Gaussian noise and in the absence of other signals, the detection process that minimizes the probability of error in the detection of s, feeds the received signal to a linear filter with impulse response $cy(T-t)$, where c is a constant, and samples the output signal from the linear filter at time T. The sample value is compared with the appropriate threshold (or thresholds) to give the detected value of s (Chapter 3). The filter maximizes the signal/noise ratio at its output at time T, and is known as a *matched filter*. The whole detection process is one of *matched-filter detection*, and no other detection process can give a lower probability of error in the detection of s.[10-14]

In the detection of a received serial stream of signal elements, the ith of which is $s_i y(t - iT)$, where the $\{s_i\}$ are statistically independent and equally likely to have any of their possible values, and where the signal-elements do not overlap in time, the arrangement of matched-filter detection can be used for each individual received signal element, in turn, and gives the best available tolerance to additive white Gaussian noise.[10] Practical detectors are often designed to approximate to the arrangement just described, even though the additive noise may not be Gaussian, and they usually give a good performance.

The effect of distortion in the attenuation-frequency and group-delay-frequency characteristics of the transmission path is to spread out the individual transmitted signal-elements in time, so that the individual signal-elements at the receiver input overlap each other. Thus, in the detection of a received signal-element by a matched-filter detector, the output signal from the matched filter contains, in addition to the wanted signal and the noise, components that originate from the neighbouring signal-elements. These interfere with the detection of the wanted signal-element and reduce the tolerance of the system to noise. They may even prevent the correct detection of the received signal in the complete absence of noise. This type of interference is known as *intersymbol interference*.

Over a channel that passes frequency components from 0 to B Hz at a significant (usable) level but no frequency components outside this band, the maximum element rate of a digital signal, for no intersymbol interference in a linear receiver, is $2B$ elements/s.[1,8] With m-level signal elements that are statistically independent and equally likely to have any of the m values, the maximum transmission (information) rate for no intersymbol interference is $2B \log_2 m$ bits/s. If now it is required to double the transmission rate, without however introducing intersymbol interference, m must be replaced by m^2. This increases the equipment complexity of the data-transmission system and, for a given transmitted power level, greatly reduces the tolerance of the system to additive noise, the reduction increasing very rapidly with m.[10] Clearly, only a limited increase in transmission rate can be achieved by increasing m.

An alternative or additional approach, which has recently been developed, is to introduce a large but controlled level of intersymbol interference between neighbouring signal-elements, thus permitting a higher signal-element rate over a given bandwidth.[3] The resultant transmission path is known as a *partial-response channel*. The intersymbol interference is corrected or compensated for in one of two ways: either through the appropriate signal coding at the transmitter, or through the appropriate detection process at the receiver.[3] The technique permits a limited but very useful increase in transmission rate to be achieved, for only a small reduction in tolerance to additive noise and a small increase in equipment complexity.[10]

The technique just mentioned is of further importance because it demonstrates the fact that severe intersymbol-interference need not itself prevent the satisfactory transmission of data, just so long as suitable signal processing is used at the transmitter or a suitable detection process is used at the receiver. Thus, even when the intersymbol interference is introduced by the transmission path rather than by the equipment filters, it should be possible to compensate or correct for the distortion, either at the transmitter or at the receiver. In this way, satisfactory transmission can be achieved, even in the presence of severe signal distortion, provided only that the distortion is known or can be estimated reasonably accurately from the received signal. The whole of this book is in fact concerned with techniques of this kind. These make it possible to increase the transmission rate over a given channel by increasing the signal-element rate rather than the number of signal levels (possible element values) and thus without the very serious reduction in tolerance to additive noise that results when the latter technique is used on its own.[10]

A most important result from the well-known sampling theorem is as follows.[11-14] If a received baseband signal, bandlimited to the frequency band 0–B Hz is sampled at regular time intervals of $1/2B$ seconds over the whole of its duration, the signal can be accurately reconstructed from the sample values and without the use of any further knowledge of the signal. This necessarily implies that all the information in the received signal waveform is contained also in the sample values. In theory, a strictly bandlimited signal has infinite duration, but the signal at the output of a linear channel with finite bandwidth, resulting from a signal of finite duration at its input, has for practical purposes also a finite duration, and so need only be sampled over this period in order to be able to reconstruct the received waveform and therefore in order to extract all the information from this waveform. Hence, if the detector operates in an appropriate manner on the *sample values* rather than on the received waveform, there need be no loss in tolerance to noise and distortion. The recent developments in microprocessors and microcomputers mean that it is now often much simpler to handle signals as sample values rather than as continuous waveforms, so that powerful detection processes can be implemented in this way without excessive equipment complexity. Thus, following the sampling of the received signal, the receiver operates on a sequence of numbers and therefore it becomes a special purpose computer. This book is concerned entirely with such techniques.

It has been mentioned before that over telephone circuits and HF radio links there are various types of additive and multiplicative noise. The theoretical analysis of the different types of noise and their effects on a data-transmission system can be very difficult. Nevertheless, it is quite easy to make a qualitative assessment of the effects of the more important types of multiplicative noise such as transient interruptions, sudden signal-level changes, sudden carrier-phase changes and small frequency-offsets.[10] The latter are effectively eliminated in a suitably designed coherent detector and, for a given transmitted signal, the effect of any one of the others on the detected signal is not greatly influenced by the particular detection process used, just so long as this is not grossly inferior. However, the different detection processes are affected very differently by additive noise. Thus the various available detection processes are best compared on the basis of their tolerances to additive noise, for a given signal distortion introduced in transmission. The noise most easily analysed theoretically and simulated on a computer is additive white Gaussian noise. This noise also has the important property that it is the least predictable of the different types of noise and is therefore, in this sense, the most 'random'.[2] Furthermore, tests with different data-transmission systems have suggested that if one data-transmission system has a better tolerance to additive white Gaussian noise than another, then it usually also has a better tolerance to the additive noise actually experienced over voice-frequency channels such as telephone circuits and HF radio links.[10] Over transmission paths that are themselves baseband channels, the noise tends to be largely additive, and the principle just mentioned again tends to apply. These considerations suggest that a good measure of the relative performances of two different detection processes in practice is given by their relative tolerances to both signal distortion and additive white Gaussian noise. Thus it is assumed throughout this book that the transmission path introduces only signal distortion and additive white Gaussian noise.

REFERENCES

1. Nyquist, H. 'Certain topics in telegraph transmission theory', *AIEE Trans.*, **47**, 617–644 (1928)
2. Shannon, C. E. 'A mathematical theory of communication', *Bell Syst. Tech. J.*, **27**, 379–423 and 623–656 (1948)
3. Kretzmer, E. R. 'Generalization of a technique for binary data transmission', *IEEE Trans. Commun. Technol.*, **COM-14**, 67–68 (1966)
4. Jones, B. J. and Teacher, V. 'Modem at 2400 bits-per-second for data transmission over telephone lines', *Elect. Commun.*, **44**, 66–71 (1969)
5. Mattsson, O. 'Modem ZAT2400 for data transmission', *Ericsson Rev.*, **50**, 101–107 (1973)
6. Pugh, A. R. 'The latest modem for the Datel 600 service', *Post Office Elect. Eng. J.*, **67**, 95–101 (1974)
7. Lucky, R. W. 'A survey of the communication theory literature: 1968–1973', *IEEE Trans. Inform. Theory*, IT-19, 725–739 (1973)
8. Bennett, W. R. and Davey, J. R. *Data Transmission*, McGraw-Hill, New York (1965)
9. Lucky, R. W., Salz, J. and Weldon, E. J. *Principles of Data Communication*, McGraw-Hill, New York (1968)
10. Clark, A. P. *Principles of Digital Data Transmission*, Pentech Press, London (1976)
11. Panter, P. F. *Modulation, Noise and Spectral Analysis*, McGraw-Hill, New York (1965)
12. Lathi, B. P. *Random Signals and Communication Theory*, Intertext Books, London (1968)
13. Schwartz, M. *Information Transmission, Modulation and Noise*, McGraw-Hill, Kogakusha, Tokyo (1970)
14. Taub, H. and Schilling, D. L. *Principles of Communication Systems*, McGraw-Hill, New York (1971)

2 Signal distortion and the discrete Fourier transform

2.1 INTRODUCTION

2.1.1 Basic assumptions

Before it is possible to decide which of the many possible detection processes is the most suitable for any given application of digital data transmission, it is necessary first to be able to classify and grade the different types of signal distortion that are introduced by the transmission path or channel. Unfortunately practical transmission paths introduce a number of different types of distortion, including both time varying and nonlinear effects, and some of these are difficult to analyse. However, over both telephone circuits and HF radio links there are two predominant types of signal distortion, known as amplitude distortion and phase (or delay) distortion.[5] These are both linear effects and the data-transmission system having the best tolerance to them will normally also have the best overall tolerance to all the different types of signal distortion present.

It is assumed that the transmission path is a linear baseband channel or else that it is a linear bandpass channel and the modulation and demodulation processes used in the data-transmission system are both linear. In the latter case, the modulator (at the transmitter) and the demodulator (at the receiver) are both considered to be part of the transmission path, which is therefore always a baseband channel. Furthermore, the filter at the output of the transmitter, that limits the transmitted signal spectrum to the available frequency band of the transmission path, and the filter at the input to the receiver, that removes the noise frequency components outside the signal frequency band, are now always low-pass filters that operate on a baseband signal. The transmitter filter, transmission path and receiver filter together form a *linear baseband channel*, as shown in Figure 2.1. It is assumed that this channel is time invariant, so that its impulse response does not vary with time. The baseband

11

channel can now introduce only two types of signal distortion known as *amplitude* distortion and *phase* distortion.

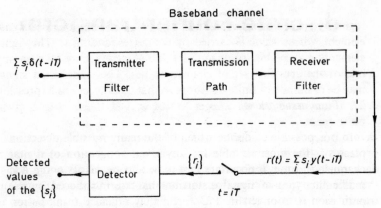

Fig. 2.1 Model of the data-transmission system

Signal distortion always appears as a change from the ideal form of the sampled impulse-reponse of the channel, which is the sequence of regularly spaced samples of the channel impulse-response, with the appropriate time interval between adjacent samples. A valuable technique for the classification and study of signal distortion is the discrete Fourier transform (DFT). The DFT has, of course, many other important applications and is of great interest in its own right.[4]

In the following discussion the DFT will be derived from first principles, in order to bring out the true significance of the transform. The DFT will then be used to define both amplitude distortion and phase distortion, in the sampled impulse-response of a linear baseband channel. By means of this analysis it is possible not only to estimate the nature and severity of the signal distortion represented by any given sampled impulse-response, but also to determine which general type of detection process is likely to be the most cost-effective for a data signal transmitted over the baseband channel. The different possible detection processes are not, however, studied in this chapter.

The model of the data-transmission system will be assumed to be as shown in Figure 2.1. This is a synchronous serial digital system, in which the receiver is held synchronized in time with the received stream of data elements.

The input signal to the baseband channel is a sequence of impulses

$$\sum_i s_i \delta(t - iT) \tag{2.1.1}$$

regularly spaced at intervals of T seconds. Each impulse is a signal element, whose value is carried by the corresponding s_i. The signal elements may be binary or multilevel, and each s_i may take on any one of the specified set of element values. The transmitted signal may be of infinite or finite duration, so that the integer i in Expression 2.1.1 may take on all integer values or just those over a given finite range.

In practice, a rectangular or rounded waveform would be used in place of the sequence of impulses at the input to the baseband channel, and the appropriate change would be made to the transmitter filter to give the same signal at the input to the transmission path as that obtained in Figure 2.1.[5] A sequence of impulses is assumed because this greatly simplifies the theoretical analysis of the system. The value s_i of a signal element is, of course, carried by the *area* of the corresponding impulse.

Since we are not here concerned with the effects of noise, the baseband channel will be assumed to be noise free, having an impulse response $y(t)$. Thus the signal at the output of the baseband channel is

$$r(t) = \sum_i s_i y(t - iT) \tag{2.1.2}$$

because, of course, the impulse $s_i \delta(t - iT)$ at the input to the baseband channel becomes the waveform $s_i y(t - iT)$ at its output.

It is assumed that the waveform $r(t)$ at the output of the baseband channel is sampled once per signal element, at the time instants $\{iT\}$, to give corresponding sample values $\{r_i\}$, where $r_i = r(iT)$. These are fed to the detector which operates on the $\{r_i\}$ to give the detected values of the $\{s_i\}$.

Suppose now that $\delta(t)$, a unit impulse at the time instant $t = 0$, is fed to the baseband channel in the absence of any other signals. The output signal from the baseband channel is $y(t)$, the impulse response of the channel. This waveform is sampled at the time instants $\{iT\}$, for all integer values of i, to give the sequence of impulses

$$y_0 \delta(t) + y_1 \delta(t - T) + y_2 \delta(t - 2T) + \ldots + y_g \delta(t - gT) \tag{2.1.3}$$

These may be considered more simply as the corresponding sequence of real sample values

$$y_0 \ y_1 \ y_2 \ldots y_g \qquad (2.1.4)$$

where $y_i = y(iT)$. It is assumed here that $y_i = 0$ when $i < 0$ and $i > g$, so that the $\{y_i\}$ for which $i < 0$ and $i > g$ can be ignored. The sequence given by Expression 2.1.4 is said to be the *sampled impulse-response* of the baseband channel.

In practice, the sampler in Figure 2.1 determines the sample value of the received signal $r(t)$ at the time instant $t = iT$, for each integer value of i, and feeds each sample value to the detector where it is stored for use in the detection process. The sampled impulse-response of the channel is therefore in fact just a sequence of numbers or values, which are represented more realistically by Expression 2.1.4 than by Expression 2.1.3. However, it is theoretically correct and often mathematically convenient (as is shown in Section 2.2.1) to consider a sequence of sample values in terms of the corresponding sequence of *impulses*, although it must now be borne in mind that this is an idealized representation of the signal at the output of the sampler and, except in the *values* of the samples, it does not correspond exactly or even approximately to the signal waveform normally obtained in practice.

The sampled impulse-response is physically realizable and of finite duration. Furthermore, if the delay in transmission is neglected, $y_0 \neq 0$. In the sampled impulse-response of any practical channel, the sample values become negligible after a certain time delay, so that, with the appropriate value of g, Expression 2.1.4 can be considered to be the sampled impulse-response of *any* given practical baseband channel.

If only a single impulse of value (area) s_h, at the time instant $t = hT$, is fed to the baseband channel in Figure 2.1, then the sample value r_i, at the detector input at the time instant $t = iT$, is $s_h y_{i-h}$, where again $y_i = 0$ when $i < 0$ and $i > g$.

With a sequence of an infinite number of signal elements at the input to the baseband channel and with the sampled impulse-response of the channel as given by Expression 2.1.4, it can be seen that the sample value at the detector input, at the time instant $t = iT$, is

$$r_i = \sum_{h=0}^{g} s_{i-h} y_h \qquad (2.1.5)$$

for any given integer value of i. Thus it may well not be possible to detect s_i from r_i, due to the presence of the other signal components. Nor may it be possible to detect any of the other element values, s_{i-1} to s_{i-g}, from r_i. If it is required to detect s_i from r_i then, in Equation 2.1.5, $s_i y_0$ is the wanted signal component and the terms $s_{i-1}y_1$ to $s_{i-g}y_g$ are the *intersymbol-interference* components. Similarly, if it is required to detect s_{i-h} from r_i, $s_{i-h}y_h$ is the wanted signal component and the terms $\{s_{i-l}y_l\}$, for $0 \leqslant l \leqslant g$ and $l \neq h$, are the intersymbol-interference components.

When $y_0 = 1$ and $y_h = 0$ for $1 \leqslant h \leqslant g$ in Expression 2.1.4, the baseband channel introduces no attenuation, delay or distortion, and from Equation 2.1.5, $r_i = s_i$, so that s_i is readily detected from r_i.

When $y_l = 1$ for an integer l in the range 1 to g, and $y_h = 0$ for $0 \leqslant h \leqslant g$ and $h \neq l$, the baseband channel introduces a delay of lT seconds but no attenuation or distortion. Now $r_i = s_{i-l}$, so that s_i is detected from r_{i+l}.

It can be seen from Equation 2.1.5 that in general the ratio of the magnitude of the wanted signal component to the number and magnitudes of the intersymbol-interference components, gives some idea of the signal distortion introduced by the transmission path. However, it can be shown that the ease with which the distortion may be eliminated and the reduction in tolerance to additive noise resulting from the signal distortion, are not necessarily determined by just the number and magnitudes of the $\{y_i\}$ in Expression 2.1.4.

One of the aims of this study is to express the signal distortion in terms of two parameters, known as amplitude distortion and phase distortion. The magnitudes of these two parameters give a much better measure of the effects of the signal distortion on the complexity and performance of the detector for the received digital signal, than do the number and magnitudes of the $\{y_i\}$ in Expression 2.1.4. Another aim of this study is to explain some of the interesting properties of the discrete Fourier transform and to show some of the ways in which it may be used in the analysis of digital signals.

2.1.2   Vectors

Before proceeding further it is necessary to summarize briefly some of the basic properties of vectors that are repeatedly used in this and the following chapters.[1-3]

For our purposes, a vector is an *ordered sequence of numbers*. These numbers are the *components* of the vector. A *real* vector is one whose components are all real.

Consider the two *n*-component row vectors

$$A = a_1 \ a_2 \ldots a_n \tag{2.1.6}$$

and

$$B = b_1 \ b_2 \ldots b_n \tag{2.1.7}$$

which are both real.

An *n*-component real vector may be represented as a *point* in an *n*-dimensional Euclidean vector space. This is illustrated in Figure 2.2 for the vectors *A* and *B*, in the simple case where $n = 2$.

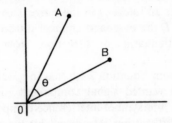

Fig. 2.2   Vectors A and B in a two-dimensional Euclidean vector space

The axes of the *n*-dimensional Euclidean vector space are orthogonal (at right angles) and such that the orthogonal projection of the *point* A on to the *i*th axis is the component a_i of the corresponding *vector* A. The *n* axes meet at the *origin* O of the vector space, and this point corresponds to the vector whose components are all zero.

The vectors *A* and *B* in Figure 2.2 can alternatively be considered as the *lines* OA and OB, so that each vector has a *direction* in the vector space. The *angle* θ between the vectors *A* and *B* is simply the angle between the lines OA and OB.

Again, the *row vector A* can be considered as the $1 \times n$ *matrix A* whose component in the *i*th column is a_i. The corresponding *column vector* is the *transpose* of the row vector *A*. It can be considered as the transpose of the $1 \times n$ matrix *A* and therefore as the $n \times 1$ matrix A^T whose component in the *i*th row is a_i. In the mathematical analysis involving vectors, they are treated as the corresponding matrices, using the normal rules of matrix algebra.

The *sum* of the two vectors *A* and *B* is the *n*-component row vector

$$A + B = a_1 + b_1 \ a_2 + b_2 \ldots a_n + b_n \tag{2.1.8}$$

The *inner product* of the vectors A and B is the component of the 1×1 matrix

$$AB^T = a_1 b_1 + a_2 b_2 + \ldots + a_n b_n \qquad (2.1.9)$$

When the vectors A and B are complex, B^T is here replaced by the conjugate transpose of B, so that b_i is replaced by the complex conjugate of b_i for each i.

The *length* of the real vector A is the distance from the origin to the corresponding point in the n-dimensional Euclidean vector space, and is

$$|A| = \left(\sum_{i=1}^{n} a_i^2 \right)^{\frac{1}{2}} = (AA^T)^{\frac{1}{2}} \qquad (2.1.10)$$

The *distance* between the real vectors A and B is the distance between the corresponding points in the vector space and is

$$|A-B| = \left(\sum_{i=1}^{n} (a_i - b_i)^2 \right)^{\frac{1}{2}} = [(A-B)(A-B)^T]^{\frac{1}{2}} \qquad (2.1.11)$$

Clearly it is also the *length* of the vector $A-B$ or $B-A$.

The *angle* θ between the real vectors A and B is given by

$$\cos \theta = \frac{AB^T}{|A| \, |B|} \qquad (2.1.12)$$

It can be seen that when $AB^T = 0$, $\theta = 90°$ or $270°$, so that the vectors are now *orthogonal* (at right angles).

It also follows from Equation 2.1.12 that

$$|AB^T| \leqslant |A| \, |B| \qquad (2.1.13)$$

and this is a particular case of the Schwarz inequality (see Reference 140 of Chapter 4).

A set of m n-component real vectors are said to be *linearly independent* if no one of the vectors can be expressed as a linear combination of one or more of the others. In other words, the m vectors $\{A_i\}$ for $i = 1, 2, \ldots, m$ are linearly independent so long as there is *no* set of m scalar quantities $\{c_i\}$ for $i = 1, 2, \ldots, m$, not all zero, such that

$$c_1 A_1 + c_2 A_2 + \ldots + c_m A_m = 0 \qquad (2.1.14)$$

The set of *all* n-component vectors that are linear combinations of m linearly independent vectors, where $m \leqslant n$, form an m-dimensional *subspace* of the *n*-dimensional Euclidean vector space. The subspace is said to be *spanned* by the *m* linearly independent vectors. Any vector lying in the subspace *must* be a linear combination of these vectors, and conversely, any vector not in the subspace *cannot* be a linear combination of the vectors. The zero vector, or origin of the *n*-dimensional vector space, lies in the subspace. Any vector that is orthogonal to the *m* linearly independent vectors must be orthogonal to every other vector lying in the subspace spanned by these and is therefore orthogonal to the subspace itself. The orthogonal projection of any vector *A* in the *n*-dimensional Euclidean vector space on to the *m*-dimensional subspace is the vector *B* in the subspace, such that the vector $A - B$ is orthogonal to the subspace. *B* is uniquely determined by *A*, and, if *A* lies in the subspace, $A = B$. Orthogonal vectors are necessarily also linearly independent, so that if *A* does not lie in the subspace, the vector $A - B$ cannot be expressed as a linear combination of any set of vectors in the subspace. The *distance* of the vector *A* from the *m*-dimensional subspace is the *length* of the vector $A - B$.

2.2 LINEAR TRANSFORMATION OF SEQUENCES

2.2.1 Z transforms

Assume that the signal at the input to the baseband channel in Figure 2.1 is the sequence of impulses

$$\sum_i s_i \delta(t - iT) \tag{2.2.1}$$

where *i* takes on all integer values. The Fourier transform of this signal is

$$\sum_i s_i \exp(-j2\pi f iT) = \sum_i s_i z^{-i} \tag{2.2.2}$$

where
$$z = \exp(j2\pi fT) \tag{2.2.3}$$

and
$$j = \sqrt{-1} \tag{2.2.4}$$

The *z* transform of the signal in Expression 2.2.1 is defined to be

$$S(z) = \sum_i s_i z^{-i} \tag{2.2.5}$$

The sampled impulse-response of the baseband channel, expressed as the corresponding sequence of impulses, is

$$\sum_{i=0}^{g} y_i \delta(t - iT) \tag{2.2.6}$$

so that the z transform of the sampled impulse-response of the channel is

$$Y(z) = \sum_{i=0}^{g} y_i z^{-i} \tag{2.2.7}$$

$Y(z)$ is, of course, the z transform of the *sequence* of sample values

$$y_0 \ y_1 \ y_2 \cdots y_g \tag{2.2.8}$$

which represent the sampled impulse-response of the channel. $Y(z)$ is often referred to simply as the z transform of the channel.

Again, the z transform of the sample values $\{r_i\}$ of the received signal in Figure 2.1 is

$$R(z) = \sum_i r_i z^{-i} \tag{2.2.9}$$

In taking the z transform of a sequence of sample values, these are treated as the corresponding sequence of *impulses*, as for instance in Expression 2.2.6. This may be done so long as the sample values are taken at regular intervals of T seconds, at the time instants $\{iT\}$, regardless of whether or not the waveform of the sampled signal in practice approximates to the corresponding sequence of impulses.

The value of the *transfer function* of the baseband channel and sampler in Figure 2.1, at a frequency f Hz, is the value of the Fourier transform of the sequence of impulses in Expression 2.2.6, at f Hz. It can be seen that this is simply the value of $Y(z)$ when $z = \exp(j2\pi fT)$. Clearly, z must here lie on the *unit circle* (the circle of unit radius with its centre at the origin) in the complex number plane. Thus, at f Hz, the transfer function has the value $Y(z)$, where z lies on the unit circle and at an angle of $2\pi fT$ radians with the positive real axis. Furthermore, the value of z is the same at all frequencies $f + (i/T)$, where i takes on all integer values, and z moves once round the unit circle in the complex number plane containing all possible values of z (the z plane) as f increases from i/T to $(i+1)/T$, for any integer i. This means that the transfer function of the baseband channel and sampler repeats itself cyclically over successive frequency intervals of $1/T$ Hz.

Suppose next that $z = \exp[(\sigma + j2\pi f)T]$ where σ may have any positive or negative real value. z can now have any complex value and so can lie anywhere in the z plane. It can be seen that just as any value of z on the unit circle in the z plane may be obtained via the Fourier transform of a sampled signal, so any value of z not on the unit circle may be obtained via the Laplace transform of the sampled signal. The transformation $z = \exp[(\sigma + j2\pi f)T]$ corresponds to the mapping of the s plane on to the z plane. All points to the left of the imaginary axis in the s plane are transformed (mapped) into points inside the unit circle in the z plane, and all points to the right of the imaginary axis are transformed into points outside the unit circle. This gives some idea of the significance of any given value of z. However, unless specifically considering values of z that do not lie on the unit circle in the z plane, it will now be assumed that $z = \exp(j2\pi fT)$.

It can be seen from Equations 2.2.5, 2.2.7 and 2.2.9 that the z transform of any sample value x_i, at the time instant $t = iT$, is $x_i z^{-i}$. Thus z^{-i} corresponds to the time instant $t = iT$ seconds, and its *coefficient* x_i, in the z transform, is the *sample value* at this time instant. Furthermore, if a set of samples is advanced in time by iT seconds, where i is an integer, the corresponding z transform is multiplied by z^i. Similarly, if the samples are delayed by iT seconds, the z transform is multiplied by z^{-i}.

Now,

$$S(z)Y(z) = \sum_i s_i z^{-i} \cdot \sum_{h=0}^{g} y_h z^{-h} \qquad (2.2.10)$$

so that the coefficient of z^{-i} in $S(z)Y(z)$ is

$$\sum_{h=0}^{g} s_{i-h} y_h = r_i \qquad (2.2.11)$$

from Equation 2.1.5. Thus

$$S(z)Y(z) = R(z) \qquad (2.2.12)$$

Consider next a finite sequence of m signal elements, at the input to the baseband channel in Figure 2.1, starting with the element $s_0 \delta(t)$ at the time instant $t = 0$. The sequence of element values is

$$s_0 \; s_1 \; s_2 \ldots s_{m-1} \qquad (2.2.13)$$

The corresponding sequence of sample values at the detector input is

$$r_0 \; r_1 \; r_2 \ldots r_{m+g-1} \tag{2.2.14}$$

The z transform of the sequence of the $\{s_i\}$ in Expression 2.2.13 is

$$S(z) = s_0 + s_1 z^{-1} + \ldots + s_{m-1} z^{-m+1} \tag{2.2.15}$$

and the z transform of the sequence of the $\{r_i\}$ in Expression 2.2.14 is

$$R(z) = r_0 + r_1 z^{-1} + \ldots + r_{m+g-1} z^{-m-g+1} \tag{2.2.16}$$

It can be seen from Equation 2.1.5 that since $s_i = 0$ for $i < 0$ and $i > m-1$, Equation 2.2.12 is again satisfied. Thus it is clear that whether the sequence of the $\{s_i\}$ is finite or infinite, the z transform of the corresponding sequence of the $\{r_i\}$ is the *product* of the z transforms of the sequences of the $\{s_i\}$ and $\{y_i\}$. On the other hand, the *sequence* of the $\{r_i\}$ itself is the *convolution* of the sequences of the $\{s_i\}$ and $\{y_i\}$, whether the sequence of the $\{s_i\}$ is finite or infinite. The great value of z transforms is that the *convolution* of two sequences becomes the *product* of their respective z transforms, which is often easier to handle and evaluate mathematically.

The z transform of a sequence of samples can be considered both as a convenient method of representing the *values* and *times of occurrence* of the samples, in terms of a polynomial in z, and also as a means of evaluating the Fourier transform of the sequence, for any value of frequency, using the fact that $z = \exp(j2\pi fT)$.

Equation 2.2.12 obviously holds also in the case where $R(z)$ is the z transform of the sequence of sample values obtained at the output of a *digital filter*, whose sampled impulse-response has the z transform $Y(z)$, and whose input signal is the sequence of sample values with z-transform $S(z)$. Indeed the whole of this and the following analysis applies also in the case where the baseband channel in Figure 2.1 is replaced by the corresponding digital filter.

The values of the *roots* of the z transform of a finite sequence are those values of z, *not* necessarily satisfying $z = \exp(j2\pi fT)$, for which the corresponding polynomial in z goes to zero.

2.2.2 Convolution matrices

With a sequence of m signal elements at the input to the baseband channel, the convolution of the sequence of the m $\{s_i\}$ with the sequence of the $g+1$ $\{y_i\}$, that gives the corresponding sequence of the $m+g$ $\{r_i\}$, can be expressed in terms of matrices, as follows.

Let S and R be the n-component row vectors

$$S = s_0 \ s_1 \ldots s_{m-1} \ 0 \ldots 0 \tag{2.2.17}$$

and

$$R = r_0 \ r_1 \ldots r_{n-1} \tag{2.2.18}$$

where

$$n = m+g \tag{2.2.19}$$

The vectors S and R are, of course, both matrices having one row and n columns, and each represents the corresponding sequence of sample values. Let Y be the $n \times n$ matrix

$$Y = \begin{bmatrix} y_0 & y_1 & . & . & . & y_{g-2} & y_{g-1} & y_g & 0 & 0 & . & . & 0 \\ 0 & y_0 & . & . & . & y_{g-3} & y_{g-2} & y_{g-1} & y_g & 0 & . & . & 0 \\ . & . & . & . & . & . & . & . & . & . & . & . & . \\ . & . & . & . & . & . & . & . & . & . & . & . & . \\ y_2 & y_3 & . & . & . & y_g & 0 & 0 & . & . & . & y_0 & y_1 \\ y_1 & y_2 & . & . & . & y_{g-1} & y_g & 0 & . & . & . & 0 & y_0 \end{bmatrix} \tag{2.2.20}$$

whose mth row is

$$Y_{m-1} = 0 \ldots 0 \ y_0 \ y_1 \ldots y_g \tag{2.2.21}$$

The matrix Y is not to be confused with $Y(z)$ which is the z transform of the sequence (or vector) Y_0 given by the first row of Y. Both of these, however, perform the same linear transformation on the signal elements involved.

Y is a circulant matrix,[1] which is a square matrix such that the $(i+1)$th row is obtained by a cyclic shift of the components of the first row by i places to the right. A cyclic shift of one place to the right of a component at the right-hand end of a row, transfers this component to the left-hand end of the row. Clearly, each row of Y contains the same components and in the same cyclic order, and all the components along the main diagonal of Y are y_0.

It can now be seen from Equation 2.1.5 that

$$R = SY \tag{2.2.22}$$

bearing in mind that $s_i = 0$ for $i<0$ and $i>m-1$. Again, R is obtained by the *convolution* of the sequences of the $\{s_i\}$ and $\{y_i\}$, and the sequences are here represented by the n-component row vectors S and Y_0, where Y_0 is the first row of the matrix Y. Y may therefore be considered as the *convolution matrix* that transforms S to R.

A further technique for evaluating the convolution of two sequences or vectors is via the *discrete Fourier transform* (DFT) which will now be studied in some detail.

2.3 THE DISCRETE FOURIER TRANSFORM

2.3.1 Definition

Consider a sequence of n sample values $\{r_i\}$ of the continuous waveform $r(t)$ at the output of the baseband channel in Figure 2.1. The samples are regularly spaced at intervals of T seconds and the first sample r_0 is at time $t = 0$. The samples may be represented as impulses whose areas are equal to the sample values. The $(i+1)$th impulse occurs at time $t = iT$ and has a value (area) r_i, for $i = 0, 1, \ldots, n-1$. Thus the sequence of impulses is

$$\sum_{i=0}^{n-1} r_i \delta(t - iT) \tag{2.3.1}$$

The Fourier transform of this sequence is

$$\sum_{i=0}^{n-1} r_i \exp(-j2\pi f iT) \tag{2.3.2}$$

where $j = \sqrt{-1}$.

The value of the Fourier transform at the discrete frequency h/nT Hz, where h is any integer, is

$$G\left(\frac{h}{nT}\right) = \sum_{i=0}^{n-1} r_i \exp\left(-j2\pi \frac{h}{nT} iT\right)$$

$$= \sum_{i=0}^{n-1} r_i \exp\left(-j2\pi \frac{hi}{n}\right)$$

$$= \sum_{i=0}^{n-1} r_i \omega^{-hi} \tag{2.3.3}$$

where

$$\omega = \exp\left(j\frac{2\pi}{n}\right) \tag{2.3.4}$$

Clearly,
$$\omega^i = \exp\left(j\frac{2\pi i}{n}\right) \qquad (2.3.5)$$

so that ω^i is a unit vector in the complex number plane, at an angle of $2\pi i/n$ radians with the positive real axis. ω^i, for $i = 0, 1, \ldots, n-1$, gives each of the n distinct nth roots of unity, which means, of course, that
$$(\omega^i)^n = \omega^{in} = 1 \qquad (2.3.6)$$

Furthermore, for any integer i,
$$\omega^i = \omega^{(i \text{ modulo-}n)} \qquad (2.3.7)$$

where
$$0 \leqslant i \text{ modulo-}n \leqslant n-1 \qquad (2.3.8)$$

and
$$i \text{ modulo-}n = i - ln \qquad (2.3.9)$$

l being the most positive integer such that $i - ln$ is nonnegative. In particular, when $i < 0$, l is the least negative integer such that $i - ln \geqslant 0$, and ω^{-i}, for $i = 0, 1, \ldots, n-1$ again gives each of the n distinct nth roots of unity.

It is evident that in Equation 2.3.3
$$G\left(\frac{h}{nT}\right) = G\left(\frac{h+in}{nT}\right) \qquad (2.3.10)$$

for any integer i, so that there are at most n different values of $G(h/nt)$ for all integer values of h.

At zero frequency (dc), $h = 0$ and the Fourier transform becomes
$$G(0) = \sum_{i=0}^{n-1} r_i \qquad (2.3.11)$$

Let F_h be the n-component row vector
$$F_h = 1 \;\; \omega^{-h} \;\; \omega^{-2h} \ldots \omega^{-(n-1)h} \qquad (2.3.12)$$

where $h = 0, 1, \ldots, n-1$, and ω is given by Equation 2.3.4. Let R be the n-component row vector
$$R = r_0 \;\; r_1 \ldots r_{n-1} \qquad (2.3.13)$$

Then, from Equation 2.3.3, the Fourier transform of the sequence of sample values $r_0, r_1, \ldots, r_{n-1}$, at the discrete frequency h/nT, is
$$G\left(\frac{h}{nT}\right) = RF_h^T \qquad (2.3.14)$$

where $F_h{}^T$ is the column vector ($n \times 1$ matrix) formed by the transpose of the row vector ($1 \times n$ matrix) F_h.

Let F be the $n \times n$ matrix whose $(h+1)$th row is F_h, so that

$$F = \begin{bmatrix} 1 & 1 & 1 & . & . & . & 1 \\ 1 & \omega^{-1} & \omega^{-2} & & & & \omega^{-(n-1)} \\ 1 & \omega^{-2} & \omega^{-4} & . & . & . & \omega^{-2(n-1)} \\ . & . & . & & & & . \\ . & . & . & & & & . \\ 1 & \omega^{-(n-1)} & \omega^{-2(n-1)} & . & . & . & \omega^{-(n-1)^2} \end{bmatrix} \quad (2.3.15)$$

Clearly, F is a symmetric matrix and the $(h+1)$th column of F is $F_h{}^T$.

$G(h/nT)$ is the $(h+1)$th component of the n-component row vector RF, for $h = 0, 1, \ldots, n-1$, so that the n discrete values of $G(h/nt)$ are given by the n components of RF. The vector RF is said to be the *discrete Fourier transform* (DFT) of R, and the linear transformation matrix F is itself known as the DFT.[4]

Whereas the components of R are real, the components of RF are often complex.

From Equations 2.3.10 and 2.3.14 it can be seen that the n components of the row vector RF contain the values of $G(h/nt)$ for *all* values of the integer h. The z transform of the sequence R is $R(z)$ in Equation 2.2.16, where $n = m+g$. But $G(h/nT)$ is the value of $R(z)$ when $z = \exp(j2\pi h/n)$. Thus the n different values of $G(h/nT)$ for $h = 0, 1, \ldots, n-1$, as given by the n components of RF, are the n values of $R(z)$ for values of z equally spaced on the unit circle in the z plane at angles of $2\pi h/n$ radians with the positive real axis.

The inverse of the matrix F is

$$F^{-1} = n^{-1} \begin{bmatrix} 1 & 1 & 1 & . & . & . & 1 \\ 1 & \omega & \omega^2 & & & & \omega^{n-1} \\ 1 & \omega^2 & \omega^4 & . & . & . & \omega^{2(n-1)} \\ . & . & . & & & & . \\ . & . & . & & & & . \\ 1 & \omega^{n-1} & \omega^{2(n-1)} & . & . & . & \omega^{(n-1)^2} \end{bmatrix} \quad (2.3.16)$$

whose $(h+1)$th row is the n-component row vector $n^{-1}E_h$, where

$$E_h = 1 \ \omega^h \ \omega^{2h} \ldots \omega^{(n-1)h} \quad (2.3.17)$$

The $(h+1)$th column of F^{-1} is $n^{-1}E_h{}^T$.

The n-component row vector RF^{-1} is said to be the *inverse discrete*

Fourier transform (inverse DFT) of R, and the linear transformation matrix F^{-1} is itself known as the inverse DFT.[4]

2.3.2   Basic properties

The $n \times n$ matrices F and F^{-1}, in Equations 2.3.15 and 2.3.16, are nonsingular symmetric matrices each of whose components has a modulus (magnitude) or unity.

If F_h is the n-component row vector formed by the $(h+1)$th row of F,

$$F_h F_i^T = 1 + \omega^{-h}\omega^{-i} + \omega^{-2h}\omega^{-2i} + \ldots + \omega^{-(n-1)h}\omega^{-(n-1)i}$$

$$= 1 + \omega^{-(h+i)} + \omega^{-2(h+i)} + \ldots + \omega^{-(n-1)(h+i)} \qquad (2.3.18)$$

When $h + i \neq 0$ or n, $F_h F_i^T = 0$

and when $h + i = 0$ or n, $F_h F_i^T = n$

as can be seen from Equation 2.3.18. It follows that the $n \times n$ matrix obtained by multiplying F by itself is

$$F^2 = n \begin{bmatrix} 1 & 0 & 0 & . & . & . & 0 & 0 \\ 0 & 0 & 0 & . & . & . & 0 & 1 \\ 0 & 0 & 0 & . & . & . & 1 & 0 \\ . & . & . & . & . & . & . & . \\ . & . & . & . & . & . & . & . \\ . & . & . & . & . & . & . & . \\ 0 & 0 & 1 & . & . & . & 0 & 0 \\ 0 & 1 & 0 & . & . & . & 0 & 0 \end{bmatrix} \qquad (2.3.19)$$

which is, of course, nonsingular.

From Equations 2.3.13 and 2.3.19,

$$n^{-1}RF^2 = r_0 \ r_{n-1} \ r_{n-2} \ldots r_1 \qquad (2.3.20)$$

and the postmultiplication of R by $n^{-1}F^2$ merely rearranges the order of the components of R. The matrix $n^{-1}F^2$ is therefore known as a *permutation* matrix.

It is clear from Equation 2.3.19 that the matrix $n^{-1}F^2$ is orthogonal and symmetric. Furthermore, the $n \times n$ matrix formed by multiplying $n^{-1}F^2$ by itself is

$$n^{-2}F^4 = I \qquad (2.3.21)$$

where I is an $n \times n$ identity matrix. Thus $n^{-\frac{1}{4}}F$ is a fourth root of the identity matrix.

From the previous analysis it can be seen that

$$F^{-2} = n^{-1} \begin{bmatrix} 1 & 0 & 0 & . & . & . & 0 & 0 \\ 0 & 0 & 0 & . & . & . & 0 & 1 \\ 0 & 0 & 0 & . & . & . & 1 & 0 \\ . & . & . & . & . & . & . & . \\ . & . & . & . & . & . & . & . \\ 0 & 0 & 1 & . & . & . & 0 & 0 \\ 0 & 1 & 0 & . & . & . & 0 & 0 \end{bmatrix} \quad (2.3.22)$$

so that nF^{-2} is a permutation matrix and is both orthogonal and symmetric. The $n \times n$ matrix formed by multiplying nF^{-2} by itself is

$$n^2 F^{-4} = I \quad (2.3.23)$$

and $n^{\frac{1}{4}}F^{-1}$ is a fourth root of the identity matrix. It is clear from Equations 2.3.19 and 2.3.22 that $n^{-1}F^2$ and nF^{-2} are the *same* permutation matrix.

It is now of interest to consider the relationships that exist between the matrices F and Y, where Y is the circulant matrix given by Equation 2.2.20.

Let C_i be the n-component row vector formed from the $(i+1)$th column of the $n \times n$ matrix Y in Equation 2.2.20. Thus

$$C_i = y_i \; y_{i-1} \; y_{i-2} \cdots y_{i-n+1} \quad (2.3.24)$$

where $y_i = 0$ for $i > g$, and $y_{-i} = y_{n-i}$ for $0 < i < n$. Now, from Equation 2.3.17,

$$E_h C_i^T = y_i + y_{i-1}\omega^h + y_{i-2}\omega^{2h} + \ldots + y_{i-n+1}\omega^{(n-1)h}$$

$$= \omega^{ih}(y_i\omega^{-ih} + y_{i-1}\omega^{-(i-1)h} + y_{i-2}\omega^{-(i-2)h} + \ldots$$

$$+ y_{i-n+1}\omega^{-(i-n+1)h})$$

$$= \omega^{ih}(y_i\omega^{-ih} + y_{i-1}\omega^{-(i-1)h} + \ldots + y_1\omega^{-h} + y_0$$

$$+ y_{n-1}\omega^{-(n-1)h} + \ldots + y_{i+1}\omega^{-(i+1)h}) \quad (2.3.25)$$

since $y_{-i} = y_{n-i}$ for $0 < i < n$, and $\omega^{-(i-n)h} = \omega^{-ih}$ for any integer i.

Equation 2.3.25 assumes, for convenience, that $2 < i < n-2$, but it can readily be seen from Equation 2.3.25 that, in general,

$$E_h C_i^T = \omega^{ih}(y_{n-1}\omega^{-(n-1)h} + y_{n-2}\omega^{-(n-2)h} + \ldots + y_1\omega^{-h} + y_0)$$

$$= \omega^{ih}(y_0 + y_1\omega^{-h} + y_2\omega^{-2h} + \ldots + y_g\omega^{-gh}) \qquad (2.3.26)$$

since $y_i = 0$ for $i > g$.

Clearly,

$$y_0 + y_1\omega^{-h} + y_2\omega^{-2h} + \ldots + y_g\omega^{-gh} = \lambda_h \qquad (2.3.27)$$

where λ_h is a function of h, for $h = 0, 1, \ldots, n-1$, but λ_h is independent of i.

It now follows that

$$E_h C_i^T = \lambda_h \omega^{ih} \qquad (2.3.28)$$

where ω^{ih} is the $(i+1)$th component of the vector E_h. But C_i^T is the n-component column vector given by the $(i+1)$th column of the $n \times n$ matrix Y, so that

$$E_h Y = \lambda_h E_h \qquad (2.3.29)$$

Thus, for $h = 0, 1, \ldots, n-1$, E_h is an *eigenvector* of the matrix Y in Equation 2.2.20 and λ_h is the corresponding *eigenvalue*.[2]

Since F^{-1} is the $n \times n$ matrix whose $(h+1)$th row is $n^{-1}E_h$, it follows from Equation 2.3.29 that $F^{-1}Y$ is the $n \times n$ matrix whose $(h+1)$th row is $\lambda_h n^{-1}E_h$. Thus

$$F^{-1}Y = DF^{-1} \qquad (2.3.30)$$

where D is the $n \times n$ diagonal matrix whose $(h+1)$th component along the main diagonal has the value λ_h in Equation 2.3.27. It follows that

$$F^{-1}YF = D \qquad (2.3.31)$$

which shows that D is obtained from Y by a *similarity transformation*, so that Y is *similar* to the diagonal matrix D.

Clearly, Y has n linearly independent eigenvectors $\{E_h\}$, given by Equation 2.3.17, and n eigenvalues $\{\lambda_h\}$, given by Equation 2.3.27. Furthermore, for *any* circulant matrix Y, the eigenvectors of Y are *independent* of the component values of Y. This is a most important result which means that the *same* similarity transformation converts any $n \times n$ circulant matrix into the corresponding diagonal matrix.[1]

Again, the n eigenvalues of Y are completely determined by the n-component row vector

$$Y_0 = y_0 \; y_1 \ldots y_g \; 0 \ldots 0 \qquad (2.3.32)$$

which is the first row of Y and, from Expression 2.2.8, is also the sampled impulse-response of the baseband channel in Figure 2.1.

It can be seen from Equations 2.3.15 and 2.3.27 that the DFT of Y_0 is

$$Y_0 F = \lambda_0 \; \lambda_1 \ldots \lambda_{n-1} \qquad (2.3.33)$$

so that the n components of $Y_0 F$ are the *eigenvalues* of Y. In other words, the n components of the DFT of the sampled impulse-response of the channel are the n eigenvalues of the $n \times n$ convolution matrix corresponding to this sampled impulse-response.

From Equations 2.3.14 and 2.3.15 it follows that

$$\lambda_h = Y_0 F_h{}^T = J\left(\frac{h}{nT}\right), \quad h = 0, 1, \ldots, n-1 \qquad (2.3.34)$$

where $J(h/nT)$ is the value of the transfer function of the baseband channel and sampler, in Figure 2.1, at the frequency h/nT Hz, and $F_h{}^T$ is the $(h+1)$th column of F. Clearly, the n components of the DFT of the sampled impulse-response of the channel are the values of the transfer function of the baseband channel and sampler, at the frequencies $\{h/nT\}$ for $h = 0, 1, \ldots, n-1$. $J(h/nT)$ has of course the same property as $G(h/nT)$ in Equation 2.3.10.

2.3.3 *Convolution of two vectors*

Consider again the finite sequence of m signal elements

$$\sum_{i=0}^{m-1} s_i \delta(t - iT)$$

at the input to the baseband channel in Figure 2.1, where the values of the signal elements, given by the areas of the corresponding impulses, are the first m components $\{s_i\}$ of the n-component row vector

$$S = s_0 \; s_1 \ldots s_{m-1} \; 0 \ldots 0 \qquad (2.3.35)$$

The corresponding sequence of n sample values at the detector input are the components $\{r_i\}$ of the n-component row vector

$$R = r_0 \ r_1 \dots r_{n-1} \qquad (2.3.36)$$

where

$$n = m+g \qquad (2.3.37)$$

The sampled impulse-response of the baseband channel in Figure 2.1 is given by the n-component row vector Y_0 in Equation 2.3.32, which is the first row of the matrix Y in Equation 2.2.20. Thus, from Equation 2.2.22,

$$R = SY \qquad (2.3.38)$$

and R is the *convolution* of the vectors S and Y_0.

Let the n-component row vectors

$$P = p_0 \ p_1 \dots p_{n-1} \qquad (2.3.39)$$

and

$$Q = q_0 \ q_1 \dots q_{n-1} \qquad (2.3.40)$$

be the discrete Fourier transforms of R and S, respectively, so that

$$P = RF \qquad (2.3.41)$$

and

$$Q = SF \qquad (2.3.42)$$

Thus

$$R = PF^{-1} \qquad (2.3.43)$$

and

$$S = QF^{-1} \qquad (2.3.44)$$

so that, from Equation 2.3.38,

$$PF^{-1} = QF^{-1}Y \qquad (2.3.45)$$

or

$$P = QF^{-1}YF \qquad (2.3.46)$$

It follows immediately from Equation 2.3.31 that

$$P = QD \qquad (2.3.47)$$

where D is the $n \times n$ diagonal matrix whose $(h+1)$th component along the main diagonal is λ_h in Equations 2.3.27 and 2.3.33. Thus

$$p_i = q_i \lambda_i \qquad (2.3.48)$$

for $i = 0, 1, \dots, n-1$.

It is clear from Equation 2.3.48 that each component of the DFT of the *convolution* of two vectors is simply the *product* of the corresponding components of the DFT's of the two vectors. This is a most important property of DFT's. It holds for the convolution of *any* two vectors provided only that the same number of components, n, is used for each DFT and n is large enough to include all the nonzero components in the *convolution* of the two vectors. The result holds, of course, in the general case where a sequence of values is fed to a digital filter, whose output sequence is the convolution of the input sequence and the sampled impulse-response of the filter.

Since each component of P is simply the *product* of the corresponding components of Q and Y_0F, P is very easily determined from Q and Y_0F. On the other hand, since R is the *convolution* of S and Y_0, the evaluation of R from S and Y_0 is considerably more complex than is the evaluation of P from Q and Y_0F. Clearly, R could be evaluated from S and Y_0 by forming Q and Y_0F, multiplying the corresponding pairs of components to give P and then forming $PF^{-1} = R$. By suitably simplifying and speeding up the implementation of the DFT into one of the possible forms known as fast Fourier transforms, it is often simpler to evaluate R by the method just described than by the direct convolution of S and Y_0.[4] We are, however, not here concerned with the method of evaluating R but rather with the nature and severity of the signal distortion corresponding to any given vector Y_0. We shall not therefore consider further the actual implementation of a DFT.

The relationships derived in the following sections are of very general application and are not just confined to an arrangement of the type shown in Figure 2.1. In order to emphasize this fact, the two vectors (or sequences) under consideration are renamed W and X.

2.3.4 Linearity

Let W and X be two n-component row vectors and let U and V be their respective DFT's, such that

$$U = WF \tag{2.3.49}$$

and
$$V = XF \tag{2.3.50}$$

where F is the $n \times n$ matrix given by Equation 2.3.15.

Since F is a linear transformation matrix,

$$aU + bV = (aW + bX)F \tag{2.3.51}$$

where a and b are two arbitrary scalars. In particular, when $a = b = 1$,

$$U + V = (W + X)F \qquad (2.3.52)$$

so that the DFT of the sum of two vectors is the sum of the individual DFT's. Similarly, the inverse DFT of the sum of two vectors is the sum of the individual inverse DFT's.

2.3.5 Shift theorem

Assume that the n components of the row vector

$$X = x_0 \ x_1 \dots x_{n-1} \qquad (2.3.53)$$

are sample values of a received signal waveform, the samples being spaced at regular intervals of T seconds, with x_0 at time $t = 0$. This is, of course, the same as the assumption made for the vector R in Section 2.3.1, as can be seen from Expression 2.3.1. Thus the set of sample values $\{x_i\}$ may be represented by the sequence of impulses

$$\sum_{i=0}^{n-1} x_i \delta(t - iT)$$

Suppose now that the sample values $\{x_i\}$ are each delayed by lT seconds (shifted to the right by l places), where l is an integer $0 < l < n$, so that the sample value x_i occurs at the time instant $t = (i+l)T$. The set of sample values may be represented by the sequence of impulses

$$\sum_{i=0}^{n-1} x_i \delta(t - (l+i)T)$$

The Fourier transform of these impulses, at a frequency h/nT Hz, is

$$G\left(\frac{h}{nT}\right) = \sum_{i=0}^{n-1} x_i \exp\left(-j2\pi \frac{h}{nT}(l+i)T\right)$$

$$= \sum_{i=0}^{n-1} x_i \omega^{-h(l+i)}$$

$$= X_{-l} F_h^T \qquad (2.3.54)$$

where ω is given by Equation 2.3.4, F_h^T is the n-component column vector formed by the $(h+1)$th column of the matrix F in Equation 2.3.15, and

$$X_{-l} = x_{n-l} \; x_{n-l+1} \ldots x_{n-1} \; x_0 \ldots x_{n-l-1} \qquad (2.3.55)$$

But $G(h/nT)$ is the $(h+1)$th component of the n-component DFT of the given sequence, and $X_{-l}F_h^T$ is the $(h+1)$th component of the n-component vector $X_{-l}F$. Thus the DFT of the sequence of sample values delayed by lT seconds is $X_{-l}F$.

Similarly, when the sample values $\{x_i\}$ are each advanced by lT seconds (shifted to the left by l places) relative to their original positions in time, the DFT of the sequence of sample values becomes X_lF, where

$$X_l = x_l \; x_{l+1} \ldots x_{n-1} \; x_0 \ldots x_{l-1} \qquad (2.3.56)$$

Clearly, a shift in time of the sequence of sample values, by a whole number of sampling intervals, appears in the DFT of the sequence as though the sequence had in fact been shifted *cyclically* in the same direction and by the same number of places. Thus a *time* shift of a sequence by a whole number of sampling intervals can be considered as the corresponding *cyclic* shift of the sample values, without in any way affecting the DFT of the sequence.

Suppose now that a sequence of sample values given by the n-component vector

$$X = x_0 \; x_1 \ldots x_{m-1} \; 0 \ldots 0 \qquad (2.3.57)$$

is fed to a digital filter whose sampled impulse-response is given by the n-component vector

$$W = w_0 \; w_1 \ldots w_g \; 0 \ldots 0 \qquad (2.3.58)$$

where
$$n = m + g \qquad (2.3.59)$$

and w_i is the sample value at time $t = iT$, for $i = 0, 1, \ldots, g$.

If M is the $n \times n$ circulant matrix whose first row is given by W, so that M is the convolution matrix corresponding to the vector W, then the output sequence from the digital filter is given by the n-component row vector XM.

Consider now the $n \times n$ circulant matrix M_l whose first row is the n-component vector

$$W_l = w_l \ldots w_g \; 0 \ldots 0 \; w_0 \ldots w_{l-1} \qquad (2.3.60)$$

W_l is derived from W by a cyclic shift of l places to the left in its components. Clearly M_l is obtained from M by a cyclic shift of l places to the left in the *columns* of M, so that XM_l is obtained from XM by a cyclic shift of l places to the left in the *components* of XM. Similarly, if M_{-l} is the $n \times n$ circulant matrix whose first row is obtained from W by a cyclic shift of l places to the right in the components of W, then XM_{-l} is obtained from XM by a cyclic shift of l places to the right in the components of XM.

W_l in Equation 2.3.60 can be considered to represent an advance in time of lT seconds in the sampled impulse-response of the digital filter, so that, in Equation 2.3.58, w_i is now the sample value at time $t = (i-l)T$ instead of at time $t = iT$. Hence, in Equation 2.3.60, a *time* shift is represented by the corresponding *cyclic* shift. This cyclic shift gives the *same* cyclic shift in the output sequence from the digital filter, where a time shift is represented by the corresponding cyclic shift, since a shift in *time* of the sampled impulse-response gives the *same* time shift in the output sequence. Thus the output sequence can here be represented by XM_l.

It is clear that an advance (or delay) in *time* of lT seconds in a sequence of samples having a sampling interval of T seconds, normally represented as a shift of l places to the left (or right) in the sequence of sample values, can alternatively be represented by a *cyclic* shift of l places to the left (or right) in the components of the sequence, whether the sequence represents a signal or the sampled impulse-response of a filter or channel. This important relationship is known here as the Shift theorem, and it holds also when $l \geqslant n$.

The representation of a time shift as a cyclic shift is of course not necessarily unique, since a cyclic shift of l places to the left (or right) corresponds to a shift in time of $l + hn$ places to the left (or right), where h is any integer. However, in all cases studied here, the time shifts in either direction are always less than $\frac{1}{2}n$ places, so that the corresponding cyclic shifts are always less than $\frac{1}{2}n$ places to the left or right, and there is now a unique relationship between the time shift and the corresponding cyclic shift. Under these conditions, a time shift is represented unambiguously by the corresponding cyclic shift, the resultant sequence *always* starting at time $t = 0$.

An immediate and sometimes useful consequence of the shift theorem is that the components $\{x_{n-i}\}$ of the vector X in Equation 2.3.53, where n is odd and $i = 1, 2, \ldots, \frac{1}{2}(n-1)$, can be taken as the samples at the time instants $\{-iT\}$ so that $x_{\frac{1}{2}(n+1)}$ is the sample value at time $t = -\frac{1}{2}(n-1)T$ and is the first component (in time) of the sequence, and $x_{\frac{1}{2}(n-1)}$ is the sample value at time $t = \frac{1}{2}(n-1)T$ and is the last component (in time) of the sequence. Furthermore, if the last l components of X are all zero, where $l \geqslant \frac{1}{2}(n-1)$, all the important results and relationships to be developed here, involving a shift in *time* of the sequence X, apply also to the corresponding *cyclic* shift, which can therefore be considered to be *identically equivalent* to the time shift. Extensive use is made of this fact in Section 2.4.

It is interesting now to evaluate the relationship between the DFT's of two sequences, where one of the sequences is given by a cyclic shift of the components of the other.

The n-component DFT of the vector X in Equation 2.3.53 is

$$XF = \sum_{i=0}^{n-1} x_i(1 \quad \omega^{-i} \quad \omega^{-2i} \ldots \omega^{-(n-1)i}) \qquad (2.3.61)$$

Suppose that the components of X are shifted cyclically l places to the right. The vector now becomes X_{-l} in Equation 2.3.55 and the DFT of X_{-l} is

$$X_{-l}F = \sum_{i=0}^{n-1} x_i(1 \quad \omega^{-(i+l)} \quad \omega^{-2(i+l)} \ldots \omega^{-(n-1)(i+l)})$$

$$= \sum_{i=0}^{n-1} x_i(1 \quad \omega^{-i}\omega^{-l} \quad \omega^{-2i}\omega^{-2l} \ldots \omega^{-(n-1)i}\omega^{-(n-1)l})$$

$$= XFD_{-l} \qquad (2.3.62)$$

where D_{-l} is the diagonal matrix whose $(i+1)$th component along the main diagonal is ω^{-il}.

Suppose next that the components of X are shifted cyclically l places to the left. The vector now becomes X_l in Equation 2.3.56 and the DFT of X_l is

$$X_lF = \sum_{i=0}^{n-1} x_i(1 \quad \omega^{-(i-l)} \quad \omega^{-2(i-l)} \ldots \omega^{-(n-1)(i-l)})$$

$$= \sum_{i=0}^{n-1} x_i(1 \quad \omega^{-i}\omega^l \quad \omega^{-2i}\omega^{2l} \ldots \omega^{-(n-1)i}\omega^{(n-1)l})$$

$$= XFD_l \qquad (2.3.63)$$

where D_l is the diagonal matrix whose $(i+1)$th component along the main diagonal is ω^{il}.

In the derivation of the *first* expression for the DFT, in both

Equations 2.3.62 and 2.3.63, use is made of the fact that $\omega^{-h(i\pm n)} = \omega^{-hi}$, where h is any integer.

In Equations 2.3.62 and 2.3.63, l may be any positive integer. However, from the definition of D_l and D_{-l}, it is evident that $D_l = D_{l+hn}$ for any integers l and h (positive or negative or zero), so that D_l has only n different values. In particular, when $0<l<n$, $D_{-l} = D_{n-l}$. Similar relationships of course also hold for X_l and X_{-l}.

Clearly, a cyclic shift of l places to the left (or right) in the n components of the vector X results in the multiplication of the corresponding DFT by D_l (or D_{-l}). Furthermore, it can be seen that

$$D_{-l} = D_l^{-1} \tag{2.3.64}$$

It follows from the Shift theorem that if a sequence of sample values is advanced or delayed in *time* by l places, the DFT of the sequence is postmultiplied by the diagonal matrix D_l or D_{-l}, respectively. Again, if the DFT of the sequence is postmultiplied by D_l or D_{-l}, where now $0<l<n$, the sequence is advanced or delayed, respectively, by $l+hn$ places in time, where h is any positive or negative integer or zero.

2.3.6 Real and imaginary components of a DFT

From Equation 2.3.15,

$$F = \begin{bmatrix} 1 & 1 & 1 & . & . & . & 1 \\ 1 & \omega^{-1} & \omega^{-2} & . & . & . & \omega^{-(n-1)} \\ 1 & \omega^{-2} & \omega^{-4} & . & . & . & \omega^{-2(n-1)} \\ . & . & . & & . & . & . \\ . & . & . & . & . & . & . \\ 1 & \omega^{-(n-1)} & \omega^{-2(n-1)} & . & . & . & \omega^{-(n-1)^2} \end{bmatrix} \tag{2.3.65}$$

where

$$\omega^{-i} = \exp\left(-j\frac{2\pi i}{n}\right)$$

$$= \cos\frac{2\pi i}{n} - j\sin\frac{2\pi i}{n} \tag{2.3.66}$$

so that

$$F = F_a - jF_b \tag{2.3.67}$$

where

$$F_a = \begin{bmatrix} 1 & 1 & 1 & \ldots & 1 \\ 1 & \cos\dfrac{2\pi}{n} & \cos\dfrac{2\pi2}{n} & \ldots & \cos\dfrac{2\pi(n-1)}{n} \\ 1 & \cos\dfrac{2\pi2}{n} & \cos\dfrac{2\pi4}{n} & \ldots & \cos\dfrac{2\pi2(n-1)}{n} \\ . & . & . & \ldots & . \\ . & . & . & \ldots & . \\ . & . & . & \ldots & . \\ 1 & \cos\dfrac{2\pi(n-1)}{n} & \cos\dfrac{2\pi2(n-1)}{n} & \ldots & \cos\dfrac{2\pi(n-1)^2}{n} \end{bmatrix} \quad (2.3.68)$$

and

$$F_b = \begin{bmatrix} 0 & 0 & 0 & \ldots & 0 \\ 0 & \sin\dfrac{2\pi}{n} & \sin\dfrac{2\pi2}{n} & \ldots & \sin\dfrac{2\pi(n-1)}{n} \\ 0 & \sin\dfrac{2\pi2}{n} & \sin\dfrac{2\pi4}{n} & \ldots & \sin\dfrac{2\pi2(n-1)}{n} \\ . & . & . & \ldots & . \\ . & . & . & \ldots & . \\ . & . & . & \ldots & . \\ 0 & \sin\dfrac{2\pi(n-1)}{n} & \sin\dfrac{2\pi2(n-1)}{n} & \ldots & \sin\dfrac{2\pi(n-1)^2}{n} \end{bmatrix} \quad (2.3.69)$$

Consider the n-component row vector X in Equation 2.3.53, all of whose components are *real*. Clearly, the DFT of X is

$$XF = XF_a - jXF_b \qquad (2.3.70)$$

Since

$$\cos\frac{2\pi hi}{n} = \cos\frac{2\pi h(n-i)}{n} \qquad (2.3.71)$$

for integers h and i having values in the range 0 to $n-1$, it can be seen that the n-component vectors formed by the 2nd and nth columns of F_a are the same. So are the vectors formed by the 3rd and $(n-1)$th columns, the 4th and $(n-2)$th columns, and so on. But for values of the integer i in the range 0 to $n-1$, the *real* part of the $(i+1)$th component of XF is

$$\sum_{h=0}^{n-1} x_h \cos\frac{2\pi hi}{n}$$

where $\cos(2\pi hi/n)$ is the $(h+1)$th component of the $(i+1)$th column of F_a. It follows that for any real n-component vector X, the real parts of the 2nd and nth components of XF are the same, as are the real parts of the 3rd and $(n-1)$th components, the 4th and $(n-2)$th components, and so on.

Since

$$\sin\frac{2\pi hi}{n} = -\sin\frac{2\pi h(n-i)}{n} \tag{2.3.72}$$

for integers h and i having values in the range 0 to $n-1$, the n-component vectors formed by the 2nd and nth columns of F_b are the negatives of each other. So are the vectors formed by the 3rd and $(n-1)$th columns of F_b, the 4th and $(n-2)$th columns, and so on. But for values of the integer i in the range 0 to $n-1$, the *imaginary* part of the $(i+1)$th component of XF is

$$-j\sum_{h=0}^{n-1} x_h \sin\frac{2\pi hi}{n}$$

where $\sin(2\pi hi/n)$ is the $(h+1)$th component of the $(i+1)$th column of F_b. Thus, for any real n-component vector X, the imaginary parts of the 2nd and nth components of XF are the negatives of each other, as are the imaginary parts of the 3rd and $(n-1)$th components, the 4th and $(n-2)$th components, and so on.

Clearly, for $i = 1, 2, \ldots, n-1$, the n-component vectors formed by the $(i+1)$th and $(n-i+1)$th columns of F are *complex conjugates* so that, for any *real* n-component vector X, the $(i+1)$th and $(n-i+1)$th *components* of XF are themselves *complex conjugates*. It can be seen from Equation 2.3.65 that the *first* component of XF must be *real*.

Since F is a symmetric (not Hermitian) matrix, the same relationships hold for the rows of F as for the columns. Thus the n-component vectors formed by the $(i+1)$th and $(n-i+1)$th rows of F are complex conjugates. Also, for any real n-component column vector X^T, the $(i+1)$th and $(n-i+1)$th components of FX^T are complex conjugates. Again, the first component of FX^T must be real.

It can be seen from Equations 2.3.16, 2.3.68 and 2.3.69 that

$$nF^{-1} = F_a + jF_b \tag{2.3.73}$$

so that, from Equation 2.3.67,

$$nF^{-1} = F^* \tag{2.3.74}$$

where F^* is the complex conjugate of the matrix F. This can in fact be seen directly from Equations 2.3.15, 2.3.16 and 2.3.5. Thus, for any real n-component row vector X,

$$nXF^{-1} = XF^* = (XF)^* \qquad (2.3.75)$$

where $(XF)^*$ is, of course, the complex conjugate of the DFT of X. This is an important result which is used in the derivation of several useful relationships.

Since F is a symmetric matrix, it follows from Equation 2.3.74 that the conjugate transpose of F is nF^{-1}, so that the conjugate transpose of $n^{-\frac{1}{2}}F$ is $n^{\frac{1}{2}}F^{-1}$ which is also the inverse of $n^{-\frac{1}{2}}F$. Thus $n^{-\frac{1}{2}}F$ is a *unitary* matrix, as is $n^{\frac{1}{2}}F^{-1}$. As has previously been shown, each of these unitary matrices when raised to the fourth power becomes the $n \times n$ identity matrix.

It is of interest now to consider the DFT's of *symmetric* and *skew-symmetric* sequences whose components are all *real*. A *symmetric* sequence is defined to be a sequence with an odd number of components, such that the central component has any real value and, for each i, the component i places to the left of the central component is the same as the component i places to the right. A *skew-symmetric* sequence is defined to be a sequence with an odd number of components, such that the central component is zero and, for each i, the component i places to the left of the central component is the negative of the component i places to the right.

When considering the DFT of a symmetric or skew-symmetric sequence, it is convenient to shift the sequence in time by the appropriate number of places (sampling intervals) so that the central component (sample) occurs at time $t = 0$. By the Shift theorem, the resultant n-component vector X (Equation 2.3.53) is such that its first component x_0 is the *central* component of the original sequence. Thus in the resultant vector, x_i and x_{n-i} (for $i = 1, 2, \ldots, \frac{1}{2}(n-1)$ and n odd) are the components i places to the right and left, respectively, of the central component in the original sequence. The cyclic shift in the components of X eliminates the delay terms in its DFT without otherwise changing the DFT, so that the basic properties of the DFT are more easily identified.

Suppose first that X is *real* and *symmetric*, such that

$$x_i = x_{n-i}, \quad i = 1, 2, \ldots, n-1 \qquad (2.3.76)$$

which means that the $(i+1)$th and $(n-i+1)$th components of X are the same.

The $(i+1)$th component of XF is

$$\sum_{h=0}^{n-1} x_h \omega^{-hi} = \sum_{h=0}^{n-1} x_h \left(\cos\frac{2\pi hi}{n} - \mathrm{j}\sin\frac{2\pi hi}{n} \right) \qquad (2.3.77)$$

It can be seen from Equations 2.3.76 and 2.3.77 that the imaginary components in F now cancel out to give a *real* vector XF. Since the $(i+1)$th and $(n-i+1)$th components of XF are complex conjugates of each other, it follows that XF is not only *real* but is also *symmetric*, such that its $(i+1)$th component equals its $(n-i+1)$th component, for $i = 1, 2, \ldots, n-1$. Similarly, it may be shown that if XF is real and symmetric, then so is X.

Suppose next that X is *real* and *skew-symmetric*, such that

$$x_0 = 0 \qquad (2.3.78)$$

and $\qquad x_i = -x_{n-i}, \quad i = 1, 2, \ldots, n-1 \qquad (2.3.79)$

which means that the $(i+1)$th and $(n-i+1)$th components of X are the negatives of each other. It can be seen from Equations 2.3.77, 2.3.78 and 2.3.79 that all the real components of F now cancel out to give an *imaginary* vector XF. Since the $(i+1)$th and $(n-i+1)$th components of XF are complex conjugates and its first component is the sum of the $\{x_i\}$, XF is not only *imaginary* but is also *skew-symmetric*, such that its first component is zero and its $(i+1)$th component is the negative of its $(n-i+1)$th component, for $i = 1, 2, \ldots, n-1$. Similarly, it may be shown that if XF is imaginary and skew-symmetric, then X is real and skew-symmetric.

It must be emphasized again, that for both the symmetric and skew-symmetric sequences X, the components of X have been shifted cyclically to the left to make x_0 the *central* component of the original sequence. This corresponds to a shift in time of the sequence of samples, to bring the central sample of the sequence to the time instant $t = 0$.

2.3.7   Reversal theorem

Let X be the n-component real row vector

$$X = x_0 \ x_1 \ldots x_g \ 0 \ldots 0 \qquad (2.3.80)$$

and let W be the n-component row vector

$$W = x_g \; x_{g-1} \ldots x_0 \; 0 \ldots 0 \qquad (2.3.81)$$

formed from X by reversing the order of its first $g+1$ components, where g is any positive integer less than n.

The DFT of X is

$$XF = \sum_{i=0}^{g} x_i (1 \; \omega^{-i} \; \omega^{-2i} \ldots \omega^{-(n-1)i}) \qquad (2.3.82)$$

and the inverse DFT of X is

$$XF^{-1} = n^{-1} \sum_{i=0}^{g} x_i (1 \; \omega^{i} \; \omega^{2i} \ldots \omega^{(n-1)i}) \qquad (2.3.83)$$

The DFT of W is

$$
\begin{aligned}
WF &= \sum_{i=0}^{g} x_{g-i} (1 \; \omega^{-i} \; \omega^{-2i} \ldots \omega^{-(n-1)i}) \\
&= \sum_{i=0}^{g} x_i (1 \; \omega^{i-g} \; \omega^{2(i-g)} \ldots \omega^{(n-1)(i-g)}) \\
&= \sum_{i=0}^{g} x_i (1 \; \omega^{i} \omega^{-g} \; \omega^{2i} \omega^{-2g} \ldots \omega^{(n-1)i} \omega^{-(n-1)g}) \\
&= nXF^{-1} D_{-g} \qquad (2.3.84)
\end{aligned}
$$

where D_{-g} is the diagonal matrix whose $(i+1)$th component along the main diagonal is ω^{-ig}. But, from Equation 2.3.75, nXF^{-1} is the *complex conjugate* of the vector XF, so that, from Equation 2.3.84,

$$WF = (XF)^* D_{-g} \qquad (2.3.85)$$

If now the components of W are shifted *cyclically* g places to the left, the vector becomes

$$W_g = x_0 \; 0 \ldots 0 \; x_g \; x_{g-1} \ldots x_1 \qquad (2.3.86)$$

and its DFT is

$$W_g F = WF D_g \qquad (2.3.87)$$

from Equation 2.3.63. Thus, from Equations 2.3.85 and 2.3.64,

$$W_g F = (XF)^* \qquad (2.3.88)$$

which means that the DFT of W_g is the *complex conjugate* of the DFT of X.

From the Shift theorem, a cyclic shift of g places to the left of the components of W corresponds to an advance in *time* of g places (sampling intervals) of the original sequence W, such that the shifted sequence has its first nonzero component (the sample value x_g) at the time instant $t = -gT$, and its last nonzero component (the sample value x_0) at the time instant $t = 0$. The sequence X has its first nonzero component (the sample value x_0) at the time instant $t = 0$, and its last nonzero component (the sample value x_g) at the time instant $t = gT$. It is clear that a reversal in time of the latter sequence, pivoted about the time instant $t = 0$, gives the former sequence. Thus the sequence W_g is in fact the true *reversal in time* of the sequence X, and, from Equation 2.3.88, its DFT is the *complex conjugate* of the DFT of X.

The Reversal theorem therefore states that if a sequence is reversed in time, the reversal being pivoted about the component at time $t = 0$, its DFT is replaced by its complex conjugate. The converse, of course, also holds.

2.3.8   *Aperiodic autocorrelation function and discrete energy-density spectrum*

Consider the sequence given by the n-component vector X in Equation 2.3.80, where now

$$n \geqslant 2g + 1 \qquad (2.3.89)$$

Assume that the sequence X is fed to a digital filter with a sampled impulse-response given by the n-component vector W in Equation 2.3.81.

The z-transforms of the vectors X and W are, respectively,

$$X(z) = x_0 + x_1 z^{-1} + \ldots + x_g z^{-g} \qquad (2.3.90)$$

and

$$W(z) = x_g + x_{g-1} z^{-1} + \ldots + x_0 z^{-g} \qquad (2.3.91)$$

If the output sequence from the filter is given by the n-component vector

$$A = a_0\, a_1 \ldots a_{n-1} \qquad (2.3.92)$$

the z transform of the output sequence is

$$
\begin{aligned}
A(z) &= X(z)W(z) \\
&= x_o x_g + (x_0 x_{g-1} + x_1 x_g) z^{-1} + \ldots \\
&\quad + (x_{g-1} x_0 + x_g x_1) z^{-2g+1} + x_g x_0 z^{-2g} \qquad (2.3.93)
\end{aligned}
$$

where the coefficient of z^{-i} in $A(z)$ is a_i, and $A(z)$ clearly has $2g+1$ terms.

The $(i+1)$th component of the n-component output vector A is evidently

$$a_i = \sum_{h=-\infty}^{\infty} x_h x_{i-g+h} \qquad (2.3.94)$$

where $x_i = 0$ for $i < 0$ and $i > g$. The n components of the output vector A contain the $2g+1$ potentially nonzero components of the discrete *aperiodic autocorrelation function* of the sequence X, given by the coefficients in $A(z)$ in Equation 2.3.93.

The value of the autocorrelation function of X for a time interval of $i-g$ places (sampling intervals) is given by a_i, the $(i+1)$th component of A. The value of the autocorrelation function for a time interval of zero is

$$a_g = \sum_{h=0}^{g} x_h^2 \qquad (2.3.95)$$

which is the squared length of the vector X. a_g is clearly positive and has a greater magnitude than any other component of A. The components of A are *real* and the sequence A is *symmetric* about the $(g+1)$th component a_g. Also $a_i = 0$ for $i < 0$ and $i > 2g$.

It can be seen from Equation 2.3.81 that the sampled impulse-response of the digital filter is

$$\sum_{i=0}^{g} x_{g-i} \delta(t - iT) \qquad (2.3.96)$$

If the time response of the filter is advanced by gT seconds, that is, by g sampling intervals, the sampled impulse-response of the digital filter becomes

$$\sum_{i=0}^{g} x_{g-i} \delta(t + (g - i)T) \qquad (2.3.97)$$

which is, of course, non-physical, since the output signal from the filter now *precedes* the input signal from which it originates. The study of this hypothetical filter is, nevertheless, of great interest.

From the Shift theorem, the n-component vector W_g, representing the sampled impulse-response in Equation 2.3.97, is obtained from the vector W, representing the sampled impulse-response in Equation 2.3.96, by shifting the n-components of W *cyclically* to the left, by g places, to give

$$W_g = x_0 \; 0 \ldots 0 \; x_g \; x_{g-1} \ldots x_1 \qquad (2.3.98)$$

as in Equation 2.3.86. But, as is shown in Section 2.3.5, a cyclic shift of g places to the left in the components of W, to give W_g, results in a cyclic shift of g places to the left in the components of A, which now becomes

$$A_g = a_g \ a_{g+1} \ldots a_{n-1} \ a_0 \ldots a_{g-1} \qquad (2.3.99)$$

The sequence of n terms given by the n-component vector A_g is defined to be the *aperiodic autocorrelation function* of the sequence given by the vector X in Equation 2.3.80, where now $n \geqslant 2g+1$. As before, a_g is the squared length of the vector X, and is also the largest component. The components of A_g are *real* and *symmetric*, such that

$$a_{g+i} = a_{g-i}, \quad i = 1, 2, \ldots, g$$
and $\qquad\qquad a_i = 0, \qquad i > 2g \qquad\qquad (2.3.100)$

So long as $n \geqslant 2g+1$, A_g contains *all* the coefficients in $A(z)$ in Equation 2.3.93 and is the aperiodic autocorrelation function of X, as just defined. However, when $n < 2g+1$, the correct autocorrelation function cannot be obtained, because A_g now has an insufficient number of terms.

It can be seen from Equation 2.3.94 that the component a_i of A_g is formed from two identical sequences X, one of which is shifted by $i-g$ places (sampling intervals) with respect to the other, each $(g+1)$-component sequence being preceded and followed by zeros to make it an infinite sequence. a_i is now given by the sum of the products of all coincident pairs of components (sample values), one from each of the two sequences. There is no question here of one of the two sequences being shifted *cyclically* with respect to the other.

Let the n-component DFT's of W_g and X be

$$W_g F = u_0 \ u_1 \ldots u_{n-1} \qquad (2.3.101)$$
and $\qquad\qquad XF = v_0 \ v_1 \ldots v_{n-1} \qquad (2.3.102)$

where $n \geqslant 2g+1$. From Equation 2.3.48, the n-component DFT of A_g is

$$A_g F = u_0 v_0 \ u_1 v_1 \ldots u_{n-1} v_{n-1} \qquad (2.3.103)$$

so that the $(i+1)$th component of $A_g F$ is $u_i v_i$. This follows because A is formed by the *convolution* of W and X, as can be seen from Equation 2.3.93, and because $W_g F$ and $A_g F$ are the DFT's of the sequences formed from W and A through shifting them to the left by g places, so that A_gF is the DFT of the *convolution* of X and the *shifted* sequence W_g.

But, from Equation 2.3.88, W_gF is the complex conjugate of XF, so that

$$u_i = v_i^* \tag{2.3.104}$$

and

$$u_iv_i = |v_i|^2 \tag{2.3.105}$$

for $i = 0, 1, \ldots, n-1$. Thus the DFT of A_g is

$$A_gF = |v_0|^2 \ |v_1|^2 \ldots |v_{n-1}|^2 \tag{2.3.106}$$

the $(i+1)$th component of A_gF being $|v_i|^2$, where v_i is the $(i+1)$th component of XF and $|v_i|$ is the modulus of v_i.

The n-component vector A_gF gives the *discrete energy-density spectrum* of the sequence X. All components of A_gF are real and nonnegative. Furthermore, since X is real, v_i is the complex conjugate of v_{n-i}, for $i = 1, 2, \ldots, n-1$. Thus

$$|v_i| = |v_{n-i}| \tag{2.3.107}$$

for $i = 1, 2, \ldots, n-1$, and A_gF is *symmetric* in the sense that its $(i+1)$th component is equal to its $(n-i+1)$th component, for each i. Alternatively, it can be seen that A_gF is a symmetric sequence since it is the DFT of the real symmetric sequence A_g.

To summarize, the discrete energy-density spectrum of the sequence X is the DFT of the aperiodic autocorrelation function of this sequence. The discrete energy-density spectrum and the aperiodic autocorrelation function are both real and symmetric sequences, and the components of the discrete energy-density spectrum are nonnegative.

2.3.9 Multiplication theorem

Let W and X be two n-component row vectors

$$W = w_0 \ w_1 \ldots w_g \ 0 \ldots 0 \tag{2.3.108}$$

and

$$X = x_0 \ x_1 \ldots x_g \ 0 \ldots 0 \tag{2.3.109}$$

where g is any nonnegative integer less than n. The $\{w_i\}$ and $\{x_i\}$ are real, and W is *not* now a function of X.

Let the n-component DFT's of W and X be

$$WF = u_0 \ u_1 \ldots u_{n-1} \tag{2.3.110}$$

and
$$XF = v_0 \ v_1 \ldots v_{n-1} \tag{2.3.111}$$

From Equation 2.3.75, nXF^{-1} is the complex conjugate of XF, so that
$$nXF^{-1} = v_0{}^* \ v_1{}^* \ldots v_{n-1}{}^* \tag{2.3.112}$$

It now follows that
$$\sum_{i=0}^{n-1} u_i v_i{}^* = WF(nXF^{-1})^T$$
$$= nWFF^{-1}X^T$$
$$= nWX^T \tag{2.3.113}$$

so that
$$\sum_{i=0}^{n-1} u_i v_i{}^* = n\sum_{i=0}^{g} w_i x_i \tag{2.3.114}$$

Taking the complex conjugate of each side of Equation 2.3.114,
$$\sum_{i=0}^{n-1} u_i{}^* v_i = n\sum_{i=0}^{g} w_i x_i \tag{2.3.115}$$

The relationships given by Equations 2.3.114 and 2.3.115 hold for *all* real vectors W and X satisfying Equations 2.3.108 and 2.3.109.

2.3.10   Signal energy

The *energy* of the real sequence given by the vector X in Equation 2.3.109 is defined as the *squared length* of X, and is given by a_g in Equation 2.3.95. The *energy* of the complex sequence XF in Equation 2.3.111 is defined as
$$\sum_{i=0}^{n-1} |v_i|^2 = \sum_{i=0}^{n-1} v_i v_i{}^* \tag{2.3.116}$$

and Equation 2.3.116 may be taken as the general definition of the energy of an n-component sequence whose $(i+1)$th component is v_i, whether the sequence is real or complex.

If the vector W in Equation 2.3.108 is set equal to X, then from Equation 2.3.114,
$$\sum_{i=0}^{n-1} |v_i|^2 = n\sum_{i=0}^{g} x_i{}^2 \tag{2.3.117}$$

so that the energy of XF is equal to n times the energy of X.

It is evident from Equations 2.3.95 and 2.3.99, that when $n \geqslant 2g + 1$, the energy of X is equal to the *first term* of A_g, the aperiodic autocorrelation function of X. Also, from Equations 2.3.106 and 2.3.116, the energy of XF is equal to the *sum* of the n components of $A_g F$, the discrete energy-density spectrum of X. Clearly, if $|v_i|^2 = 1$ for $i = 0, 1, \ldots, n-1$, then XF has n units of energy and X has one unit of energy.

2.4   SIGNAL DISTORTION

2.4.1   *Introduction*

In the synchronous serial digital data-transmission system shown in Figure 2.1, the linear baseband channel is said to introduce *signal distortion* when the sampled impulse-response of the channel has more than one nonzero component (sample). Unfortunately, the number and relative values of the nonzero components of the sampled impulse-response do not themselves directly indicate either the ease with which the distortion may be eliminated or the reduction in tolerance to additive noise resulting from the signal distortion. The purpose of this analysis is to describe the sampled impulse-response of the channel in terms of both *amplitude* distortion and *phase* distortion, since the magnitudes of these two parameters give a useful guide as to the severity of the signal distortion and the ease with which it may be eliminated. Since our prime concern here is with the analysis of the signal distortion, we shall consider the principles of *linear equalization* for each of the two types of distortion, only as far as is necessary to bring out some of the important properties of each type of distortion. We shall not, however, be concerned with the effects of noise on the received signal or with the various techniques for detecting the received signal in the presence of noise.

Suppose now that the sampled impulse-response of the baseband channel in Figure 2.1 is given by the n-component real row vector

$$Y_0 = y_0 \; y_1 \ldots y_g \; 0 \ldots 0 \qquad (2.4.1)$$

where
$$n \geqslant 2g + 1 \qquad (2.4.2)$$

Under these conditions, the aperiodic autocorrelation function of

Y_0 involves a sequence of no more than n components, and a *time* shift in the sequence Y_0, of up to g places (sampling intervals) in either direction, is identically equivalent to the corresponding *cyclic* shift.

The DFT of Y_0 is the n-component row vector

$$Y_0F = \lambda_0 \; \lambda_1 \ldots \lambda_{n-1} \tag{2.4.3}$$

where the $\{\lambda_i\}$ are the n eigenvalues of the $n \times n$ circulant matrix whose first row is Y_0 in Equation 2.4.1. The circulant matrix is, of course, the convolution matrix Y in Equation 2.2.20, where n is now given by Equation 2.4.2 and not Equation 2.2.19. From the analysis in Section 2.3.6, λ_0 is real, and

$$\lambda_i = \lambda_{n-i}^*, \quad i = 1, 2, \ldots, n-1 \tag{2.4.4}$$

where λ_{n-i}^* is the complex conjugate of λ_{n-i}.

The aperiodic autocorrelation function of the sequence Y_0 is the n-component row vector

$$B_g = b_g \; b_{g+1} \ldots b_{n-1} \; b_0 \ldots b_{g-1} \tag{2.4.5}$$

where

$$b_i = \sum_{h=-\infty}^{\infty} y_h y_{i-g+h} \tag{2.4.6}$$

and $y_i = 0$ for $i < 0$ and $i > g$. All the $\{b_i\}$ are real, b_g is positive, $b_i = 0$ for $i > 2g$, and the remaining $\{b_i\}$ may be positive or negative with magnitudes less than b_g. B_g is symmetric in the sense that

$$b_{g+i} = b_{g-i}, \quad i = 1, 2, \ldots, g \tag{2.4.7}$$

The discrete energy-density spectrum of the sequence Y_0 is the n-component row vector

$$B_gF = |\lambda_0|^2 \; |\lambda_1|^2 \ldots |\lambda_{n-1}|^2 \tag{2.4.8}$$

where, of course, the components of B_gF are all real and non-negative, and B_gF is symmetric in the sense that

$$|\lambda_i|^2 = |\lambda_{n-i}|^2, \quad i = 1, 2, \ldots, n-1 \tag{2.4.9}$$

It is clear from the previous analysis that none of the relationships summarized in Equations 2.4.1 to 2.4.9 is dependent upon the particular relationship between n and g, provided only that $n > 2g$. n may, of course, be odd or even.

A linear baseband channel, such as that in Figure 2.1, introduces four different effects on the transmitted signal. These are signal attenuation (or gain), signal delay, amplitude distortion and phase distortion. The first two of these are not distortion effects and it is convenient to consider signal attenuation together with amplitude distortion, and signal delay together with phase distortion, because of the obvious relationship between each pair of effects. Signal attenuation will be assumed to be $\geqslant 0$ dB or $\leqslant 0$ dB, so that this includes signal gain.

Both amplitude distortion and phase distortion will be *defined* in terms of the n components of $Y_0 F$, these being the values of the transfer function of the baseband channel and sampler at the n discrete frequencies $\{i/nT\}$, for $i = 0, 1, \ldots, n-1$. The values of the transfer function at other frequencies will be *ignored*. Y_0 is, of course, uniquely determined by $Y_0 F$, and vice versa, since F is nonsingular. This method of describing signal distortion leads to a simple and powerful technique for analysing its properties.

2.4.2   Phase distortion

When the baseband channel and sampler in Figure 2.1 introduce amplitude distortion, it means that the n discrete values of their transfer function, $J(i/nT)$ for $i = 0, 1, \ldots, n-1$, do not all have the same amplitude or modulus. But the n values of $J(i/nT)$ are the n components $\{\lambda_i\}$ of $Y_0 F$, as can be seen from Equation 2.3.34. Thus the presence of amplitude distortion implies that the $n\{|\lambda_i|\}$ do not all have the same value.

Pure phase distortion is here defined as signal distortion that includes no amplitude distortion and no signal attenuation. Suppose therefore that the baseband channel and sampler in Figure 2.1 introduce *no amplitude distortion* and *no signal attenuation*. The absence of amplitude distortion implies that, in Equation 2.4.3,

$$|\lambda_i| = c, \quad i = 0, 1, \ldots, n-1 \qquad (2.4.10)$$

where c is a positive constant. The absence of signal attenuation implies that the energy (or squared length) of Y_0 is equal to unity, so that

$$\sum_{i=0}^{g} y_i^2 = 1 \qquad (2.4.11)$$

It follows from Equations 2.4.10 and 2.3.117 that $c = 1$, so that now

$$|\lambda_i| = 1, \quad i = 0, 1, \ldots, n-1 \tag{2.4.12}$$

Also, since λ_0 is real, $\lambda_0 = \pm 1$.

Equation 2.4.12 is the condition that *must* be satisfied when the baseband channel introduces no amplitude distortion and no signal attenuation. It follows that, in the absence of both amplitude distortion and signal attenuation, each component $|\lambda_i|^2$ of the discrete energy-density spectrum in Equation 2.4.8 has the value unity. It is interesting to observe that a unit impulse has an energy spectral-density of unity over all frequencies (positive and negative), the unit impulse having, of course, infinite energy.

Since the $\{\lambda_i\}$ are the eigenvalues of the real $n \times n$ convolution matrix Y, it is clear that all the eigenvalues of Y are of modulus unity. From Equation 2.3.31,

$$Y = FDF^{-1} \tag{2.4.13}$$

where D is the $n \times n$ diagonal matrix whose $(i+1)$th component along the main diagonal is λ_i. Also

$$Y^T = F^{-1}DF \tag{2.4.14}$$

since the matrices F, F^{-1} and D are all symmetric. It follows from Equation 2.3.22 that

$$YY^T = FDF^{-2}DF = n^{-1}FDKDF \tag{2.4.15}$$

where K is the $n \times n$ permutation matrix

$$K = \begin{bmatrix} 1 & 0 & 0 & . & . & . & 0 & 0 \\ 0 & 0 & 0 & . & . & . & 0 & 1 \\ 0 & 0 & 0 & . & . & . & 1 & 0 \\ . & . & . & . & . & . & . & . \\ . & . & . & . & . & . & . & . \\ 0 & 0 & 1 & . & . & . & 0 & 0 \\ 0 & 1 & 0 & . & . & . & 0 & 0 \end{bmatrix} \tag{2.4.16}$$

But, for each i, $|\lambda_i| = 1$ and $\lambda_i = \lambda_{n-i}^*$. Also $\lambda_0 = \pm 1$. Thus

$$DKD = K \tag{2.4.17}$$

and, from Equations 2.4.15 and 2.3.22,

$$YY^T = n^{-1}FKF$$

$$= FF^{-2}F = I \tag{2.4.18}$$

where I is an $n \times n$ identity matrix. Thus

$$Y^T = Y^{-1} \qquad (2.4.19)$$

and Y is a *real orthogonal* matrix. The baseband channel and sampler in Figure 2.1 therefore perform an *orthogonal transformation* on the transmitted data signal.

Conversely, if Y is a real orthogonal matrix, all its eigenvalues have a modulus of unity, which, of course, means that all components of $Y_0 F$ have a modulus of unity and the baseband channel and sampler in Figure 2.1 introduce no amplitude distortion and no attenuation. It follows that a necessary and sufficient condition for no amplitude distortion and no attenuation is that the real convolution matrix Y is an *orthogonal* matrix.

Assume next that, in the synchronous serial data-transmission system of Figure 2.1, a sequence of m signal elements, in the form of the m impulses

$$\sum_{i=0}^{m-1} s_i \delta(t - iT) \qquad (2.4.20)$$

is fed to the input of the baseband channel. The sequence of element values may be represented more simply by the n-component row vector

$$S = s_0 \ s_1 \ldots s_{m-1} \ 0 \ldots 0 \qquad (2.4.21)$$

where now $n = m + 2g$. The sequence could be represented by just the first m components of S, but the reason for choosing the given value of n will become apparent later.

After transmission over the baseband channel and sampling at the channel output, the sequence (or vector) S is transformed into the sequence given by the n-component row vector

$$SY = \sum_{i=0}^{m-1} s_i Y_i \qquad (2.4.22)$$

where Y is the $n \times n$ circulant matrix, whose first row is the n-component row vector Y_0 in Equation 2.4.1, and whose $(i+1)$th row (for $i = 0, 1, \ldots, m+g-1$) is the n-component row vector

$$Y_i = \overbrace{0 \ldots 0}^{i} \ y_0 \ y_1 \ldots y_g \ \overbrace{0 \ldots 0}^{m+g-i-1} \qquad (2.4.23)$$

The last g of the n components of SY are all zero.

Since Y is an orthogonal matrix, the $n\{Y_i\}$ are orthogonal vectors, which means that

$$Y_h Y_i^T = 0 \qquad (2.4.24)$$

whenever $h \neq i$. Clearly, the m signal-elements $\{s_i Y_i\}$ at the output of the channel are orthogonal, so that the channel uses an orthogonal transformation to convert the m orthogonal signal-elements $\{s_i\}$ at the input of the channel into the m orthogonal signal-elements $\{s_i Y_i\}$ at its output.

For exact orthogonality to hold between the received signal-elements $\{s_i Y_i\}$, it is necessary that Y_0 has an infinite number of nonzero components, so that $g \to \infty$. In practice, however, a reasonable approximation to true orthogonality is usually obtained with values of $g > 8$. For convenience, exact orthogonality will be assumed.

Suppose now that the n sample values given by SY in Equation 2.4.22 are fed to a digital filter whose sampled impulse-response is given by the n-component row vector

$$W = y_g \ y_{g-1} \ldots y_0 \ 0 \ldots 0 \qquad (2.4.25)$$

W is derived from Y_0 by reversing the order of its first $g+1$ components. The sampled impulse-response of the filter is clearly the *reverse* of that of the baseband channel in Figure 2.1, so that the filter is *matched* to the channel and is therefore also *matched* to each individual signal-element $s_i Y_i$.

It can be seen that the sequence at the input to the digital filter has up to $m+g$ nonzero components, whereas the output sequence from the digital filter has up to $m+2g$ nonzero components. It is for this reason that n has been given the value $m+2g$.

Let M be the $n \times n$ circulant matrix whose first row is the vector W, so that M is the convolution matrix corresponding to the sampled impulse-response W. But

$$M = Y^T L \qquad (2.4.26)$$

where L is the $n \times n$ permutation matrix obtained from the identity matrix by a cyclic shift of g places to the right of its columns, so that $Y^T L$ is the $n \times n$ matrix obtained from Y^T by shifting its columns cyclically by g places to the right.

The received signal at the input to the digital filter is the set of n sample values given by the vector SY, and the output signal from the digital filter is therefore the set of sample values given by the n-component vector

$$SYM = SYY^TL = SL \qquad (2.4.27)$$

since, of course, Y is an orthogonal matrix. This means that the output signal from the digital filter is the sequence S shifted to the right by g places and so delayed in time by gT seconds.

From Equations 2.4.19 and 2.4.26,

$$M = Y^{-1}L \qquad (2.4.28)$$

where Y^{-1} is an orthogonal matrix. Thus the digital filter performs an orthogonal transformation on the received signal which is the *inverse* of that introduced by the channel, and the filter also delays the signal by gT seconds. Thus the data signal is restored to its original form, but delayed by g places. Clearly, the digital filter is not only *matched* to the channel but is also the *inverse* of the channel.

It is now of interest to consider in more detail the aperiodic autocorrelation function B_g and the discrete energy-density spectrum B_gF of the sampled impulse-response of the baseband channel, where the channel introduces no amplitude distortion and no signal attenuation.

The discrete energy-density spectrum B_gF in Equation 2.4.8 is here such that

$$|\lambda_i|^2 = 1, \quad i = 0, 1, \ldots, n-1 \qquad (2.4.29)$$

Thus it can be seen from Equation 2.3.34 that the sequence Y_0 (the sampled impulse-response of the baseband channel in Figure 2.1) now has a Fourier transform of modulus unity at each of the n discrete frequencies $\{i/nT\}$ and therefore unit energy spectral density at each of these frequencies. This implies that the baseband channel has unit gain (or no attenuation) at each frequency, which is to be expected in the absence of both amplitude distortion and signal attenuation.

Clearly, the aperiodic autocorrelation function B_g of the sequence Y_0 is the inverse DFT of the n-component sequence whose $(i+1)$th component is $|\lambda_i|^2$ in Equation 2.4.29, so that from Equations 2.4.5 and 2.3.16,

$$b_{g+i} = n^{-1}(1+\omega^i+\omega^{2i}+ \ldots +\omega^{(n-1)i}) \qquad (2.4.30)$$

for $-g \leqslant i \leqslant n-g-1$, bearing in mind that $\omega^{-hi} = \omega^{h(n-i)}$. When

$i \neq 0$, but assuming still that $-g \leqslant i \leqslant n-g-1$,

$$1+\omega^i+\omega^{2i}+ \ldots +\omega^{(n-1)i} = 0 \qquad (2.4.31)$$

since ω^i is here one of the nth roots of unity not including unity itself. Thus, when $i \neq 0$,

$$b_{g+i} = 0 \qquad (2.4.32)$$

When $i = 0$, b_{g+i} is the first component of B_g and has the value

$$b_g = 1 \qquad (2.4.33)$$

as can be seen from Equation 2.4.30. Hence the aperiodic auto-correlation function of the sequence Y_0 becomes

$$B_g = 1 \ 0 \ 0 \ldots 0 \qquad (2.4.34)$$

Since the $(i+1)$th component of B_g, for $i = 0, 1, \ldots, n-g-1$, is the value of the aperiodic autocorrelation function of the sequence Y_0, for a time interval (relative time delay) of i places, it follows from Equations 2.4.5, 2.4.6 and 2.4.34 that

$$\sum_{h=-\infty}^{\infty} y_h y_{i+h} = 0 \qquad (2.4.35)$$

when $i \neq 0$, assuming that $y_i = 0$ for $i < 0$ and $i > g$. This means that the sequence Y_0 is *orthogonal* to the sequence obtained from Y_0 by shifting its components one or more places to the left or right. Y_0 is here assumed to be shifted in *time* by a multiple of T seconds, where T is the sampling interval used for the sampled impulse-response of the baseband channel in Figure 2.1. Finally, Equation 2.4.33 shows that Equation 2.4.11 must hold. The result given by Equation 2.4.35 is entirely consistent with that given by Equation 2.4.24, and confirms the fact that the baseband channel performs an orthogonal transformation on the transmitted data signal.

It is interesting now to compare the previous analysis by matrix algebra, of the transmission of a finite sequence of m signal elements, with an analysis of the transmission of an *infinite* sequence of signal elements, in terms of the z transforms of the transmitted signals. Suppose therefore that, in the synchronous serial data-transmission system of Figure 2.1, a *continuous* sequence of signal elements is transmitted, such that the z transform of the sequence of impulses at the input to the baseband channel is

$$S(z) = \sum_i s_i z^{-i} \qquad (2.4.36)$$

The z transform of the sampled impulse-response of the baseband channel is

$$Y(z) = \sum_{i=0}^{g} y_i z^{-i} \qquad (2.4.37)$$

and the z transform of the sequence obtained after sampling the output signal from the baseband channel, at the time instants $\{iT\}$, is

$$S(z)Y(z) = \sum_{i} s_i z^{-i} Y(z) \qquad (2.4.38)$$

It is evident from Equation 2.4.35 that each individual signal-element in this sampled signal is orthogonal with respect to every other signal-element.

Suppose now that the sequence given by Equation 2.4.38 is fed to a digital filter whose sampled impulse-response is given by W in Equation 2.4.25 and whose z transform is

$$W(z) = \sum_{i=0}^{g} y_{g-i} z^{-i} \qquad (2.4.39)$$

The filter is *matched* to the channel, and, from Equation 2.4.37, $W(z)$ is obtained from $Y(z)$ by *reversing* the order of its coefficients.

It can be seen from Equations 2.3.92 and 2.3.93 that the coefficient of z^{-i} in $Y(z)W(z)$ is b_i, so that from Equation 2.4.34,

$$Y(z)W(z) = z^{-g} \qquad (2.4.40)$$

Thus the z transform of the output sequence from the filter is

$$\sum_{i} s_i z^{-i} Y(z)W(z) = \sum_{i} s_i z^{-i-g} \qquad (2.4.41)$$

This means that the sample value at the filter output, at time $t = (h+g)T$, is s_h, so that s_h can be detected from this sample value without interference from any of the other $\{s_i\}$.

From Equation 2.4.40,

$$W(z) = z^{-g} Y^{-1}(z) \qquad (2.4.42)$$

where $Y(z)Y^{-1}(z) = 1$. As before, the matched filter introduces a transformation on the received signal that is the *inverse* of that introduced by the channel and the filter also *delays* the signal by gT seconds (g places). The filter gives a sequence of orthogonal signal-elements at its output that are a delayed copy of the signal elements at the input to the channel, as can be seen from Equation 2.4.41.

With a finite value of g, Equations 2.4.40 to 2.4.42 hold only approximately, $Y^{-1}(z)$ being an infinite series (Section 4·3.)

It is clear from the preceding discussion that $W(z)$ represents an *orthogonal* transformation and is both the *reverse* and *inverse* of $Y(z)$, neglecting the delay introduced. Similarly, the *orthogonal* transformation $Y(z)$ is both the *reverse* and *inverse* of $W(z)$, again neglecting the signal delay. Thus it seems that just as an orthogonal matrix is one whose inverse is also its transpose, so the z transform corresponding to an orthogonal transformation is one whose inverse is also its reverse.

The important points emerging from the preceding discussion can now be summarized as follows. If the baseband channel and sampler in Figure 2.1 introduce no amplitude distortion and no signal attenuation, then $|\lambda_i| = 1$, for $i = 0, 1, \ldots, n-1$ where $n \geqslant 2g+1$, and every component $|\lambda_i|^2$ of the discrete energy-density spectrum of the sampled impulse-response of the channel has the value unity. The channel (including the sampler) may also introduce no phase distortion and no signal delay, in which case the sampled impulse-response of the channel is

$$Y_0 = 1 \; 0 \ldots 0 \qquad (2.4.43)$$

which represents perfect transmission. Alternatively, the channel may introduce a signal delay of iT seconds (i places) in which case

$$Y_0 = \overbrace{0 \ldots 0}^{i} \; 1 \; 0 \ldots 0 \qquad (2.4.44)$$

On the other hand, the channel may introduce *pure phase distortion* in which case the sequence Y_0 has *more* than one nonzero component (ideally an infinite number) but its discrete energy-density spectrum again has all its components equal to unity. The combination of these two conditions may be taken as a definition of pure phase distortion. The length of the vector Y_0 must here be equal to unity, to satisfy the condition of no signal attenuation.

Pure phase distortion is always an orthogonal transformation of the transmitted signal, and vice versa. The corresponding z transform is such that the reversal of the order of its coefficients gives the inverse orthogonal transformation together with the appropriate delay. A linear equalizer for a baseband channel introducing pure phase distortion is a filter matched to the channel. Its z transform is both the reverse and inverse of that of the channel (neglecting the delay involved), and it introduces itself both the reverse and inverse of the phase distortion introduced by the channel.

2.4.3 Representation of phase distortion by the poles and zeros of the z-transform

The baseband channel in Figure 2.1 has a sampled impulse-response Y_0 whose z-transform is

$$Y(z) = y_0 + y_1 z^{-1} + \ldots + y_g z^{-g}$$
$$= y_0 (1 - \alpha_1 z^{-1})(1 - \alpha_2 z^{-1}) \ldots (1 - \alpha_g z^{-1}) \quad (2.4.45)$$

where $\alpha_1, \alpha_2, \ldots, \alpha_g$ are the g *roots* of $Y(z)$, and may, of course, be real or complex. The channel can evidently be considered as g separate channels (or segments of the whole channel) connected in sequence and separated by samplers, the ith segment having the sampled impulse-response

$$X_i = 1 \quad -\alpha_i \ 0 \ldots 0 \quad (2.4.46)$$

with z-transform

$$X_i(z) = 1 - \alpha_i z^{-1} \quad (2.4.47)$$

The physical nature of this channel, when α_i is complex, is considered in Section 2.4.8. The n-component DFT of X_i is

$$X_i F = 1 - \alpha_i \ 1 - \alpha_i \omega^{-1} \ 1 - \alpha_i \omega^{-2} \ldots 1 - \alpha_i \omega^{-(n-1)} \quad (2.4.48)$$

whose $(h+1)$th component is $1 - \alpha_i \omega^{-h}$. As before, $n \geqslant 2g + 1$.

The $(h+1)$th component of the n-component DFT for the *whole* channel is the product of y_0 and the $(h+1)$th components of the DFT's of the g individual segments of the channel, and is

$$y_0 (1 - \alpha_1 \omega^{-h})(1 - \alpha_2 \omega^{-h}) \ldots (1 - \alpha_g \omega^{-h}) \quad (2.4.49)$$

for $h = 0, 1, \ldots, n-1$. From Equation 2.4.8, the $(h+1)$th component of the n-component discrete energy-density spectrum of Y_0 is

$$y_0^2 \ |1 - \alpha_1 \omega^{-h}|^2 \ |1 - \alpha_2 \omega^{-h}|^2 \ldots |1 - \alpha_g \omega^{-h}|^2 \quad (2.4.50)$$

Suppose now that the ith segment of the channel is *replaced* by one having the sampled impulse-response

$$W_i = -\alpha_i^* \ 1 \ 0 \ldots 0 \quad (2.4.51)$$

with z-transform

$$W_i(z) = -\alpha_i^* + z^{-1} \qquad (2.4.52)$$

where α_i^* is the complex conjugate of α_i. The n-component DFT of W_i is

$$W_iF = -\alpha_i^* + 1 \quad -\alpha_i^* + \omega^{-1} \quad -\alpha_i^* + \omega^{-2} \ldots -\alpha_i^* + \omega^{-(n-1)} \qquad (2.4.53)$$

and it can be seen that

$$|-\alpha_i^* + \omega^{-h}| = |1 - \alpha_i\omega^{-h}| \qquad (2.4.54)$$

for $h = 0, 1, \ldots, n-1$. Evidently the modulus (magnitude) of any component of W_iF is the *same* as that of the corresponding component of X_iF, and, from Expression 2.4.50, the change in the ith segment of the channel does not change the discrete energy-density spectrum of Y_0.

Now the *root* of $-\alpha_i^* + z^{-1}$ is the complex conjugate of the *reciprocal* of the root of $1 - \alpha_i z^{-1}$. Thus the replacement of one or more of the roots of $Y(z)$ by the complex conjugates of their reciprocals, with the corresponding change in Y_0, does not change the discrete energy-density spectrum of Y_0, so long as there is no change in the value of y_0 in Equations 2.4.45 and 2.4.50, which implies no change in the signal attenuation. The transformation causing this change must therefore be such that the n-component DFT of its sampled impulse-response has all its components with unit modulus, as in Equation 2.4.12. Thus the transformation is *orthogonal* and introduces *pure phase distortion*. It can be seen from Equation 2.4.45 that, in practice, the replacement of a *root* of $Y(z)$ by the complex conjugate of its reciprocal may or may not be accompanied by a change in the value of y_0 and therefore by a change in signal attenuation, since the value of y_0 does not affect the roots of $Y(z)$.

If m of the g roots of $Y(z)$, α_1 to α_m, are replaced by the complex conjugates of their reciprocals, the corresponding factor of $Y(z)$, which is

$$U(z) = (1 - \alpha_1 z^{-1})(1 - \alpha_2 z^{-1}) \ldots (1 - \alpha_m z^{-1}) \qquad (2.4.55)$$

is replaced by

$$V(z) = (-\alpha_1^* + z^{-1})(-\alpha_2^* + z^{-1}) \ldots (-\alpha_m^* + z^{-1}) \qquad (2.4.56)$$

$V(z)$ is thus obtained from $U(z)$ by first *reversing* the order of its coefficients and then replacing these by their complex conjugates.

Since we are here concerned with *real* signals, that is, with signal samples whose values have no imaginary components, the coefficients of $U(z)$ are all real. $V(z)$ is therefore obtained from $U(z)$ simply by reversing the order of the coefficients, and the roots of $V(z)$ are just the reciprocals of the roots of $U(z)$, complex roots of $U(z)$ or $V(z)$ occurring in complex-conjugate pairs. It now follows directly from the Reversal theorem (Section 2.3.7) that the discrete energy-density spectrum of the sequence given by $V(z)$ is the same as that of the sequence given by $U(z)$, so that $V(z)$ is obtained from $U(z)$ by means of a linear transformation that introduces pure phase distortion.

If $U(z)$ has a root α_i on the unit circle in the z plane, $|\alpha_i| = 1$, and, since $z = \exp(j2\pi fT)$, the factor $1 - \alpha_i z^{-1}$ of $U(z)$ has the value zero at a frequency f Hz such that $\exp(j2\pi fT) = \alpha_i$. In other words, a channel or filter with z-transform $1 - \alpha_i z^{-1}$ has infinite attenuation at the given frequency f Hz. The inverse of $1 - \alpha_i z^{-1}$ is $(1 - \alpha_i z^{-1})^{-1}$, which is such that

$$(1 - \alpha_i z^{-1})(1 - \alpha_i z^{-1})^{-1} = 1$$

Thus a channel or filter with z-transform $1 - \alpha_i z^{-1}$ in cascade with a channel or filter with z-transform $(1 - \alpha_i z^{-1})^{-1}$ have a resultant z transform of unity and therefore no attenuation or gain at any frequency. This implies that the channel or filter with z-transform $(1 - \alpha_i z^{-1})^{-1}$ has infinite gain at the given frequency f Hz, which is, of course, not physically realizable. Clearly, if $U(z)$ has one or more roots on the unit circle in the z plane, its inverse $U^{-1}(z)$, which is such that $U(z)U^{-1}(z) = 1$, is not physically realizable.

The z transform of a linear channel or filter that converts an input sequence of sample values with z-transform $U(z)$ into an output sequence with z-transform $V(z)$, and at the same time introduces a delay of hT seconds, where h is an appropriate nonnegative integer such as to make this physically realizable (or at least approximately so), is $z^{-h}U^{-1}(z)V(z)$, since

$$U(z)\ z^{-h}U^{-1}(z)V(z) = z^{-h}V(z) \qquad (2.4.57)$$

It can be seen from Equation 2.4.55 that $U^{-1}(z)$ is the product of the m factors $\{(1 - \alpha_i z^{-1})^{-1}\}$, for $i = 1, 2, \ldots, m$. Now, if $U(z)$ has any roots on the unit circle in the z plane, these same roots appear also in $V(z)$. But each of the linear factors of $V(z)$ (on the right-hand side of Equation 2.4.56) that contains one of these roots is accompanied by the *inverse* factor in $U^{-1}(z)$, so that, in $U^{-1}(z)V(z)$, the two factors cancel out. Clearly, $z^{-h}U^{-1}(z)V(z)$ does not imply infinite gain at any frequency and so is physically realizable. This can alternatively be seen from the fact that the transformation from $U(z)$ to $V(z)$ does not change any roots of $U(z)$ that lie on the unit circle, so that these roots are not, strictly speaking, involved in the transformation.

For reasons that will shortly be considered in more detail, the physically realizable inverse of a given z transform and the physically realizable inverse of the z transform obtained by reversing the order of the coefficients of the given z transform, are themselves such that one is obtained from the other by reversing the order of the coefficients, so long as each has the same number of terms. Thus, because $V(z)$ is obtained from $U(z)$ by reversing the order of its coefficients, a reversal in the order of the coefficients of $z^{-h}U^{-1}(z)$ gives $z^{-l}V^{-1}(z)$, and vice versa, where $z^{-h}U^{-1}(z)$ and $z^{-l}V^{-1}(z)$ are the physically realizable inverses of $U(z)$ and $V(z)$, respectively, (or at least approximate to them) and l is the appropriate non-negative integer.

It follows that a reversal in the order of the coefficients of $z^{-h}U^{-1}(z)V(z)$ gives $z^{-l}V^{-1}(z)U(z)$, and vice versa. But

$$z^{-h}U^{-1}(z)V(z)\,.\,z^{-l}V^{-1}(z)U(z) = z^{-h-l} \qquad (2.4.58)$$

so that, neglecting signal delay, $z^{-l}V^{-1}(z)U(z)$ is both the *reverse* and *inverse* of $z^{-h}U^{-1}(z)V(z)$, and $z^{-h}U^{-1}(z)V(z)$ is both the *reverse* and *inverse* of $z^{-l}V^{-1}(z)U(z)$. It can therefore be seen that the z-transform $z^{-h}U^{-1}(z)V(z)$ represents an *orthogonal transformation*, as does the z-transform $z^{-l}V^{-1}(z)U(z)$, and both z transforms represent *pure phase distortion*.

Furthermore, in order that the aperiodic autocorrelation function of Y_0 should satisfy Equation 2.4.34, which is the condition for the channel to introduce an orthogonal transformation on the transmitted signal and so to introduce pure phase distortion, it is *necessary* that the z transform formed by reversing the order of the coefficients of $Y(z)$, the z transform of Y_0, is also the inverse z-transform suitably delayed. But, as will presently be shown, this can *only* be satisfied if the z transform of Y_0 is

$$Y(z) = \pm z^{-h}X^{-1}(z)W(z) \qquad (2.4.59)$$

where $X(z)$ is an arbitrary z-transform with real coefficients, $W(z)$

is the z transform formed from $X(z)$ by reversing the order of its coefficients, and h is a nonnegative integer such as to make $Y(z)$ physically realizable (or at least approximately so). Thus *any* practical example of pure phase distortion can be represented by $Y(z)$ in Equation 2.4.59, $X^{-1}(z)$ now being taken as a finite series.

It can be seen that the modulus of the ith component of the DFT of the sequence with z-transform $z^{-h}X^{-1}(z)$, for each i, is the *reciprocal* of the modulus of the ith component of the sequence with z-transform $X(z)$. This follows because the ith component of the DFT of the sequence with z-transform

$$z^{-h}X^{-1}(z)X(z) = z^{-h}$$

is the product of the other two ith components and must clearly have a modulus of unity. Also, from the Reversal theorem (Section 2.3.7), the modulus of the ith component of the DFT of the sequence with z-transform $W(z)$ is *equal* to the modulus of the ith component of the DFT of the sequence with z-transform $X(z)$, for each i. But the ith component of the DFT of the sequence with z-transform $Y(z)$ is plus or minus the product of the ith components of the DFT's of the sequences with z-transforms $z^{-h}X^{-1}(z)$ and $W(z)$, as can be seen from Equation 2.4.59. Thus each component of the DFT of the sequence with z-transform $Y(z)$ has a modulus of unity, and $Y(z)$ represents pure phase distortion. Equation 2.4.59 represents a *sufficient* condition that must be satisfied by $Y(z)$ to ensure that it represents pure phase distortion. The fact that it is also a *necessary* condition can be seen most clearly by considering the location of the poles and zeros of $Y(z)$.

The values of z such that $Y(z) = 0$ are said to be the *zeros* of $Y(z)$ in the z plane, and the values of z such that $Y(z) \rightarrow \infty$ are said to be the *poles* of $Y(z)$. Clearly a *root* of $Y(z)$ is also a zero of $Y(z)$. Thus, if $1 - \alpha z^{-1}$ is a factor of $Y(z)$, $Y(z)$ has a zero at $z = \alpha$, and if $(1 - \alpha z^{-1})^{-1}$ is a factor of $Y(z)$, $Y(z)$ has a pole at $z = \alpha$. Obviously, a pole and zero at the same value of z cancel out, since the corresponding factors are the inverses of each other. When $Y(z)$ has one or more poles, it has ideally an infinite number of terms, but a good approximation to the ideal is obtained with a finite number of terms, except where there are one or more poles on the unit circle in the z plane. Again, if $(1 - \alpha z^{-1})^m$ is a factor of $Y(z)$, $Y(z)$ has m zeros at $z = \alpha$, and if $(1 - \alpha z^{-1})^{-m}$ is a factor of $Y(z)$, $Y(z)$ has m poles at $z = \alpha$. Hence, just as $Y(z)$ may have repeated roots, so there may be more than one pole or zero at any one value of z. A pole at the origin together with a zero at infinity correspond to a *delay* of one sampling interval of T seconds, since their presence means that z^{-1} is a factor of $Y(z)$. Similarly, a zero at the origin together with a pole at infinity correspond to an *advance* in time of one sampling interval, since their presence means that z is a factor of $Y(z)$. Because these poles and zeros at the origin and infinity represent simple shifts in time, they are neglected here when considering signal distortion.

It is well known that if the coefficients of a polynomial in z are reversed in order, without otherwise changing the polynomial, then the zeros (roots) of the polynomial, other than any at the origin or infinity, are transferred to the corresponding reciprocal values of z. Suppose for the moment that, in Equation 2.4.59, $X(z)$ has a finite number of terms and $W(z)$ is the polynomial formed from $X(z)$ by replacing each component z^{-i} by z^{i}, without however changing the coefficients. $W(z)$ is now obtained from $X(z)$ by reversing the order of its coefficients, the reversal being pivoted about the coefficient of z^0. This means, of course, that the sequence W (with z-transform $W(z)$) is obtained from the sequence X (with z-transform $X(z)$) by reversing the order of the latter sequence, the reversal being pivoted about the component in the sequence X at time $t = 0$. Since $X(z)$ has a finite number of terms, it has no poles, other than any at the origin or infinity, which are of course neglected here. Clearly, the zeros of $W(z)$ are given by transferring the zeros of $X(z)$ to the respective *reciprocal* values of z.

The *inverse* of $X(z)$ is $X^{-1}(z)$, which is such that $X(z)X^{-1}(z) = 1$. $X^{-1}(z)$ has as many poles as $X(z)$ has zeros, the poles of $X^{-1}(z)$ occurring at the same values of z as the zeros of $X(z)$. $X^{-1}(z)$ has no zeros, other than any at the origin or infinity. $W^{-1}(z)$ has also the same basic properties. It follows that the transfer of the poles of $X^{-1}(z)$ to the respective reciprocal values of z gives the poles of $W^{-1}(z)$. Furthermore, if there is a reversal in the order of the coefficients of $X(z)$, then, because $X(z)X^{-1}(z) = 1$, there must also be a reversal in the order of the coefficients of $X^{-1}(z)$. This means that $W^{-1}(z)$ is obtained from $X^{-1}(z)$ by reversing the order of its coefficients. Thus the reversal of the order of the coefficients of a polynomial containing only poles (other than any at the origin or infinity) transfers these poles to the corresponding reciprocal values of z. Since this applies also to zeros, the result may be extended to the more general case where a polynomial contains both poles and zeros. Hence, the *reversal* of the order of the coefficients of a z transform, transfers its poles and zeros to the corresponding reciprocal values of z. For comparison, the *inversion* of the z transform replaces its poles and zeros by zeros and poles, respectively.

It is clear now that in the general case, where $W(z)$ and $X(z)$ each contain both poles and zeros, $z^{-h}X^{-1}(z)W(z)$ in Equation 2.4.59 represents an orthogonal transformation and therefore pure phase distortion. Nevertheless, any poles in $W(z)$ are always accompanied by zeros in $X^{-1}(z)$ at the corresponding reciprocal values of z. $W(z)$ and $X^{-1}(z)$ can therefore always be redefined so that $W(z)$ contains only zeros and $X^{-1}(z)$ contains only poles. Any poles of $X^{-1}(z)$ on the unit circle always coincide with zeros of $W(z)$ and so do not appear in $Y(z)$.

The value of the transfer function $J(f)$ of the baseband channel and sampler in Figure 2.1, at a frequency f Hz, is the value of $Y(z)$ when $z = \exp(\mathrm{j}2\pi fT)$, and so is the value of $Y(z)$ at the point on the unit circle in the z plane making an angle of $2\pi fT$ radians with the positive real axis. Thus a pole or zero *near* the unit circle results in an increase or decrease, respectively, in the magnitudes of $J(f)$, at values of f that correspond to points on the unit circle close to the pole or zero. Furthermore, the closer the pole or zero is to the unit circle in the z plane, the greater is the increase or decrease in the magnitudes of $J(f)$ at these values of f. When the pole or zero lies *on* the unit circle, at $z = \mathrm{e}$ where of course $|\mathrm{e}| = 1$, then at the corresponding values of f, $J(f)$ is infinite or zero, respectively. The corresponding values of f are given by $\exp(\mathrm{j}2\pi fT) = \mathrm{e}$. A pole on the unit circle is not physically realizable, since it implies infinite gain at the corresponding frequencies.

It can be seen that $Y(z)$ in Equation 2.4.59 must have at least one *pole* not at the origin or infinity (except in the degenerate case where $Y(z)$ represents no distortion), so that $Y(z)$ is not a polynomial of finite degree. Thus pure phase distortion can only be represented exactly by an infinite sequence Y_0 and is only approximately realized by the finite sequence assumed in Expression 2.4.1. The approximation to the ideal infinite sequence is achieved through the use of the following simple relationships.

A pole may be represented approximately by a set of zeros, as follows. Consider a pole at $z = \alpha$, represented by the z transform $(1 - \alpha z^{-1})^{-1}$. If $|\alpha| < 1$,

$$(1-\alpha z^{-1})^{-1} \simeq 1+\alpha z^{-1}+\alpha^2 z^{-2}+ \ldots +\alpha^{n-1}z^{-(n-1)} \quad (2.4.60)$$

which has zeros (roots) at

$$z = \alpha\omega^i, \quad i = 1, 2, \ldots, n-1, \quad (2.4.61)$$

where

$$\omega^i = \exp\left(j\frac{2\pi i}{n}\right) \quad (2.4.62)$$

If $|\alpha| > 1$,

$$(1-\alpha z^{-1})^{-1} = -\alpha^{-1}z(1-\alpha^{-1}z)^{-1}$$

$$\simeq -\alpha^{-1}z(1+\alpha^{-1}z+\alpha^{-2}z^2+ \ldots$$

$$+\alpha^{-(n-1)}z^{n-1}) \quad (2.4.63)$$

The fact that the expansion in Equation 2.4.60 or 2.4.63 holds for $z = \pm 1$ and for any real value of α satisfying the given inequality,[6] implies that the expansion holds also for any value of z on the unit circle in the z plane and for any complex value of α satisfying the given inequality.

Since z^i is the z transform of an advance in time of iT seconds, and since we are here considering the z transform of the sampled impulse-response of the channel, the z transform given by Equation 2.4.63 is not physically realizable. However, if a delay of nT seconds is included in the response given by Equation 2.4.63, this has the z transform

$$z^{-n}(1-\alpha z^{-1})^{-1} \simeq -\alpha^{-1}z^{-(n-1)}(1+\alpha^{-1}z+\alpha^{-2}z^2+ \ldots$$

$$+\alpha^{-(n-1)}z^{n-1})$$

$$= -\alpha^{-n}(1+\alpha z^{-1}+\alpha^2 z^{-2}+ \ldots$$

$$+\alpha^{n-1}z^{-(n-1)}) \quad (2.4.64)$$

which is physically realizable. Furthermore, the zeros of this z transform are given by Equation 2.4.61, as before.

When $|\alpha| = 1$, the z transform $(1-\alpha z^{-1})^{-1}$ represents a pole on the unit circle in the z plane, and this is, of course, not physically realizable.

Evidently, a pole at $z = \alpha$ (where $|\alpha| \neq 1$), together with a delay of nT seconds when $|\alpha| > 1$, may be approximately represented by $n-1$ zeros, equally spaced on the circle of radius $|\alpha|$ in the z plane at the $n-1$ of the n possible values of $\alpha\omega^i$ that exclude α.

A z transform having one or more poles *inside* the unit circle in the z plane but no poles outside, can be realized exactly in practice by means of the appropriate feedback circuit, which is quite stable (Section 4.3.5). A z transform having one or more poles *outside* the unit circle, however, can only be realized exactly by means of an unstable feedback circuit, and to be physically realizable the z transform must here be accompanied by infinite signal delay. Nevertheless, as can be seen from Equations 2.4.60 and 2.4.64, in a z transform having poles both inside and outside the unit circle in the z plane, each factor representing a pole can be expressed approximately by the appropriate polynomial in z^{-1}, with a finite number of terms and of finite degree, so long as the necessary signal delay is introduced into the original (ideal) z transform, for each pole outside the unit circle. The approximation to the ideal z transform represents a system that is both stable and physically realizable, and the approximation may be made as close to the ideal as required by taking a sufficient number of terms in each polynomial that approximately represents a pole.

Although the z transform of any *finite* sequence has no poles but only zeros, the basic structure of the z transform of a sequence exhibiting phase distortion is more simply represented in terms of both poles and zeros, assuming therefore that the sequence is infinite. Where the z transform of the ideal infinite sequence has one or more poles, these appear in the z transform of the finite sequence, that approximates to it, as the corresponding sets of zeros given by Equation 2.4.61.

It has been shown that Equation 2.4.59 represents a *sufficient* condition to ensure that $Y(z)$ represents pure phase distortion. It remains now to consider whether or not the equation also represents a *necessary* condition. In $z^{-h}Y^{-1}(z)$, which is the inverse of $Y(z)$ suitably delayed, each pole and zero of $Y(z)$ (other than any at the origin and infinity) is replaced by a zero or pole, respectively. But, if $Y(z)$ represents pure phase distortion, it must be such that the z transform formed from $Y(z)$ by reversing the order of its coefficients is also $z^{-h}Y^{-1}(z)$. Furthermore, the reversal of the order of the coefficients of $Y(z)$ transfers each pole and zero to the corresponding reciprocal value of z. For these conditions to be satisfied, each pole and zero of $Y(z)$ must be accompanied by a zero or pole, respectively, at the reciprocal value of z, and for this to apply, $Y(z)$ must satisfy Equation 2.4.59 modified by the inclusion of an arbitrary multiplying real constant c on the right-hand side. However, since it is assumed here that $Y(z)$ introduces no attenuation, $c = 1$. Thus Equation 2.4.59 is a condition that must *necessarily* be satisfied by $Y(z)$.

Since pure phase distortion is represented by the z-transform $Y(z)$ in Equation 2.4.59, it is such that each pole and zero in the z transform is accompanied by a zero or pole, respectively, at the reciprocal value of z. Since the coefficients of the z transform are all *real*, all poles and all zeros at complex values of z occur in *complex conjugate pairs*, so that each pole and zero is also accompanied by a zero or pole, respectively, at the *complex conjugate* of the reciprocal value of z. This latter condition applies to $Y(z)$, when it represents pure phase distortion and when its coefficients may be real or complex. It therefore applies to both *real* and *complex* signals and so is of more general application than the condition previously given, which of course applies only to real signals.

In conclusion, when a filter introducing pure phase distortion is connected in cascade with a channel, and when the poles of the z transform of the filter coincide with zeros of the z transform of the channel, then the action of the filter is to replace these zeros by the corresponding set of zeros at the reciprocal values of z. Thus the replacement of one or more zeros in a z transform, by zeros at the reciprocal values of z, is pure phase distortion and is therefore an orthogonal transformation, as has already been shown.

When the filter is matched to the baseband channel introducing pure phase distortion, the poles and zeros of the filter z-transform coincide with the zeros and poles, respectively, of the channel z-transform, so that the channel and filter in cascade have a z transform with no poles or zeros (other than poles at the origin and zeros at infinity, corresponding to the delay in transmission). Thus the matched filter is also a *linear equalizer* for the channel, since it equalizes or corrects the channel response. The equalizer has a z transform with an infinite number of terms and usually it can only be realized approximately in practice. Furthermore, the equalizer always introduces some signal delay.

2.4.4   Amplitude distortion

Pure amplitude distortion is here defined as signal distortion that includes no phase distortion and no signal delay. Suppose therefore that the baseband channel and sampler in Figure 2.1 introduce *no phase distortion* and *no signal delay*.

When the transfer function $J(i/nT)$ of the baseband channel and sampler has a *complex* value at one or more of the n discrete frequencies $\{i/nT\}$, this necessarily implies phase shifts at the corresponding frequencies and therefore *some* combination of phase distortion and signal delay in the transmitted signal. Thus for no phase distortion and no signal delay, the n discrete values of $J(i/nT)$, for $i = 0, 1, \ldots, n-1$, must all be *real*. But the n values of $J(i/nT)$ are the n components $\{\lambda_i\}$ of Y_0F, as can be seen from Equation 2.3.34, where of course Y_0F is the DFT of the sampled impulse-response Y_0 of the baseband channel in Figure 2.1. Thus in the absence of both phase distortion and signal delay, the $\{\lambda_i\}$ must all be real and so must be the vector Y_0F. But from Equation 2.4.4, λ_i and λ_{n-i} are complex conjugates, so that now $\lambda_i = \lambda_{n-i}$, for each i. This means that Y_0F is real and symmetric. It is evident from the analysis in Section 2.3.6 that, under these conditions, if the sampled impulse-response of the baseband channel in Figure 2.1 is given by the n-component row vector

$$Y_0 = y_0 \ y_1 \ldots y_{n-1} \qquad (2.4.65)$$

then $$y_i = y_{n-i}, \quad i = 1, 2, \ldots, n-1 \qquad (2.4.66)$$

and the vector Y_0 is *real* and *symmetric*. Conversely, as is shown in Section 2.3.6, if Y_0 is real and symmetric, then so also is Y_0F.

It is assumed here that in the symmetric sequence Y_0 the components have been shifted cyclically to the left to make y_0 the *central* component of the original sequence. This corresponds to an advance in time of the sampled impulse-response of the channel, to bring the central sample of the symmetric sequence to the time instant $t = 0$. Clearly, this is not physically realizable, and in practice, amplitude distortion must always be accompanied by the corresponding signal delay, to make the distortion physically realizable. It is however simpler to study amplitude distortion in the absence of signal delay, since the delay terms are now eliminated from Y_0F. The signal delay will therefore be ignored, bearing in mind that the system can always be made physically realizable by reintroducing the appropriate delay.

It is clear from Equations 2.4.1 and 2.4.65 that the sampled impulse-response of the baseband channel in Figure 2.1 can now be rewritten as the n-component vector

$$Y_0 = y_{\frac{1}{2}g} \ldots y_g \; 0 \ldots 0 \; y_0 \ldots y_{\frac{1}{2}g-1} \qquad (2.4.67)$$

where $n \geqslant 2g + 1$, g is even, and

$$y_{\frac{1}{2}g+i} = y_{\frac{1}{2}g-i}, \quad i = 1, 2, \ldots, \tfrac{1}{2}g \qquad (2.4.68)$$

The $n \times n$ convolution matrix Y corresponding to the sampled impulse-response Y_0 in Equation 2.4.67 is the $n \times n$ circulant matrix whose first row is the vector Y_0. It is evident from Equation 2.4.68 that Y is now a *real symmetric* matrix, and this may be shown more rigorously as follows.

In the presence of no phase distortion and no delay, the $\{\lambda_i\}$, which are the eigenvalues of Y, must all be real. Thus, from Equations 2.3.31 and 2.3.74,

$$Y = FDF^{-1} = n^{-1}FDF^* \qquad (2.4.69)$$

where D is the real diagonal matrix whose $(i+1)$th component along the main diagonal is λ_i, and F^* is the complex conjugate of F. Since F, D and F^* are symmetric matrices, it follows that

$$Y^T = n^{-1}F^*DF \qquad (2.4.70)$$

But, since D is real,

$$F^*DF = (FDF^*)^* \qquad (2.4.71)$$

so that

$$Y^T = Y^* = Y \qquad (2.4.72)$$

since Y is real. Thus Y is a symmetric matrix.

Conversely, if Y is a real symmetric matrix, all its eigenvalues are real and the baseband channel and sampler in Figure 2.1 introduce no phase distortion and no delay. It follows that a necessary and sufficient condition for no phase distortion and no delay is that the real convolution matrix Y is a *symmetric* matrix. This means that the sequence Y_0 is symmetric about its component at time $t = 0$, so that a reversal in time of the sequence, pivoted about this component, does not change the sequence.

For comparison, a necessary and sufficient condition for no amplitude distortion and no attenuation is that the real convolution matrix Y is an *orthogonal* matrix, which means that a reversal in time of the sequence Y_0, pivoted about its component at time $t = 0$, gives the *inverse* sequence.

Another important distinction between amplitude and phase effects is that amplitude distortion and signal attenuation are always accompanied by a change both in the discrete energy-density spectrum and in the aperiodic autocorrelation function of the sampled impulse-response of the channel, from their ideal values in Equations 2.4.29 and 2.4.34, respectively, whereas phase distortion and signal delay are not accompanied by a change either in the discrete energy-density spectrum or in the aperiodic autocorrelation function.

As before, it is instructive to consider the transmission of a continuous sequence of signal elements. Assume therefore that, in the synchronous serial data-transmission system of Figure 2.1, a continuous sequence of signal elements is transmitted, such that the z transform of the sequence of impulses at the input to the baseband channel is $S(z)$ in Equation 2.4.36. The z transform of the sequence obtained after sampling the output signal from the baseband channel, is $S(z) Y(z)$ in Equation 2.4.38, where $Y(z)$ in Equation 2.4.37 is the z transform of the sampled impulse-response of the channel. This is assumed to include sufficient delay to be physically realizable.

Suppose now that the received sequence with z-transform $S(z) Y(z)$ is fed to a digital filter with z-transform $z^{-h} Y^{-1}(z)$, where z^{-h} is the z transform of the delay of hT seconds needed to make the filter response physically realizable. The output sequence from the digital filter has the z-transform

$$S(z)Y(z)z^{-h}Y^{-1}(z) = z^{-h}S(z)$$

$$= \sum_i s_i z^{-i-h} \qquad (2.4.73)$$

so that this is simply a delayed version of the transmitted sequence of values, and s_l can be detected from the sample value at time $t = (l+h)T$, without interference from the other $\{s_i\}$.

The digital filter with z-transform $z^{-h} Y^{-1}(z)$ is a *linear equalizer* for the channel, since it equalizes or corrects the channel response. Contrary to the case where the channel introduces pure phase distortion, the sampled impulse-response of the linear equalizer here is not very simply related to the sampled impulse-response of the channel. The equalizer is *not* matched to the channel.

2.4.5   *Representation of amplitude distortion by the poles and zeros of the z transform*

Assume that the baseband channel introduces pure amplitude distortion (with no delay), so that it is, of course, not physically realizable. The sampled impulse-response Y_0 of the channel is now symmetric about its central component which occurs at time $t = 0$, and $Y(z)$ contains some positive powers of z and only real coefficients. It is clear that the reversal of the order of the coefficients of $Y(z)$, without changing the position of the coefficient of z^0 so that the reversal is pivoted about this coefficient, gives $Y(z)$ again. As will presently be shown, this necessarily implies that

$$Y(z) = \pm X(z)W(z) \qquad (2.4.74)$$

where $W(z)$ is obtained from $X(z)$ by reversing the order of its coefficients. The first term of $X(z)$ is assumed to contain the component z^0, and $W(z)$ is obtained from $X(z)$ through replacing each component z^{-i} by z^i without however changing the coefficient. Both $W(z)$ and $X(z)$ have the same number of terms and only real coefficients, and the values of z at the zeros of $W(z)$ are the *reciprocals* of those at the zeros of $X(z)$. Thus all zeros of $Y(z)$ occur in *reciprocal pairs*, and $X(z)$ can always be selected to have all its zeros inside the unit circle in the z plane and $W(z)$ to have all its zeros outside the unit circle. If $W(z)$ and $X(z)$ have an infinite number of terms, then they may have poles as well as zeros. In this case the values of z at the poles of $W(z)$ are the reciprocals of those at the poles of $X(z)$, and again $X(z)$ can always be selected to have all its poles inside the unit circle and $W(z)$ to have all its poles outside. Since all coefficients of $Y(z)$ are *real*, all its poles and all its zeros at complex values of z occur not only in reciprocal pairs but also in complex-conjugate pairs. Clearly, each zero is accompanied by another zero at the complex conjugate of the reciprocal value of z, and similarly for each pole. This condition is in fact satisfied by both *real* and *complex* signals and is therefore of more general application than the simpler condition previously described, which applies only to real signals.

Obviously, the sequence W (with z-transform $W(z)$) is the time reversal of the sequence X (with z-transform $X(z)$), the sequence W being reversed with respect to time $t = 0$ by pivoting the sequence X about its first component. $Y(z)$ is now the z transform of the sequence Y_0 which is symmetric about its component at time $t = 0$. It follows from the Reversal theorem (Section 2.3.7) that the ith component of the DFT of the sequence W is the complex conjugate of the ith component of the DFT of the sequence X, for each i. But the ith component of the DFT of the sequence Y_0 is plus or minus the product of the ith components of the DFT's of the sequences X and W, as can be seen from Equation 2.4.74. Thus each component of the DFT of the sequence Y_0 is real, and $Y(z)$ represents pure amplitude distortion. All sequences here are finite.

Equation 2.4.74 clearly represents a *sufficient* condition that must be satisfied by $Y(z)$ to ensure that it represents pure amplitude distortion. A *necessary* condition for pure amplitude distortion, that must be satisfied by the real sequence Y_0, is that it is symmetric about its central component which occurs at time $t = 0$. Under this condition, the reversal of the order of the coefficients of $Y(z)$, without changing the position of the coefficient of z^0, leaves $Y(z)$ unchanged. However, this reversal must transfer each pole and zero of $Y(z)$ to the corresponding reciprocal value of z. For this not to change $Y(z)$, each pole and zero of $Y(z)$ must be accompanied by a pole or zero, respectively, at the reciprocal value of z, and, for this to be true, $Y(z)$ must satisfy Equation 2.4.74. The z-transforms $W(z)$ and $X(z)$ can now always be found to satisfy both Equation 2.4.74 and the other assumed conditions. The requirement that the sequence W be obtained from the reversal of the sequence X, by pivoting this about its component at time $t = 0$, follows from the fact that the *central* component of Y_0 is at time $t = 0$. Clearly Equation 2.4.74 is a condition that must *necessarily* be satisfied by $Y(z)$.

Any channel (and sampler) together with the appropriate matched filter have a z transform that is the product of the z transforms of the channel and matched filter. Furthermore, since the sampled impulse-response of the matched filter is the reverse of that of the channel, the values of z at the poles and zeros of the z transform of the matched filter are the reciprocals of those at the poles and zeros, respectively, of the z transform of the channel. Thus all poles and all zeros in the z transform of the channel and matched filter occur in reciprocal pairs, so that the channel and matched filter together introduce pure amplitude distortion. The resultant sampled impulse-response is symmetric and is, of course, the aperiodic autocorrelation function of the channel itself. Clearly, a matched filter eliminates all phase distortion. It is not, however, a pure phase equalizer except in the particular case where the channel introduces pure phase distortion. Whenever the channel introduces some amplitude distortion, the matched filter always increases this distortion.

It is clear that the z-transform $z^{-h}Y^{-1}(z)$ of the linear equalizer for the channel introducing pure amplitude distortion may be obtained from $Y(z)$ by replacing all zeros by poles and all poles by zeros, and, of course, introducing a delay of hT seconds. Thus in $z^{-h}Y^{-1}(z)$, all poles and all zeros occur in reciprocal pairs and the sampled impulse-response of the linear equalizer is symmetric. Also, the channel and linear equalizer connected in cascade have a z transform with no poles or zeros (other than poles at the origin and zeros at infinity, corresponding to the delay in transmission). A linear equalizer for a practical channel introducing pure amplitude distortion has a z transform with an infinite number of terms and this can only be realized approximately in practice. Furthermore the equalizer always introduces some signal delay.

2.4.6   Severe amplitude distortion

Suppose that the sampled impulse-response of the baseband channel in Figure 2.1 is given by the n-component row vector

$$Y_0 = y_0 \ y_1 \ldots y_g \ 0 \ldots 0 \tag{2.4.75}$$

with z-transform

$$Y(z) = y_0 + y_1 z^{-1} + y_2 z^{-2} + \ldots + y_g z^{-g}$$
$$= y_0(1 - \alpha_1 z^{-1})(1 - \alpha_2 z^{-1}) \ldots (1 - \alpha_g z^{-1}) \tag{2.4.76}$$

where the $\{\alpha_i\}$ are the g zeros (roots) of $Y(z)$ and $n \geqslant g + 1$. Assume also that y_0 and y_g are both nonzero. This means that the delay in transmission is neglected, so that the first nonzero component of the sampled impulse-response occurs at the time instant $t = 0$. Furthermore $Y(z)$ is physically realizable.

From Equation 2.3.34, the value of the transfer function of the baseband channel and sampler, at the frequency h/nT, for $h = 0$, $1, \ldots, n-1$, is

$$\lambda_h = Y_0 F_h^T \tag{2.4.77}$$

where λ_h is the $(h+1)$th component of the n-component DFT of Y_0, and

$$F_h = 1 \quad \omega^{-h} \quad \omega^{-2h} \ldots \omega^{-(n-1)h} \qquad (2.4.78)$$

Consider now the case where, for some integer h in the range 0 to $n-1$, $\lambda_h = 0$. Thus, from Equation 2.4.77,

$$\lambda_h = y_0 + y_1 \omega^{-h} + y_2 \omega^{-2h} + \ldots + y_g \omega^{-gh} = 0 \qquad (2.4.79)$$

This represents the total loss in transmission of the signal frequency-component at h/nT Hz and therefore corresponds to severe amplitude distortion whenever the transmitted signal has a significant frequency component at h/nT Hz.

Since the coefficients of the polynomial in Equation 2.4.79 are the same as those in Equation 2.4.76, it is clear from these equations that now

$$\lambda_h = y_0(1 - \alpha_1 \omega^{-h})(1 - \alpha_2 \omega^{-h}) \ldots (1 - \alpha_g \omega^{-h}) = 0 \qquad (2.4.80)$$

It follows that if $\lambda_h = 0$, one or more of the $\{\alpha_i\}$ must satisfy

$$1 - \alpha_i \omega^{-h} = 0 \qquad (2.4.81)$$

so that

$$\alpha_i = \omega^h = \exp\left(j\frac{2\pi h}{n}\right) \qquad (2.4.82)$$

where $j = \sqrt{-1}$. In other words, one or more of the zeros (roots) of $Y(z)$ must lie on the unit circle in the z plane, at the point $\exp\left(j\frac{2\pi h}{n}\right)$. Conversely, the latter condition ensures that $\lambda_h = 0$.

It must be borne in mind that $\omega^h = \omega^{h+ln}$ for any integer l, so that no distinction is made between ω^h and ω^{h+ln} and therefore between the frequencies h/nT and $(h+ln)/nT$ Hz. Of course, with an n-component DFT and a sampling interval of T seconds, as in Figure 2.1, the discrete frequency components of the data signal, given by the DFT, are themselves all integral multiples of $1/nT$ Hz. We shall therefore consider only the frequencies $\{h/nT\}$, for $h = 0, 1, \ldots, n-1$.

Suppose next that *none* of the $\{\alpha_i\}$ satisfy Equation 2.4.81. Under these conditions,

$$1 - \alpha_i \omega^{-h} \neq 0 \qquad (2.4.83)$$

for $i = 1, 2, \ldots, g$, and therefore $\lambda_h \neq 0$, as can be seen from Equation 2.4.80.

Thus $\lambda_h = 0$, if and only if one or more of the zeros of $Y(z)$ lie on the unit circle in the z plane, at the point $\exp\left(j\frac{2\pi h}{n}\right)$.

Since λ_h is one of the n eigenvalues of the $n \times n$ convolution matrix Y, in Equation 2.2.20, it is clear that, when $\lambda_h = 0$, Y becomes singular. For the given value of n, there is not now an inverse convolution matrix Y^{-1} by means of which the signal distortion introduced by the baseband channel in Figure 2.1 can be eliminated. Again, there is not now a physically realizable z-transform approximating to $z^{-l}Y^{-1}(z)$ for any nonnegative integer l, since this must have one or more poles on the unit circle in the z plane and therefore infinite gain at the corresponding frequencies. The fact that $\lambda_h = 0$ means that any frequency component of the transmitted signal, at h/nT Hz, is reduced to zero by the baseband channel and cannot therefore be restored again by any *linear* process at the receiver. Notice that $z^{-l}Y^{-1}(z)$ is not physically realizable whenever $Y(z)$ has one or more zeros *anywhere* on the unit circle.

If no zeros of $Y(z)$ lie on the unit circle in the z plane, $\lambda_h \neq 0$ for each integer h and any $n \geqslant g+1$, so that Y is a nonsingular matrix and therefore has an inverse. Again, $Y^{-1}(z)$ now has no poles on the unit circle so that there is a physically realizable z-transform approximating to $z^{-l}Y^{-1}(z)$ for the appropriate nonnegative integer l.

It follows that a necessary and sufficient condition for a channel with z-transform $Y(z)$ to be capable of being equalized *linearly* is that $Y(z)$ has no zeros (roots) on the unit circle in the z plane. This is a most important result.

Suppose now that $Y(z)$ has the real factor

$$X(z) = x_0 + x_1 z^{-1} + x_2 z^{-2} + \ldots + x_l z^{-l} \qquad (2.4.84)$$

all of whose l zeros lie on the unit circle, the remaining factor of $Y(z)$ having *no* zeros on the unit circle. Since the $\{x_i\}$ are real, the zeros of $X(z)$ are ± 1 or complex-conjugate pairs of unit modulus.

Suppose next that $n \geqslant 2g+1$ and $Y(z)$ represents pure amplitude distortion together with the appropriate delay to make it physically realizable. This means that g is even and $z^{\frac{1}{2}g}Y(z)$ represents pure amplitude distortion (with no delay). It is evident that $X(z)$ must now also represent pure amplitude distortion together with the appropriate delay to make it physically realizable, which means that l is even and $z^{\frac{1}{2}l}X(z)$ represents pure amplitude distortion (with no delay). Thus

$$x_{\frac{1}{2}l+i} = x_{\frac{1}{2}l-i}, \quad i = 1, 2, \ldots, \tfrac{1}{2}l \qquad (2.4.85)$$

It can be seen that for these conditions to hold, the zeros of $X(z)$ must all be in reciprocal pairs which are now also complex-conjugate pairs and may of course include one or more *pairs* of zeros at 1 or -1.

In the general case where $Y(z)$ does not represent pure amplitude distortion and delay, l need not be even and it is possible for $X(z)$ to contain an *odd* number of zeros at 1 or -1. Under these conditions the $\{x_i\}$ do not form a symmetric sequence (as defined in Section 2.3.6), and, with n even, $z^h X(z)$ necessarily introduces phase distortion in addition to amplitude distortion, *whatever* the value of the integer h. The phase distortion includes a time shift of $\frac{1}{2}T$.

Clearly, whenever $Y(z)$ represents pure amplitude distortion and delay, its zeros, whether off or on the unit circle, occur in *reciprocal pairs*, all complex zeros being at the same time in *complex-conjugate pairs*. The corresponding sampled impulse-response of the baseband channel is always symmetric about the central sample value at time $t = \frac{1}{2}gT$, g of course being even.

2.4.7 Assessment of the distortion introduced by a linear baseband channel

Consider the z transform of the sampled impulse-response of the baseband channel in Figure 2.1, where the z transform may have an infinite number of terms and is represented by means of its poles and zeros in the z plane.

In the case of pure amplitude distortion, all poles and all zeros occur in reciprocal pairs, whereas with pure phase distortion, each pole is accompanied by a zero at the reciprocal value of z, and each zero by a pole. All poles and all zeros at complex values of z occur in complex-conjugate pairs.

Any linear baseband channel whose output signal is sampled at regular intervals of time, as in Figure 2.1, can be considered to be composed of a number of separate segments, each of which may be represented by the appropriate digital filter. Since the z transform of the channel is the *product* of the z transforms of its segments, the poles and zeros of the z transform of the channel are given by the *sum* of the poles and zeros of its individual segments.

Suppose now that the channel is composed of two segments. The z transform of the first of these has just two poles or two zeros, which occur at reciprocal real values of z. This segment clearly introduces pure amplitude distortion. The z transform of the second segment has a pole at one of the two reciprocal real values of z and a zero at the other. This segment clearly introduces pure phase distortion. The z transform of the channel now has just two poles or two zeros at one of the two values of z, since the coincident pole and zero cancel out. Furthermore the channel introduces both amplitude and phase distortions. It follows that any two coincident poles or zeros at a real value of z, with no pole or zero at the reciprocal value of z, are formed by the combination of the corresponding amplitude and phase distortions. Thus any single pole or zero at a real value of z, with no pole or zero at the reciprocal value of z, represents a z transform which when multiplied by itself is a combination of amplitude and phase distortions, so that any such pole or zero itself represents a combination of amplitude and phase distortions. Although the resultant levels of the amplitude and phase distortions are determined by the distribution of *all* poles and zeros in the z plane, any single pole or zero at a real value of z, with no pole or zero at the reciprocal value of z, implies the presence of at least *some* amplitude distortion and *some* phase distortion.

Suppose now that the z transform of the first segment of the channel has four poles or zeros at two pairs of complex-conjugate values of z which also form two pairs of reciprocal values of z. The z transform of the second segment has two poles and two zeros at the same four values of z, both poles and zeros being at complex-conjugate values of z, and the values of z at the poles being the reciprocals of those at the zeros. The first segment clearly introduces pure amplitude distortion and the second introduces pure phase distortion. By an exactly similar line of reasoning to that used before, it can be seen that any pair of poles or zeros at complex-conjugate values of z, not accompanied by a pair of poles or zeros at the reciprocal values of z, imply the presence of at least *some* amplitude distortion and *some* phase distortion.

In the most general case, which applies to both real and complex signals (Section 2.4.8), a z transform having *any* poles or zeros, not accompanied by poles or zeros at the complex conjugates of the reciprocal values of z, represents a *combination* of amplitude and phase distortions. Furthermore, in the z transform of the sampled impulse-response of any linear baseband channel, any two poles or zeros at values of z, one of which is the complex conjugate of the reciprocal of the other, imply the presence of some amplitude distortion, and any pole or zero having a zero or pole, respectively, at the complex conjugate of the reciprocal value of z implies the presence of some phase distortion. If there are no poles or zeros, except at the origin and at infinity, then there is no signal distortion, but there may of course be signal attenuation and delay.

So long as the z transform of the sampled impulse-response of the channel does not have too many poles and zeros, a study of the distribution of these can indicate not only the predominant type of distortion but also its severity. Some idea of the latter can be obtained by noting the following two points. Firstly, two poles or zeros at values of z, one of which is the complex conjugate of the reciprocal of the other, each represent the same *level* of distortion. This follows from the fact that when a pole or zero is transferred from a value of z to the complex conjugate of its reciprocal, the corresponding *factor* of the z transform is replaced by the complex conjugate of the *reversed* factor, and there is no change in the discrete energy-density spectrum of the signal. Secondly, the nearer a pole or zero is to the unit circle in the z plane, the greater is the corresponding signal distortion. In particular, a pole or zero at a value of z such that $\frac{1}{2} < |z| < 2$ represents appreciable signal distortion, the distortion often increasing rapidly with the *number* of poles or zeros lying within this area. Poles in the same neighbourhood tend to reinforce each other, as do zeros, but poles and zeros in the same neighbourhood tend to cancel. Poles or zeros, regularly spaced over the circumference of a circle of any radius and with its centre at the origin, tend to cancel, the degree of cancellation increasing with the number of poles or zeros.

The signal distortion introduced by a linear baseband channel is perhaps best studied by considering the channel to be composed of two constituent channels, A and B, connected in cascade and with a sampler in between, where channel A introduces pure phase distortion and channel B introduces pure amplitude distortion.

If the n-component DFT of the channel A is

$$\phi_0 \ \phi_1 \ldots \phi_{n-1} \tag{2.4.86}$$

and the n-component DFT of the channel B is

$$\theta_0 \ \theta_1 \ldots \theta_{n-1} \tag{2.4.87}$$

then $$\phi_i \theta_i = \lambda_i, \quad i = 0, 1, \ldots, n-1 \tag{2.4.88}$$

where, of course, the n-component DFT of the whole baseband channel is

$$Y_0 F = \lambda_0 \ \lambda_1 \ldots \lambda_{n-1} \qquad (2.4.89)$$

Furthermore,

$$\theta_i = \theta_{n-i} = \pm |\lambda_i| \qquad (2.4.90)$$

and

$$\phi_i = \phi^*_{n-i} = \theta_{n-i} = \pm |\lambda_i| \frac{\lambda_i}{\theta_i} \qquad (2.4.91)$$

for each i, $|\phi_i|$ being held at unity when $\lambda_i = 0$. Ideally $n \rightarrow \infty$ here, since n must be large enough to accommodate the convolution of the sampled impulse-responses of channels A and B, channel A having a sampled impulse-response with theoretically an infinite number of components, as also often does channel B. Thus, in practice, $n \geqslant 2g + 1$.

It can be seen that the sampled impulse-response of each of the channels A and B is real, but for the purposes of the present analysis it is not necessary that either channel be physically realizable.

It will now be shown that when the baseband channel and sampler in Figure 2.1 have a sampled impulse-response with a finite number of nonzero values (all of which are real), the baseband channel and sampler can always be split into the two separate channels A and B, as just described. The analysis demonstrates some interesting properties of these channels.

Suppose that the z transform of the baseband channel and sampler is

$$Y(z) = y_0(1 - \alpha_1 z^{-1})(1 - \alpha_2 z^{-1}) \ldots (1 - \alpha_g z^{-1}) \qquad (2.4.92)$$

as in Equation 2.4.45, the coefficients in $Y(z)$ all being real. Then

$$Y^{\frac{1}{2}}(z) = y_0^{\frac{1}{2}}(1 - \alpha_1 z^{-1})^{\frac{1}{2}}(1 - \alpha_2 z^{-1})^{\frac{1}{2}} \ldots (1 - \alpha_g z^{-1})^{\frac{1}{2}} \qquad (2.4.93)$$

When $|\alpha_i| \leqslant 1$, for any $i = 1, 2, \ldots, g$,

$$(1 - \alpha_i z^{-1})^{\frac{1}{2}} = 1 - \tfrac{1}{2}\alpha_i z^{-1} - \tfrac{1}{8}\alpha_i^2 z^{-2} - \tfrac{1}{16}\alpha_i^3 z^{-3}$$
$$- \tfrac{5}{128}\alpha_i^4 z^{-4} - \ldots \qquad (2.4.94)$$

and when $|\alpha_i| \geqslant 1$,

$$(1 - \alpha_i z^{-1})^{\frac{1}{2}} = (-\alpha_i z^{-1})^{\frac{1}{2}}(1 - \alpha_i^{-1} z)^{\frac{1}{2}}$$
$$= (-\alpha_i z^{-1})^{\frac{1}{2}}(1 - \tfrac{1}{2}\alpha_i^{-1} z - \tfrac{1}{8}\alpha_i^{-2} z^2 - \tfrac{1}{16}\alpha_i^{-3} z^3$$
$$- \tfrac{5}{128}\alpha_i^{-4} z^4 - \ldots) \qquad (2.4.95)$$

The fact that the expansion in Equation 2.4.94 or 2.4.95 holds for $z = \pm 1$ and for any real value of α_i satisfying the given inequality,[6] implies that the expansion holds also for any value of z on the unit circle in the z plane and for any complex value of α_i satisfying the given inequality.

$Y^{\frac{1}{2}}(z)$ can now be evaluated from $Y(z)$, using Equations 2.4.93 to 2.4.95. An alternative and perhaps simpler method of evaluating $Y^{\frac{1}{2}}(z)$ is outlined in Section 4.6.3.

It can readily be shown that the coefficients of the powers of z in $Y^{\frac{1}{2}}(z)$ are either all real or else all imaginary, since the roots of $Y(z)$ that are not real occur in complex-conjugate pairs. The imaginary components can, if required, be replaced by real components, by considering $[-Y(z)]^{\frac{1}{2}}$ in place of $Y^{\frac{1}{2}}(z)$. This in no way affects any of the important results of the following analysis.

It can be seen from Equation 2.4.95 that $Y^{\frac{1}{2}}(z)$ may contain the components $\{z^{i+\frac{1}{2}}\}$, where the values of the integer i may be positive, negative and zero. $Y(z)$, of course, contains only integral powers of z. This means that the sampling instants of the sequence whose z transform is $Y^{\frac{1}{2}}(z)$ have been shifted in time by $\frac{1}{2}T$ seconds relative to the sampling instants $\{iT\}$, which, strictly speaking, is not physically realisable. The components $\{z^{i+\frac{1}{2}}\}$ in $Y^{\frac{1}{2}}(z)$ can, if required, be replaced by the components $\{z^i\}$, simply by introducing a delay of $\frac{1}{2}T$ seconds into the sequence with z transform $Y^{\frac{1}{2}}(z)$. This in no way affects any of the important results of the following analysis.

To simplify the analysis, $Y^{\frac{1}{2}}(z)$ will not be modified in any way, but some care is now required in the interpretation of the results. These hold whether or not $Y^{\frac{1}{2}}(z)$ has imaginary coefficients or components $\{z^{i+\frac{1}{2}}\}$ but assume that $Y(z)$ has no zeros at $z = \pm 1$.

Let $X(z)$ be the z transform obtained from $Y^{\frac{1}{2}}(z)$ by replacing each component z^{-i} by z^i (or $z^{-i-\frac{1}{2}}$ by $z^{i+\frac{1}{2}}$), without however changing the coefficients. Thus $X(z)$ is obtained from $Y^{\frac{1}{2}}(z)$ by reversing the order of its coefficients, the reversal being pivoted about the coefficient of z^0 (which may be zero). If the coefficients in $Y^{\frac{1}{2}}(z)$ are imaginary, then so also are the coefficients in $X(z)$ and $X^{-1}(z)$. This means that $X(z)Y^{\frac{1}{2}}(z)$ and $X^{-1}(z)Y^{\frac{1}{2}}(z)$ both contain only *real* coefficients, regardless of whether the coefficients in $Y^{\frac{1}{2}}(z)$ are real or imaginary. The multiplication of the coefficients of $Y^{\frac{1}{2}}(z)$ by $\sqrt{-1}$, together with the corresponding changes in $X(z)$ and $X^{-1}(z)$ such that these maintain their relationships with $Y^{\frac{1}{2}}(z)$, does not change $X^{-1}(z) Y^{\frac{1}{2}}(z)$, and simply changes the sign of $X(z) Y^{\frac{1}{2}}(z)$. If $Y^{\frac{1}{2}}(z)$ contains the components $\{z^{i+\frac{1}{2}}\}$, where i has integer values, then so also do $X(z)$ and $X^{-1}(z)$. This means that $X(z) Y^{\frac{1}{2}}(z)$ and $X^{-1}(z) Y^{\frac{1}{2}}(z)$ both contain only *integral* powers of z, regardless of whether $Y^{\frac{1}{2}}(z)$ does or not. The introduction of a time shift (multiplication by z^i) in $Y^{\frac{1}{2}}(z)$, together with the corresponding time shifts in $X(z)$ and $X^{-1}(z)$, does not change $X(z) Y^{\frac{1}{2}}(z)$, and introduces a time shift in $X^{-1}(z) Y^{\frac{1}{2}}(z)$ without otherwise changing it. None of the effects just mentioned changes the poles and zeros (other than those at the origin and infinity) of any of the z transforms.

Now, from Equation 2.4.59, $X^{-1}(z) Y^{\frac{1}{2}}(z)$ is the z transform of a channel or filter introducing *pure phase distortion*, and from Equation 2.4.74, $X(z) Y^{\frac{1}{2}}(z)$ is the z transform of a channel or filter introducing *pure amplitude distortion*. Furthermore,

$$X^{-1}(z) Y^{\frac{1}{2}}(z) . X(z) Y^{\frac{1}{2}}(z) = Y(z) \qquad (2.4.96)$$

Thus the baseband channel and sampler in Figure 2.1 can be represented by two channels in cascade: the channel A having the z-transform $X^{-1}(z) Y^{\frac{1}{2}}(z)$ and therefore introducing pure phase distortion, and the channel B having the z-transform $X(z) Y^{\frac{1}{2}}(z)$ and therefore introducing pure amplitude distortion. The baseband channel and sampler may be made physically realizable (or at least approximately so) by introducing sufficient delay.

If $Y(z)$ represents pure phase distortion, then so must $Y^{\frac{1}{2}}(z)$, which means that $X(z) = Y^{-\frac{1}{2}}(z)$ and $X^{-1}(z) = Y^{\frac{1}{2}}(z)$. From Equation 2.4.96, channel A now has the z transform $Y(z)$, and channel B has a z transform of unity, as would be expected. If $Y(z)$ represents pure amplitude distortion, then so must $Y^{\frac{1}{2}}(z)$ or $\sqrt{-1} \, Y^{\frac{1}{2}}(z)$, which means that $X(z) = Y^{\frac{1}{2}}(z)$ and $X^{-1}(z) = Y^{-\frac{1}{2}}(z)$. From Equation 2.4.96, channel A now has a z-transform of unity, and channel B has the z-transform $Y(z)$, again as would be expected.

The coefficients of $X(z) Y^{\frac{1}{2}}(z)$ give the *aperiodic autocorrelation function* of the sequence with z-transform $Y^{\frac{1}{2}}(z)$, the function being sensitive only to amplitude distortion and attenuation. The coefficients of $X^{-1}(z) Y^{\frac{1}{2}}(z)$ give the corresponding function that is sensitive only to phase distortion and delay.

It is shown in Section 4.3.10 that phase distortion can be equalized linearly with no reduction in the tolerance of the data-transmission system to additive white Gaussian noise, relative to the case where there is no signal distortion of any kind. In other words, pure phase distortion does not itself introduce any irreparable degradation into the system performance and can usually be equalized quite easily. Furthermore, it is shown in Chapter 4 that in the presence of pure phase distortion and additive white Gaussian noise, the *best* detection process is achieved by a linear equalizer, followed by the appropriate decision thresholds, no advantage being gained here through the use of any more sophisticated detection process. These statements are intuitively reasonable in view of the fact that pure phase distortion is an orthogonal transformation and a phase equalizer is a matched filter. On the other hand, pure amplitude distortion often results in a reduction in tolerance to noise that cannot be avoided, no matter what detection process is used. Again, in the presence of amplitude distortion an advantage in tolerance to additive white Gaussian noise can always be gained by replacing a linear equalizer by the appropriate and more sophisticated detection process. This is shown in Chapters 4 and 5.

It follows that with the model of the baseband channel just assumed, the channel A can be equalized linearly at the receiver, and the combination of the whole channel and linear equalizer can now be considered to be exactly equivalent to just the channel B on its own. This is, of course, only possible because the equalization of channel A gives no reduction in tolerance to additive white Gaussian noise relative to the case where the channel B is present on its own.

Thus, in assessing the severity of the signal distortion introduced by a channel, it is clearly the *amplitude distortion* that is the most important factor to be considered. The amplitude distortion can be determined by taking the DFT of the sampled impulse-response of the channel to give $Y_0 F$ in Equation 2.4.89. In the absence of amplitude distortion, $|\lambda_i|$ has the same value for each i, this value being a measure of the signal attenuation. The degree to which the $\{|\lambda_i|\}$ differ in value is a measure of the amplitude distortion.

Alternatively, it can be seen from Equation 2.4.8 that the $(i+1)$th component of the discrete energy-density spectrum is $|\lambda_i|^2$, and the inverse DFT of the discrete energy-density spectrum is the aperiodic autocorrelation function B_g in Equation 2.4.5. This means that B_g is completely determined by the amplitude distortion and signal attenuation, so that B_g depends *only* on these two parameters. In the absence of amplitude distortion and signal attenuation, B_g is given by Equation 2.4.34, regardless of the phase distortion or signal delay, so that B_g is obviously unaffected by phase distortion and signal delay.

The advantage of using B_g as a measure of amplitude distortion is that it is very simply evaluated from the sampled impulse-response of the channel and it is more closely related to the sampled impulse-response (at least conceptually) than is the modulus of the DFT of this response, or the discrete energy-density spectrum.

Signal attenuation or gain affects the magnitudes of all components of B_g equally. Furthermore, $b_{g+i} = b_{g-i}$ for $i = 1, 2, \ldots, g$, and the remaining $\{b_i\}$ are all zero. In the absence of amplitude distortion, $b_i = 0$ for each $i \neq g$. Thus the quantity

$$d = \frac{1}{b_g^2} \sum_{i=g+1}^{2g} b_i^2 \qquad (2.4.97)$$

gives a measure of the severity of the amplitude distortion and therefore of the irreducible degradation in performance likely to be experienced in the data-transmission system of Figure 2.1. The quantity d is, of course, independent of signal attenuation, signal delay and phase distortion.

When $d > \frac{1}{8}$ there is significant amplitude distortion and when $d > \frac{1}{2}$ the distortion is severe. Although this is necessarily a crude measure of the amplitude distortion, since it takes no account of the detailed structure or shape of the distortion, it does give a general guide as to the type of detection process that is likely to be the most cost-effective. If $d < \frac{1}{8}$, a simple linear equalizer is probably the most suitable arrangement at the receiver. If $d > \frac{1}{2}$, a more sophisticated detection process would be worth while and should give a useful advantage in performance (Chapter 5).

2.4.8 Signals with real and imaginary components

It is of interest finally to consider the important case where two double sideband suppressed carrier AM signals in phase quadrature are transmitted over a band-pass channel. The two signals are generated at the transmitter from two baseband signals that are in element synchronism (having coincident element-boundaries). These are used to modulate the respective signal carriers that have the *same* frequency and are at a constant phase angle of 90°, to give the two

AM signals in phase quadrature. These signals are demodulated at the receiver by two coherent demodulators, whose reference carriers are phase synchronized to the respective signal carriers.

The baseband signal fed to each modulator at the transmitter is a sequence of impulses, which are regularly spaced in time and carry the respective element values. Corresponding impulses in the two baseband signals coincide in time. In each modulator, the impulses are filtered to produce a rounded waveform which is then used to modulate the signal carrier. It is convenient for our purposes to consider the sequence of impulses as the baseband modulating waveform, and the filter as a part of the modulator. The two demodulated baseband signals at the receiver are sampled simultaneously, at regular time intervals once per signal element, to give the corresponding two sequences of sample values, that are used in the detection of the received signal-elements.

When the transmission path introduces no noise or distortion, the demodulated and sampled baseband signal at the output of each coherent demodulator is an exact copy of the corresponding baseband signal (sequence of impulses) at the transmitter. When the transmission path introduces signal distortion, this generally appears in the demodulated and sampled baseband signals as a combination of intersymbol and interchannel interference, since the data-transmission system is of course a parallel system having two channels.

To analyse the system theoretically, the modulating and demodulated baseband signals, associated with the AM signal whose carrier phase *lags* that of the other by 90°, are considered as *real* signals, and the modulating and demodulated baseband signals associated with the other AM signal, are considered as *imaginary* signals. Thus, in the simple case where each AM signal is binary coded, a *pair* of coincident elements of the two AM signals have the resultant value s_i, where s_i has one of the values $\pm k \pm jk$ (k is real and $j = \sqrt{-1}$). The transmission path, together with the two linear modulators at the transmitter and the two linear demodulators at the receiver, now form a *linear baseband channel* whose sampled impulse-response is a *complex* vector. The real and imaginary components of this vector are given by the sample values of the respective demodulated baseband signals at the receiver, in response to a unit impulse (at time $t = 0$) at the input to the transmitter modulator associated with the AM signal whose carrier phase *lags* that of the other by 90°.

Thus the transmitted signal vector and the sampled impulse-response of the baseband channel become the complex vectors that correspond to the real vectors previously assumed, and Equation 2.2.22 still holds, except that R, S and Y are now all complex.

When the baseband channel introduces *pure phase distortion*, it introduces an *orthogonal transformation* into the transmitted signal, as before, such that the received signal-elements corresponding to the resultant (complex) transmitted signal-elements are orthogonal, their inner products being zero. The sampled impulse-response of the baseband channel (Equation 2.4.1) is now such that the sequence, formed from Y_0 by reversing the order of its first $g + 1$ components and then replacing each of these by its complex conjugate, is the *inverse* of Y_0 delayed by g sampling intervals. The sequence Y_0 and its inverse are both such that the sum of the squares of the moduli of the n components is equal to unity. The sequence obtained by convolving the inverse (and delayed) sequence with Y_0 has all its components zero except for the $(g + 1)$th component which has the value unity. The matrix Y in Equation 2.4.19 now becomes a *unitary* matrix, whose inverse is therefore equal to its conjugate transpose. The eigenvalues of this matrix all have a modulus of unity, so that all components of the DFT of the sampled impulse-response of the baseband channel have a modulus of unity, which is the condition that must be satisfied in the presence of pure phase distortion. It is, of course, assumed here that the channel introduces no signal attenuation. With the appropriate detection process, pure phase distortion again introduces no reduction in tolerance to additive white Gaussian noise relative to the case where there is no distortion.

When the baseband channel introduces *pure amplitude distortion* and no delay, the sampled impulse-response of the baseband channel is given by Equation 2.4.67, where now $y_{\frac{1}{2}g + i}$ is equal to the complex conjugate of $y_{\frac{1}{2}g - i}$, for $i = 1, 2, \ldots, \frac{1}{2}g$, and $y_{\frac{1}{2}g}$ is real. The sampled impulse-response obtained by delaying this by $\frac{1}{2}g$ sampling intervals is given by Equation 2.4.1. If the first $g + 1$ components of the latter sequence are reversed in order and each component is then replaced by its complex conjugate, the sequence is left unchanged. The convolution matrix Y corresponding to the sampled impulse-response in Equation 2.4.67 is a *Hermitian* matrix and is therefore equal to its conjugate transpose. The eigenvalues of this matrix are all *real*, so that all components of the DFT of the sampled impulse-response of the baseband channel are real, which is the condition that must be satisfied in the presence of pure amplitude distortion.

The simple relationships of the type just mentioned, which exist between signals that are real and signals that have both real and imaginary components (complex signals), carry over, as might be expected, to the other properties of these signals, as well as to the various available techniques for detecting and estimating them. For instance, the comparison of the different techniques studied in Chapters 4–6 is not significantly affected by whether real or complex signals are used. Real signals are used much more often in practice than are complex signals and they are therefore of considerable importance in their own right. Furthermore, the detailed study of real signals brings out very clearly the important properties of *both* real and complex signals, and at the same time avoids the more serious conceptual difficulties involved with the latter. The remaining chapters of the book are therefore concerned solely with *real signals*. Once the principles and techniques have been mastered for these, it is a relatively simple matter to extend them to complex signals.

REFERENCES

1. Browne, E. T. *Introduction to the Theory of Determinants and Matrices*, Chapel Hill, University of North Carolina Press (1958)
2. Paige, L. J. and Swift, J. D. *Elements of Linear Algebra*, Blaisdell Publishing Co., New York (1961)
3. Ayres, F. *Matrices*, McGraw-Hill, New York (1962)
4. Gold, B. and Rader, C. M. *Digital Processing of Signals*, McGraw-Hill, New York (1969)
5. Clark, A. P. *Principles of Digital Data Transmission*, Pentech Press, London (1976)
6. Dwight, H. B. *Tables of Integrals and other Mathematical Data*, 4th edn, pp. 1–18, Macmillan, New York (1961)

3 Principles of detection and estimation

3.1 INTRODUCTION

3.1.1 Basic assumptions

We are here concerned with the principles of both the optimum detection and the optimum linear estimation of some given parameter of a sampled baseband digital signal, where this is received in the presence of noise. The parameter is a scalar or vector quantity. It is assumed that the received signal is the continuous baseband waveform

$$r(t) = s(t) + w(t) \qquad (3.1.1)$$

where $s(t)$ is the waveform of the digital signal and $w(t)$ is the noise waveform.

Two different cases are considered. In the first of these, the parameter of $s(t)$ has any one of a given set of possible values, and the receiver is required to *detect*, with the minimum probability of error, which of the possible values of the parameter of $s(t)$ is actually received. In the second case, the parameter of $s(t)$ is either as just described or else has any one of a continuous range (infinite set) of values, and it is required to *estimate* its value as accurately as possible. In every case, $s(t)$ is the resultant waveform of one or more received signal-elements, these being the individual pulses of the digital signal. The parameter of $s(t)$ to be detected or estimated itself carries the corresponding *set* of element values and is the information to be transmitted. If $s(t)$ contains one signal element, the parameter is a scalar quantity; if $s(t)$ contains m signal elements, the parameter is an m-component vector and, strictly speaking, becomes a set of m scalar parameters.

The received waveform $r(t)$ is sampled n times, at the time instants $\{jT\}$, where T is measured in seconds and $j = 1, 2, \ldots, n$. The resultant n sample values $\{r(jT)\}$ may be represented by the components of the row vector

$$R = r_1 \; r_2 \ldots r_n \qquad (3.1.2)$$

where $r_j = r(jT)$. The parameter of $s(t)$ is detected or estimated from the n sample values $\{r_j\}$, that is, from the received vector R. Thus the received signal is processed as shown in Figure 3.1.

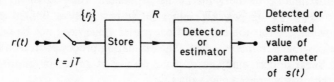

Fig. 3.1   Detection or estimation of the parameter of s(t)

A detection process is the *selection* of one of a finite set of 'symbols' or 'messages' in response to the received signal. The symbols (or messages) themselves form the information being transmitted. Ideally each symbol is associated with a different *discrete* value of the parameter of $s(t)$. In practice, however, the receiver may not have an exact prior knowledge of the parameter value associated with each symbol, so that, as far as the receiver is concerned, each symbol is associated with a range or infinite set of parameter values. So long as the sets of parameter values associated with different symbols are disjoint (nonoverlapping), the transmitted symbol may be uniquely determined from the received parameter value, in the absence of noise.

Clearly, a process of detection is a *decision* as to which of the possible symbols has been transmitted, and it is essentially *nonlinear*. The decision is based on the received signal together with the available prior knowledge of this signal. In order to be able to reach the correct decision, the receiver must obviously have prior knowledge of the *number* of different possible symbols, and it must know which symbol to associate with which range of parameter values.

In order to simplify the terminology used, no distinction is often made between the *symbol* that is transmitted and the corresponding *actual value* of the parameter of $s(t)$, which is of course one of the finite set of discrete values. Under the most favourable conditions, the receiver has prior knowledge of the different actual possible values of the parameter of $s(t)$ and it selects one of these to give the detected symbol. Under less favourable conditions, the receiver only knows that the parameter value must lie in one of a number of different ranges and it selects one of these ranges to give the detected symbol. In either case, the detected symbol is often taken to be the corresponding actual possible parameter value, even though, in the latter case, the receiver has no prior knowledge of this.

A process of *estimation* is defined to be an assessment of or approximation to the value of the parameter, using either no prior knowledge of its possible values or at best only a limited prior knowledge. Thus the estimated parameter value need not be one of the possible parameter values. The estimation processes studied here are linear and in many cases make no use of any prior knowledge of the possible parameter values.

Unless some rather special conditions are satisfied, the optimum detection or estimation of the parameter of $s(t)$, from the received vector R, does not necessarily correspond to the optimum detection or estimation of the parameter from the received waveform $r(t)$. It can never be better and it may be considerably inferior. In the particular case where $r(jT)$ is zero for all integers $j < 1$ and $j > n$ and where the bandwidth of $r(t)$ is limited to the frequency range 0 to $1/2T$ Hz (considering only positive frequencies), it can be seen from Nyquist's sampling theorem[3,4] that $r(t)$ is completely defined by R. This necessarily implies that the optimum detection or estimation of the parameter of $s(t)$ from the vector R is also the optimum detection or estimation of the parameter from $r(t)$, since of course all the information in $r(t)$ is contained also in R.

In practice, the noise waveform $w(t)$ is continuous and of effectively infinite duration, so that the conditions just assumed do not apply. However the data signal $s(t)$ is normally of finite duration, at least for practical purposes. Assume therefore that $s(jT)$ is zero for all integers $j < 1$ and $j > n$, and suppose furthermore that the samples $\{w(jT)\}$ of the noise waveform are statistically independent and also independent of $s(t)$. The latter two assumptions hold, at least approximately, in many practical systems. Under these conditions, the received samples $\{r(jT)\}$ for all integers $j < 1$ and $j > n$ are statistically independent of $s(t)$. It is now reasonable to conclude[3] that these $\{r(jT)\}$ cannot assist in any way in the detection or estimation of the parameter of $s(t)$, since they are in no way dependent either upon $s(t)$ or on the noise samples $\{w(jT)\}$ for $j = 1, 2, \ldots, n$. In other words, in the optimum detection or estimation of the parameter of $s(t)$ from the received samples $\{r(jT)\}$, it is only necessary to use the n samples $\{r(jT)\}$, for $j = 1, 2, \ldots, n$.

If now the bandwidth of $r(t)$ is limited to the frequency band from zero frequency to $1/2T$ Hz (considering only positive frequencies), then so also is the bandwidth of $s(t)$. The received signal waveform $s(t)$ is thus completely defined by the n samples $\{s(jT)\}$ for $j = 1, 2, \ldots, n$, since the remaining $\{s(jT)\}$ are all zero. Furthermore, the received waveform $r(t)$ is completely defined by the infinite set of sample values $\{r(jT)\}$, for all integer values of j, but only those samples for $j = 1, 2, \ldots, n$ contribute in any useful way to the detection or estimation of the parameter of $s(t)$. Hence the optimum detection or estimation of the parameter of $s(t)$ from R is also the optimum detection or estimation of the parameter from $r(t)$. If the bandwidth of either $s(t)$ or $w(t)$, or of both $s(t)$ and $w(t)$, exceeds $1/2T$ Hz, so that the bandwidth of $r(t)$ exceeds $1/2T$ Hz, the arrangement just described no longer gives the best detection or estimation of the parameter of $s(t)$, from the received waveform $r(t)$.

Throughout the following discussion we will be concerned with the optimum detection or estimation of the parameter of $s(t)$, from the received vector R, regardless of whether or not this corresponds to the optimum detection or estimation of the parameter from $r(t)$. No particular assumptions are therefore made concerning the bandwidth of the received waveform $r(t)$. The detection or estimation of the parameter of $s(t)$ can of course be optimum in many different senses of the word, dependent upon which particular quantity we wish to maximize or minimize. The word 'optimum' will therefore require careful definition for each of the different cases that are studied.

Let the n sample values $\{s(jT)\}$, for $j = 1, 2, \ldots, n$, be the n components of the row vector

$$S = s_1 \; s_2 \ldots s_n \tag{3.1.3}$$

where $s_j = s(jT)$, and let the n sample values $\{w(jT)\}$, for $j = 1, 2, \ldots, n$, be the n components of the row vector

$$W = w_1 \; w_2 \ldots w_n \tag{3.1.4}$$

where $w_j = w(jT)$. It will now be assumed throughout the following discussion that $w(t)$ is a stationary Gaussian random process which is such that the $\{w_j\}$ are statistically independent Gaussian random variables with zero mean and variance σ^2. Furthermore, the $\{w_j\}$ are statistically independent of the $\{s_j\}$.

From Equation 3.1.1,

$$R = S + W \tag{3.1.5}$$

and $$r_j = s_j + w_j \qquad (3.1.6)$$

for $j = 1, 2, \ldots, n$.

✗   In Equation 3.1.5, R, S, and W each constitute an *ordered set* of *n random variables* and is therefore the corresponding *random vector*. The *value* of a vector is the *ordered set* of values of its n components, and the *sample value* of a random vector is the actual value of the vector obtained in any one occurrence (transmission or reception) of the vector. A random vector itself is the ensemble (collection) of all the different possible sample values of that vector, each value being associated with the corresponding probability or probability density. A random vector with just one component ($n = 1$) becomes a random variable.

A *discrete* random vector is one having a finite number of different possible sample values, whereas a *continuous* random vector is one having a continuous range (and hence infinite number) of different possible sample values. The complete *set* of different possible sample values of a discrete random vector X is designated $\{X\}$.

In order to reduce the complexity of the notation, which would otherwise become very cumbersome in the later chapters of the book, the *same* symbol will be used for a random vector and its sample value, the correct meaning being in every case made clear in the context.

When the vector R in Equation 3.1.5 is taken to represent the vector actually received, in some particular transmission, the vectors R, S and W in Equation 3.1.5 become the *sample values* of the corresponding random vectors. Thus their values are now fixed. On the other hand, when considering the mean or mean-square value of any of the vectors R, S and W, or of any function of these vectors, or alternatively, when considering the probability of any of these vectors having a value within some given range, the vectors must be taken as *random vectors*, since we are now concerned with *all possible* sample values of the vectors under consideration, and not just with the particular sample values that happen to be received in any given transmission. Equation 3.1.5 is satisfied, regardless of whether R, S and W are random vectors or sample values of random vectors.

S normally has different values in different transmissions, so that the detected or estimated value of S is itself a random vector, normally differing from one transmission to another. In any *one* detection or estimation process, however, S becomes a *sample value* of the corresponding random vector as does the detected or estimated value of S. The detected value of S should here, if possible, be correct, and the estimated value of S should be as close as possible to its correct value.

In all the various cases to be studied here, the receiver will be assumed to be as shown in Figure 3.1, and the detection or estimation process will operate on the received vector R, given by Equation 3.1.5. Different forms of the signal waveform $s(t)$ will however be considered, so that the corresponding n-component signal vector will not necessarily be represented in terms of a single vector S.

Before proceeding further with the derivation of the optimum detection and estimation processes, it is necessary first to clarify an important mathematical relationship that is used in the analysis.

3.1.2 Bayes' theorem

If $P(A|B)$ is the conditional probability of the event A, given the event B, and $P(B|A)$ is the conditional probability of the event B, given the event A, then the two conditional probabilities are related by the following equation

$$P(A|B) = \frac{P(B|A)P(A)}{P(B)} \qquad (3.1.7)$$

where $P(A)$ is the probability of the event A and $P(B)$ is the probability of the event B. Equation 3.1.7 is known as Bayes' theorem. A and B are two *events* and the equation involves only *probabilities*.

It may be shown[3] that this relationship holds also for the case where each of the symbols A and B represents a *set* of sample values of the corresponding set of continuous random variables, so that A and B become sample values of the corresponding continuous random vectors. Equation 3.1.7 now involves probability *densities* and becomes

$$p(A|B) = \frac{p(B|A)p(A)}{p(B)} \qquad (3.1.8)$$

where $A = (a_1, a_2, \ldots, a_n)$ and $B = (b_1, b_2, \ldots, b_n)$. $p(A)$ is the value of the probability *density* function of the random vector with sample value A, when the random vector has this sample value.

$p(A|B)$ is the value of the corresponding *conditional* probability density function, given B. Similarly for $p(B)$ and $p(B|A)$.

Again, Bayes' theorem also holds where the n random variables, whose sample values are given by A, are *discrete*, so that A may have any one of a finite set of possible values and is a sample value of a *discrete* random vector. B is a sample value of a continuously variable random vector, as before. The equation now contains the appropriate *mixture* of probabilities and probability densities,[3] and becomes

$$P(A|B) = \frac{p(B|A)P(A)}{p(B)} \tag{3.1.9}$$

$P(A)$ is here the *probability* that the random vector with sample value A has this sample value. $P(A|B)$ is the corresponding *conditional* probability, given B.

Throughout the following discussion, $P(.)$ will be taken to signify a *probability* and $P(.|.)$ a *conditional probability*, whereas $p(.)$ will be taken to signify a *probability density* and $p(.|.)$ a *conditional probability density*. The quantities inside the brackets here are *sample values* of the corresponding random vectors.

3.2   DETECTION OF A SIGNAL WITH KNOWN PARAMETERS IN NOISE

3.2.1   *Simple hypothesis testing*

Suppose that the received waveform is

$$r(t) = s_i(t) + w(t) \tag{3.2.1}$$

where $i = 0$ or 1, so that one of two possible signal waveforms $s_0(t)$ and $s_1(t)$ is received. $s_i(t)$ is a binary signal element, which is assumed to extend over the time interval T to nT seconds. $s_0(t)$ and $s_1(t)$ may have any two waveforms which are continuous and of finite energy.

The receiver of Figure 3.1 is used to detect the value of i, which is of course the binary value of the received signal-element. Thus the received signal-plus-noise waveform $r(t)$ is sampled n times, at the time instants $t = jT$, for $j = 1, 2, \ldots, n$, and the n sample values form the n components of the row vector R in Equation 3.1.2. The corresponding n sample values of the noise waveform $w(t)$, form the n components of the row vector W in Equation 3.1.4.

As before, the components of W are statistically independent Gaussian random variables with zero mean and variance σ^2.

Let the n sample values of $s_i(t)$ be the n components of the row vector

$$S_i = s_{i1} \; s_{i2} \ldots s_{in} \tag{3.2.2}$$

where $s_{ij} = s_i(jT)$. Thus

$$R = S_i + W \tag{3.2.3}$$

and

$$r_j = s_{ij} + w_j \tag{3.2.4}$$

for $j = 1, 2, \ldots, n$. The *random variables* $\{w_j\}$ are statistically independent of the $\{s_{ij}\}$, so that the continuous *random vector* W is statistically independent of the signal vector S_i.

The detector is assumed to have prior knowledge of both S_0 and S_1. It uses this prior knowledge in the detection of i from R. For any received vector R it is *possible* that either S_0 or S_1 is actually received, so that the detector cannot detect the value of i with certainty. The problem is therefore how to make the best guess as to the correct value of i, from the received vector R. A detection process *starts* with the reception of a vector R, which is therefore known to the detector, in addition to S_0 and S_1, and the process *ends* with a *decision* as to whether $i = 0$ or 1.

The detector can clearly make one of two *hypotheses*:

(1) H_0 that $i = 0$ and therefore that $S_i = S_0$.
(2) H_1 that $i = 1$ and therefore that $S_i = S_1$.

For any given received vector R, the detector can either decide in favour of H_0 or H_1, and the detector clearly requires a *decision rule* to decide which values of R lead to the acceptance of H_0 and which to the acceptance of H_1.

The vectors R, S_i and W may be represented as points in an n-dimensional linear Euclidean vector space, sometimes known as *signal space*. The orthogonal projections of any point in the signal space on to the n orthogonal axes give the n components of the corresponding vector. The detector uses the *position* of the received vector R in this vector space to detect the value of i. It divides the vector space into two regions, D_0 and D_1. If R lies in D_0, the detector accepts H_0 and so detects i as 0. If R lies in D_1, the detector accepts H_1 and so detects i as 1. The two regions of the vector space are known as *decision regions*, and the boundary separating them is known as the *decision boundary*.[1] The arrangement is shown in Figure 3.2.

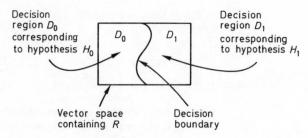

Decision region D_0 corresponding to hypothesis H_0

D_0 D_1

Decision region D_1 corresponding to hypothesis H_1

Vector space containing R

Decision boundary

Fig. 3.2 Decision regions in the n-dimensional vector space

If R lies in some region D_i of the vector space, this is written $R \in D_i$. The probability that this occurs is written $P(R \in D_i)$. Of course, R is here a *random vector* whose *sample value* lies in the region D_i.

Clearly, the *decision rule* is as follows:

(1) If $R \in D_0$, accept H_0 and detect i as 0.
(2) If $R \in D_1$, accept H_1 and detect i as 1.

It is assumed that R must lie *either* in D_0 *or* in D_1. It cannot lie in both D_0 and D_1, nor can it lie in neither of these. In other words, the two decision regions are disjoint and together they fill the whole of the n-dimensional vector space.

The problem is now to determine the *best decision boundary*.

3.2.2 Bayes' likelihood-ratio test

In the detection process for the received signal-element there are two kinds of error:

(1) Accept H_1 when H_0 is true. This has probability α.
(2) Accept H_0 when H_1 is true. This has probability β.

The remaining two possible outcomes of the detection process are:

(1) Accept H_0 when H_0 is true. This has probability $1 - \alpha$.
(2) Accept H_1 when H_1 is true. This has probability $1 - \beta$.

Now[1]
$$\alpha = P(R \in D_1|H_0) = P(R \in D_1|S_0)$$

$$= \int_{D_1} p(R|S_0)\mathrm{d}R \tag{3.2.5}$$

$$\beta = P(R \in D_0|H_1) = P(R \in D_0|S_1)$$

$$= \int_{D_0} p(R|S_1)\mathrm{d}R \tag{3.2.6}$$

$$1 - \alpha = P(R \in D_0|S_0) = \int_{D_0} p(R|S_0)\mathrm{d}R \tag{3.2.7}$$

$$1 - \beta = P(R \in D_1|S_1) = \int_{D_1} p(R|S_1)\mathrm{d}R \tag{3.2.8}$$

It is assumed here that, for $i = 0$ or 1,

$$p(R|S_i) = p(r_1, r_2, \ldots, r_n|s_{i1}, s_{i2}, \ldots, s_{in})$$

is the value of the conditional joint probability density function of the random variables with sample values $r_1, r_2, \ldots, r_n$, when the random variables have the given values $r_1, r_2, \ldots, r_n$, and given the values $s_{i1}, s_{i2}, \ldots, s_{in}$. A *conditional* probability density function has all the properties of an ordinary probability density function. Thus

$$\int_{D_j} p(R|S_i)\mathrm{d}R = \int_{D_j} \ldots \int p(r_1, r_2, \ldots, r_n|s_{i1}, s_{i2}, \ldots$$
$$\ldots, s_{in})\mathrm{d}r_1\mathrm{d}r_2 \ldots \mathrm{d}r_n$$

where $j = 0$ or 1 and the integration is taken over all values of the vector R lying in the decision region D_j. R is here a *sample value* of a continuous random vector.

Suppose now that

$$P(H_0) = P(S_0) = q_0 \tag{3.2.9}$$

$$P(H_1) = P(S_1) = q_1 \tag{3.2.10}$$

Clearly
$$q_0 + q_1 = 1 \tag{3.2.11}$$

q_0 is the *a priori* probability of S_0 and q_1 is the *a priori* probability of S_1. The detector is assumed to have prior knowledge of q_0 and q_1.

In some situations certain errors are more *costly* than others and account must be taken of this in choosing the best decision boundary.[1,2] Let

C_{00} = cost of deciding that hypothesis H_0 is true when H_0 is true.
C_{10} = cost of deciding that hypothesis H_1 is true when hypothesis H_0 is true.
C_{01} = cost of deciding that hypothesis H_0 is true when hypothesis H_1 is true.
C_{11} = cost of deciding that hypothesis H_1 is true when H_1 is true.

The four costs are taken to be nonnegative.

Given these four costs and given the *a priori* probabilities of H_0 and H_1, the decision boundary can now be chosen to *minimize the average cost* in a detection process.[1]

The conditional expected cost, given that H_0 is true, is

$$g_0 = C_{00}(1-\alpha)+C_{10}\alpha \qquad (3.2.12)$$

and the conditional expected cost, given that H_1 is true, is

$$g_1 = C_{11}(1-\beta)+C_{01}\beta \qquad (3.2.13)$$

Given the probability q_0 that H_0 is true and the probability q_1 that H_1 is true, the average or expected cost when either H_0 or H_1 may be true is

$$G = q_0 g_0 + q_1 g_1 \qquad (3.2.14)$$

Thus

$$G = q_0 C_{00} + q_1 C_{11} + q_0(C_{10}-C_{00})\alpha + q_1(C_{01}-C_{11})\beta \qquad (3.2.15)$$

G is usually called the *average risk*[1].

Let

$$C_\alpha = C_{10}-C_{00}>0 \qquad (3.2.16)$$

and

$$C_\beta = C_{01}-C_{11}>0 \qquad (3.2.17)$$

The cost of making an error is greater than the cost of making a correct decision, and in most applications

$$C_{00} = C_{11} = 0 \qquad (3.2.18)$$

From Equations 3.2.15, 3.2.16 and 3.2.17,

$$G = q_0 C_{00} + q_1 C_{11} + q_0 C_\alpha \alpha + q_1 C_\beta \beta \qquad (3.2.19)$$

Now from Equations 3.2.5 and 3.2.6,

$$G = q_0 C_{00} + q_1 C_{11} + q_0 C_\alpha \int_{D_1} p(R|S_0)dR + q_1 C_\beta \int_{D_0} p(R|S_1)dR \qquad (3.2.20)$$

But $\int_{D_0+D_1} p(R|S_0)dR = \int_{D_0} p(R|S_0)dR + \int_{D_1} p(R|S_0)dR = 1$ (3.2.21)

so that[1]

$$G = q_0C_{00}+q_1C_{11}+q_0C_\alpha\left[1-\int_{D_0} p(R|S_0)dR\right]+q_1C_\beta\int_{D_0} p(R|S_1)dR$$

$$= q_0C_{00}+q_1C_{11}+q_0C_\alpha+\int_{D_0}[-q_0C_\alpha p(R|S_0)$$
$$+q_1C_\beta p(R|S_1)]dR \quad (3.2.22)$$

Since q_0C_{00}, q_1C_{11} and q_0C_α are constants, G is minimum when the integral is minimum. But $q_0C_\alpha p(R|S_0)$ and $q_1C_\beta p(R|S_1)$ are non-negative. Thus for the integral to have its minimum value we must assign to the decision region D_0 those values of R for which

$$q_0C_\alpha p(R|S_0) > q_1C_\beta p(R|S_1) \quad (3.2.23)$$

The remaining points in the *n*-dimensional vector-space containing R satisfy

$$q_0C_\alpha p(R|S_0) \leqslant q_1C_\beta p(R|S_1) \quad (3.2.24)$$

Thus we must assign to the decision region D_1 those values of R for which

$$q_0C_\alpha p(R|S_0) < q_1C_\beta p(R|S_1) \quad (3.2.25)$$

and the decision boundary between D_0 and D_1 becomes

$$q_0C_\alpha p(R|S_0) = q_1C_\beta p(R|S_1) \quad (3.2.26)$$

It follows from the above that we must assign to D_0 those values of R for which

$$\frac{p(R|S_1)}{p(R|S_0)} < \frac{q_0C_\alpha}{q_1C_\beta} \quad (3.2.27)$$

and we must assign to D_1 those values of R for which

$$\frac{p(R|S_1)}{p(R|S_0)} > \frac{q_0C_\alpha}{q_1C_\beta} \quad (3.2.28)$$

The *decision boundary* is given by

$$\frac{p(R|S_1)}{p(R|S_0)} = \frac{q_0C_\alpha}{q_1C_\beta} \quad (3.2.29)$$

Since there is zero probability that R satisfies Equation 3.2.29 and so falls on the decision boundary, points on the decision boundary can be assigned arbitrarily to either decision region.

Let

$$\frac{p(R|S_1)}{p(R|S_0)} = L(R) \qquad (3.2.30)$$

where $L(R)$ is the *likelihood ratio*,

and let

$$\frac{q_0 C_\alpha}{q_1 C_\beta} = K \qquad (3.2.31)$$

where K is the *threshold*.

Thus the decision boundary is given by

$$L(R) = K \qquad (3.2.32)$$

If

$$L(R) < K$$

R lies in the decision region D_0 and hypothesis H_0 is accepted. S_i is now detected as S_0, so that i is detected as 0.

If

$$L(R) > K$$

R lies in the decision region D_1 and hypothesis H_1 is accepted. S_i is now detected as S_1, so that i is detected as 1.

The detection process just described is known as *Bayes' likelihood ratio test*.[1,2]

It can be seen that Bayes' likelihood ratio test holds for *any* joint probability density function of the n noise samples $\{w_j\}$ and not just for the joint Gaussian probability density function assumed here. Thus, whatever the joint probability density function of the noise samples $\{w_j\}$, Bayes' likelihood ratio test gives the best detection process for i from R. However, if the noise samples $\{w_j\}$ are not statistically independent of each other or not statistically independent of the signal samples $\{s_{ij}\}$, a better detection process for i can in general be obtained by using more than the n samples $\{r_j\}$ of the received waveform $r(t)$, assumed here. Again, if the received signal waveform $s_i(t)$ is nonzero at any time instants $\{jT\}$, for integers $j < l$ or $j > n$, or both, a better detection process for i can be obtained by including the corresponding received samples $\{r_j\}$ in the vector R.

3.2.3 Ideal observer strategy

In most data-transmission systems,

$$C_{10} = C_{01} = 1 \qquad (3.2.33)$$

and

$$C_{00} = C_{11} = 0 \qquad (3.2.34)$$

so that

$$C_\alpha = C_\beta = 1 \qquad (3.2.35)$$

From Equation 3.2.19, $G = q_0\alpha + q_1\beta = P_e$ (3.2.36)

where P_e is the *probability of error* in the detection of the element binary value.

The decision boundary is now given by

$$\frac{p(R|S_1)}{p(R|S_0)} = \frac{q_0}{q_1} \qquad (3.2.37)$$

Since the decision boundary is selected to minimize the average cost G, it is clear from Equation 3.2.36 that it also *minimizes the probability of error*. This arrangement is known as the *ideal observer detection system*.[1] It is optimum in the sense that no other receiver can operate with a lower average error rate.

If S_0 and S_1 are equally likely, so that $q_0 = q_1 = \frac{1}{2}$, the decision boundary becomes

$$\frac{p(R|S_1)}{p(R|S_0)} = 1 \qquad (3.2.38)$$

This is the optimum decision boundary for most practical data-transmission systems. The element value i is here detected as 0 or 1, depending upon whether $p(R|S_1) < p(R|S_0)$ or $p(R|S_1) > p(R|S_0)$, respectively, which is a process of *maximum likelihood* detection.

3.2.4 Neyman Pearson strategy

In some applications, such as radar, the error involved in accepting the hypothesis H_1 when H_0 is true, is more costly than the other error. The best strategy here is to fix α (the probability of the more costly error) at an acceptable level α_0 and then to minimize β. This is the *Neyman Pearson strategy*.[1] Now

$$\alpha = \int_{D_1} p(R|S_0)dR = \alpha_0 \qquad (3.2.39)$$

Usually more than one choice of D_1 results in $\alpha = \alpha_0$. The optimum decision boundary is that for which

$$\beta = \int_{D_0} p(R|S_1)dR \qquad (3.2.40)$$

is a minimum.

3.2.5 Interpretation of the likelihood-ratio test

From Equations 3.2.30 and 3.1.9, the likelihood ratio can be written

$$L(R) = \frac{p(R|S_1)}{p(R|S_0)} = \frac{P(S_1|R)\dfrac{p(R)}{P(S_1)}}{P(S_0|R)\dfrac{p(R)}{P(S_0)}} = \frac{q_0 P(S_1|R)}{q_1 P(S_0|R)} \qquad (3.2.41)$$

where $p(R)$ is the value of the joint probability density function of the random variables with sample values $r_1, r_2, \ldots, r_n$, when the random variables have the given values $r_1, r_2, \ldots, r_n$. $P(S_i|R)$, for $i = 0$ or 1, is the conditional probability of S_i, given R, so that it is the *a posteriori* probability of S_i.

From Equations 3.2.29 and 3.2.41, the decision boundary in Bayes' test is given by

$$\frac{P(S_1|R)}{P(S_0|R)} = \frac{C_\alpha}{C_\beta} \qquad (3.2.42)$$

When $C_\alpha = C_\beta$, as in most data-transmission systems, the decision boundary in Bayes' test becomes

$$\frac{P(S_1|R)}{P(S_0|R)} = 1 \qquad (3.2.43)$$

so that Bayes' test now selects S_0 or S_1, and hence detects the element value i as 0 or 1, depending upon whether S_0 or S_1 has the greater *a posteriori probability* of being correct. It can be seen from Equations

3.2.37 and 3.2.41 that this is the ideal observer strategy which, of course, *minimizes the probability of error*.

3.2.6 Bayes' strategy for one sample value

Assume that the received waveform

$$r(t) = s_i(t) + w(t) \tag{3.2.44}$$

is sampled only *once* and the element value i is detected from this sample value.

It is required to determine the decision boundary that minimizes the average cost.

Let the single sample values of $r(t)$, $s_i(t)$ and $w(t)$ be r, s_i and w, respectively, where w is a Gaussian random variable with zero mean and variance σ^2. Strictly speaking, r, s_i and w are one-component vectors, and the vector space containing r is one-dimensional. r, s_i and w can however be considered as scalar quantities. Clearly

$$r = s_i + w \tag{3.2.45}$$

From the received value of r the detector decides whether s_i is s_0 or s_1 and hence whether $i = 0$ or 1. It is required to optimize this decision process. The detector is assumed to have prior knowledge of s_0, s_1, q_0 and q_1.

According to Bayes' strategy, the one-dimensional vector space is divided into two decision regions and the decision boundary separating these is given by

$$L(r) = K \tag{3.2.46}$$

or

$$\log L(r) = \log K \tag{3.2.47}$$

where

$$L(r) = \frac{p(r|s_1)}{p(r|s_0)} \tag{3.2.48}$$

and

$$K = \frac{q_0 C_\alpha}{q_1 C_\beta} \tag{3.2.49}$$

It is permissible to express the decision boundary by Equation 3.2.47, since $\log x$ increases with x over all positive values of x, so that whenever $x_2 > x_1$, $\log x_2 > \log x_1$. The logarithms are to the base e.

In Equation 3.2.48, $p(r|s_1)$ is the probability density function of a

Gaussian random variable with mean s_1 and variance σ^2, and $p(r|s_0)$ is the probability density function of a Gaussian random variable with mean s_0 and variance σ^2. Thus

$$p(r|s_1) = \frac{1}{\sqrt{(2\pi\sigma^2)}} \exp\left[-\frac{(r-s_1)^2}{2\sigma^2} \right] \tag{3.2.50}$$

and

$$p(r|s_0) = \frac{1}{\sqrt{(2\pi\sigma^2)}} \exp\left[-\frac{(r-s_0)^2}{2\sigma^2} \right] \tag{3.2.51}$$

so that

$$L(r) = \frac{\exp\left[-\dfrac{(r-s_1)^2}{2\sigma^2} \right]}{\exp\left[-\dfrac{(r-s_0)^2}{2\sigma^2} \right]}$$

$$= \exp\left(\frac{1}{2\sigma^2}[(r-s_0)^2 - (r-s_1)^2] \right) \tag{3.2.52}$$

and

$$\log L(r) = \frac{1}{2\sigma^2}[(r-s_0)^2 - (r-s_1)^2] \tag{3.2.53}$$

From Equation 3.2.47, the decision boundary is given by

$$\frac{1}{2\sigma^2}[(r-s_0)^2 - (r-s_1)^2] = \log K \tag{3.2.54}$$

or

$$(r-s_0)^2 - (r-s_1)^2 = 2\sigma^2 \log K \tag{3.2.55}$$

or

$$|r-s_0|^2 - |r-s_1|^2 = 2\sigma^2 \log K \tag{3.2.56}$$

where $|r-s_0|$ is the *distance* between r and s_0, and $|r-s_1|$ is the *distance* between r and s_1. K is of course given by Equation 3.2.49.

When

$$|r-s_0|^2 - |r-s_1|^2 < 2\sigma^2 \log K$$

$$L(r) < K$$

and r lies in the decision region D_0. For *any* such value of r, s_i is detected as s_0, and hence i is detected as 0.

When

$$|r-s_0|^2 - |r-s_1|^2 > 2\sigma^2 \log K$$

$$L(r) > K$$

and r lies in the decision region D_1. For *any* such value of r, s_i is detected as s_1, and hence i is detected as 1.

When s_0 and s_1 are *equally likely* to be received and it is required to *minimize the error probability*, $q_0 = q_1 = \frac{1}{2}$ and $C_\alpha = C_\beta = 1$. Thus $K = 1$ and $\log K = 0$. From Equation 3.2.56, the decision boundary is now given by

$$|r-s_0|^2 - |r-s_1|^2 = 0 \qquad (3.2.57)$$

so that s_i is detected as s_0 when

$$|r-s_0|^2 - |r-s_1|^2 < 0$$

and s_i is detected as s_1 when

$$|r-s_0|^2 - |r-s_1|^2 > 0$$

Thus s_i is detected as s_0 when r is *closer* to s_0 than to s_1, and s_i is detected as s_1 when r is *closer* to s_1 than to s_0.

Since r, s_i and w in Equation 3.2.45 are, strictly speaking, one-component vectors, it is clear that r must lie in the one-dimensional vector space given by the *line* through s_0 and s_1, so that r must lie *on* this line. From Equation 3.2.57, the decision boundary separating the two decision regions in the one-dimensional vector space is given by

$$r^2 - 2rs_0 + s_0{}^2 - r^2 + 2rs_1 - s_1{}^2 = 0 \qquad (3.2.58)$$

or
$$2r(s_1 - s_0) - (s_1 + s_0)(s_1 - s_0) = 0 \qquad (3.2.59)$$

or
$$r = \tfrac{1}{2}(s_1 + s_0) \qquad (3.2.60)$$

so that the decision boundary is the *point* which *bisects* the line joining s_1 and s_0, as shown in Figure 3.3. Both s_1 and s_0 are at a distance d from the decision boundary, where

$$d = \tfrac{1}{2}|s_1 - s_0| \qquad (3.2.61)$$

The probability of error in this detection process is determined as follows, assuming for convenience that $s_1 > s_0$.

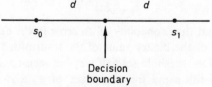

Fig. 3.3 *Decision boundary in the one-dimensional vector space*

s_0 is wrongly detected as s_1 whenever r is on the opposite side of the decision boundary to s_0, given that s_0 is in fact received. The decision boundary is the point whose value is $\frac{1}{2}(s_1 + s_0)$. Thus the probability that s_0 is detected as s_1 is

$$P(r > \tfrac{1}{2}(s_1 + s_0)|s_i = s_0)$$
$$= P(r - s_0 > \tfrac{1}{2}(s_1 - s_0)|s_i = s_0)$$
$$= P(w > d)$$
$$= \int_d^\infty \frac{1}{\sqrt{(2\pi\sigma^2)}} \exp\left(-\frac{w^2}{2\sigma^2}\right) dw \qquad (3.2.62)$$

since here $r = s_0 + w$, and w is a Gaussian random variable with zero mean and variance σ^2.

The probability that s_1 is wrongly detected as s_0 is

$$P(r < \tfrac{1}{2}(s_1 + s_0)|s_i = s_1)$$
$$= P(r - s_1 < -\tfrac{1}{2}(s_1 - s_0)|s_i = s_1)$$
$$= P(w < -d)$$
$$= \int_{-\infty}^{-d} \frac{1}{\sqrt{(2\pi\sigma^2)}} \exp\left(-\frac{w^2}{2\sigma^2}\right) dw$$
$$= \int_d^\infty \frac{1}{\sqrt{(2\pi\sigma^2)}} \exp\left(-\frac{w^2}{2\sigma^2}\right) dw \qquad (3.2.63)$$

since here $r = s_1 + w$.

Thus whether s_0 or s_1 is received, the probability of error is

$$\int_d^\infty \frac{1}{\sqrt{(2\pi\sigma^2)}} \exp\left(-\frac{w^2}{2\sigma^2}\right) dw = \int_{d/\sigma}^\infty \frac{1}{\sqrt{(2\pi)}} \exp(-\tfrac{1}{2}w^2) dw$$
$$= Q\left(\frac{d}{\sigma}\right) \qquad (3.2.64)$$

where
$$Q(y) = \int_y^\infty \frac{1}{\sqrt{(2\pi)}} \exp(-\tfrac{1}{2}x^2) dx \qquad (3.2.65)$$

It is clear that the probability of an error in the detection process is *independent* of the binary value of the transmitted signal, so that with the decision threshold set half way between s_0 and s_1, the error probability is the same for all values of q_0 and q_1. This error probability is the minimum obtainable for all decision boundaries when $q_0 = q_1 = \frac{1}{2}$. It is not, however, the minimum obtainable for any other values of q_0 and q_1.

3.2.7 Bayes' strategy for n sample values

Assume that the received signal waveform

$$r(t) = s_i(t) + w(t) \qquad (3.2.66)$$

is sampled n times and the element value i is detected from the n sample values. It is required to determine the decision boundary that minimizes the average cost.

Let the n sample values of $r(t)$ be the n components of the row vector

$$R = r_1 \; r_2 \ldots r_n \qquad (3.2.67)$$

Let the n sample values of $s_i(t)$ be the n components of the row vector

$$S_i = s_{i1} \; s_{i2} \ldots s_{in} \qquad (3.2.68)$$

Let the n sample values of $w(t)$ be the n components of the row vector

$$W = w_1 \; w_2 \ldots w_n \qquad (3.2.69)$$

The components of W are statistically independent Gaussian random variables with zero mean and variance σ^2, as before. Clearly,

$$R = S_i + W \qquad (3.2.70)$$

and

$$r_j = s_{ij} + w_j \qquad (3.2.71)$$

for $j = 1, 2, \ldots, n$.

From the value of the received vector R, the detector decides whether S_i is S_0 or S_1 and hence whether $i = 0$ or 1. It is required to optimize this decision process. The detector is assumed to have prior knowledge of S_0, S_1, q_0 and q_1.

According to Bayes' strategy, the n-dimensional vector space containing R is divided into two decision regions and the decision boundary separating these is given by

$$L(R) = K \qquad (3.2.72)$$

or

$$\log L(R) = \log K \qquad (3.2.73)$$

where
$$L(R) = \frac{p(R|S_1)}{p(R|S_0)} = \frac{p(r_1, r_2, \ldots, r_n|S_1)}{p(r_1, r_2, \ldots, r_n|S_0)} \qquad (3.2.74)$$

$p(R|S_i)$, for $i = 0$ or 1, is the value of the conditional joint probability density function of the random variables with sample values r_1, $r_2, \ldots, r_n$, when the random variables have the given values $r_1, r_2, \ldots, r_n$, and given the values $s_{i1}, s_{i2}, \ldots, s_{in}$. K is given by Equation 3.2.49, as before.

It can be seen from Equation 3.2.71 that r_j is a sample value of a Gaussian random variable with a mean value s_{ij} and a variance σ^2. Furthermore the different $\{r_j\}$ are sample values of statistically independent Gaussian random variables. Thus the conditional probability density function of the random variable with sample value r_j, given S_i, is

$$p(r_j|S_i) = \frac{1}{\sqrt{(2\pi\sigma^2)}} \exp\left[-\frac{(r_j - s_{ij})^2}{2\sigma^2} \right] \qquad (3.2.75)$$

for $j = 1, 2, \ldots, n$, and

$$p(r_1, r_2, \ldots, r_n|S_i) = p(r_1|S_i)p(r_2|S_i) \ldots p(r_n|S_i) \qquad (3.2.76)$$

Now
$$\begin{aligned}
p(R|S_i) &= p(r_1, r_2, \ldots, r_n|S_i) \\
&= \prod_{j=1}^{n} \frac{1}{\sqrt{(2\pi\sigma^2)}} \exp\left[-\frac{(r_j - s_{ij})^2}{2\sigma^2} \right] \\
&= \frac{1}{(2\pi\sigma^2)^{\frac{1}{2}n}} \exp\left[\sum_{j=1}^{n} -\frac{(r_j - s_{ij})^2}{2\sigma^2} \right] \\
&= \frac{1}{(2\pi\sigma^2)^{\frac{1}{2}n}} \exp\left[-\frac{1}{2\sigma^2} \sum_{j=1}^{n} (r_j - s_{ij})^2 \right] \qquad (3.2.77)
\end{aligned}$$

Hence
$$L(R) = \frac{\exp\left[-\dfrac{1}{2\sigma^2} \displaystyle\sum_{j=1}^{n} (r_j - s_{1j})^2 \right]}{\exp\left[-\dfrac{1}{2\sigma^2} \displaystyle\sum_{j=1}^{n} (r_j - s_{0j})^2 \right]}$$

$$= \exp\left[\frac{1}{2\sigma^2}\left(\sum_{j=1}^{n}(r_j-s_{0j})^2 - \sum_{j=1}^{n}(r_j-s_{1j})^2\right)\right] \quad (3.2.78)$$

and

$$\log L(R) = \frac{1}{2\sigma^2}\left(\sum_{j=1}^{n}(r_j-s_{0j})^2 - \sum_{j=1}^{n}(r_j-s_{1j})^2\right) \quad (3.2.79)$$

The distance between the vectors R and S_i is $|R-S_i|$, so that the square of the distance between R and S_i is

$$|R-S_i|^2 = \sum_{j=1}^{n}(r_j-s_{ij})^2 \quad (3.2.80)$$

Thus

$$\log L(R) = \frac{1}{2\sigma^2}(|R-S_0|^2 - |R-S_1|^2) \quad (3.2.81)$$

From Equation 3.2.73, the decision boundary is given by

$$\frac{1}{2\sigma^2}(|R-S_0|^2 - |R-S_1|^2) = \log K \quad (3.2.82)$$

or

$$|R-S_0|^2 - |R-S_1|^2 = 2\sigma^2 \log K \quad (3.2.83)$$

When

$$|R-S_0|^2 - |R-S_1|^2 < 2\sigma^2 \log K$$

$$L(R) < K$$

and R lies in the decision region D_0. For *any* such value of R, S_i is detected as S_0, and hence i is detected as 0.

When

$$|R-S_0|^2 - |R-S_1|^2 > 2\sigma^2 \log K$$

$$L(R) > K$$

and R lies in the decision region D_1. For *any* such value of R, S_i is detected as S_1, and hence i is detected as 1.

When S_0 and S_1 are equally likely to be received and it is required to minimize the error probability, then $q_0 = q_1 = \frac{1}{2}$ and $C_\alpha = C_\beta = 1$, so that $\log K = 0$. The decision boundary is now given by

$$|R-S_0|^2 - |R-S_1|^2 = 0 \quad (3.2.84)$$

so that S_i is detected as S_0 when

$$|R-S_0|^2 - |R-S_1|^2 < 0$$

and S_i is detected as S_1 when

$$|R - S_0|^2 - |R - S_1|^2 > 0$$

Thus S_i is detected as S_0 when R is *closer* to S_0 than to S_1, and S_i is detected as S_1 when R is *closer* to S_1 than to S_0.

The decision boundary given by Equation 3.2.84 is an $(n-1)$-dimensional hyperplane which perpendicularly bisects the line joining the vectors S_0 and S_1. Both S_0 and S_1 are at a distance d from this hyperplane, where

$$d = \tfrac{1}{2}|S_1 - S_0| \qquad (3.2.85)$$

In order to determine the probability of error in the detection of S_i from R, it is first necessary to establish an important property of the noise vector W.

Let J be a unit vector (vector of unit length) which may have any direction in the n-dimensional vector space. Since J has unit length,

$$JJ^T = 1 \qquad (3.2.86)$$

The value of the orthogonal projection of the noise vector W on to J, or more precisely, on to the one-dimensional subspace spanned by J in the n-dimensional vector space, is the inner product of the vectors W and J, given by

$$WJ^T = JW^T \qquad (3.2.87)$$

This is a Gaussian random variable with zero mean and variance given by the expected value of the square of the inner product of W and J, which is

$$E[(WJ^T)^2] = E[JW^TWJ^T]$$
$$= JHJ^T \qquad (3.2.88)$$

where $\qquad\qquad H = E[W^TW] \qquad (3.2.89)$

is the correlation or covariance matrix of the n noise samples $\{w_j\}$. From Equation 3.2.89, the component in the ith row and jth column of the $n \times n$ matrix H is

$$h_{ij} = E[w_i w_j] \qquad (3.2.90)$$

where w_i and w_j are the ith and jth components respectively of the noise vector W.

Since the different noise samples are statistically independent and therefore uncorrelated, and since they have zero mean,

$$E[w_i w_j] = 0 \qquad (3.2.91)$$

when $i \neq j$. Since the noise samples have zero mean and a variance σ^2,

$$E[w_i^2] = \sigma^2 \qquad (3.2.92)$$

Thus

$$H = \sigma^2 I \qquad (3.2.93)$$

where I is an $n \times n$ identity matrix.

From Equations 3.2.88 and 3.2.93, the *value* of the orthogonal projection of W on to the one-dimensional subspace spanned by J, and therefore the value of the noise component in the direction of J, is a Gaussian random variable with zero mean and variance

$$JHJ^T = \sigma^2 JIJ^T$$

$$= \sigma^2 JJ^T$$

$$= \sigma^2 \qquad (3.2.94)$$

from Equation 3.2.86. The orthogonal projection of W on to the one-dimensional subspace is, of course, itself the n-component row vector

$$WJ^T J = JW^T J \qquad (3.2.95)$$

Thus the value of the orthogonal projection of the noise vector W on to *any* direction in the n-dimensional vector space is a Gaussian random variable with zero mean and variance σ^2.

The probability of an error in the detection of S_i may now be deduced with the aid of Figure 3.4.

When $S_i (= S_0 \text{ or } S_1)$ is received in the presence of a noise vector

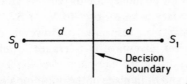

Fig. 3.4   Decision boundary in the n-dimensional vector space

W, it is wrongly detected if R is on the opposite side of the decision boundary to S_i. This occurs when the orthogonal projection of W on to the line joining S_0 and S_1 has a value greater than d in the direction from S_i towards the decision boundary. But the value of the projection of W is a Gaussian random variable with zero mean and variance σ^2. Alternatively, the noise vector W may be represented in terms of n orthogonal components whose directions are chosen as required, the value of each component being a Gaussian random variable with zero mean and variance σ^2. One of these components is now selected to have a direction parallel to the line joining S_0 and S_1, so that the remaining $n-1$ components are orthogonal to this line. The latter components of W can only shift R in a direction parallel to the decision boundary and so do not affect the distance of R from the boundary. They cannot therefore produce errors, nor can they even affect the error probability. Thus the error probability is dependent solely on the component of W that is parallel to the line joining S_0 to S_1. Hence the probability of an error in the detection of S_i, whether $i = 0$ or 1, is

$$\int_d^\infty \frac{1}{\sqrt{(2\pi\sigma^2)}} \exp\left(-\frac{w^2}{2\sigma^2}\right) dw = Q\left(\frac{d}{\sigma}\right) \tag{3.2.96}$$

where d is the distance from both S_0 and S_1 to the decision boundary. This result applies for any value of n, so long as the Gaussian noise samples are statistically independent.

When the receiver has no prior knowledge of q_0 and q_1 (the *a priori* probabilities of S_0 and S_1) it is conventional to assume that $q_0 = q_1 = \frac{1}{2}$. Under these conditions the correct decision boundary is as shown in Figure 3.4. Thus the error probability is given by Equation 3.2.96 and is independent of whether S_0 or S_1 is received. The error probability for this detector is clearly independent of the actual values of q_0 and q_1.

When $q_0 \neq q_1$ the error probability can be made smaller by using the optimum decision boundary for the correct values of q_0 and q_1. The decision boundary is here closer to S_0 or S_1, depending upon whether $q_0 < q_1$ or $q_0 > q_1$, respectively. However, unless $q_0 \ll q_1$ or $q_0 \gg q_1$, no very great advantage in tolerance to additive Gaussian noise is gained by using the optimum decision boundary for the *actual* values of q_0 and q_1 in place of the optimum decision boundary assuming that $q_0 = q_1 = \frac{1}{2}$.

3.2.8 Optimum detection process for an l-level signal

Assume, as before, that the received waveform is

$$r(t) = s_i(t) + w(t) \tag{3.2.97}$$

except that now i has one of l different values $0, 1, \ldots, l-1$ (where $l \geqslant 2$). Each of the corresponding l signals $\{s_i(t)\}$ has a different waveform, which may have any suitable shape. Clearly, $s_i(t)$ is an l-level signal element.

The received waveform $r(t)$ is sampled n times and the element value i is detected from the n sample values. These are given by the n-component row vector

$$R = S_i + W \tag{3.2.98}$$

as before.

The detector is assumed to have prior knowledge of the l vectors $\{S_i\}$, which implies that these l vectors are held in an appropriate store. The detector also has prior knowledge of the *a priori* probabilities of the different vectors $\{S_i\}$. The whole of this prior knowledge is used to optimize the detection process, that is, to *minimize the probability of error* in the detection of the element value.

From the value of the received vector R the detector selects one of the l vectors $\{S_i\}$ as the detected signal, and the corresponding value of i gives the detected element value.

In order to minimize the probability of error in a detection process[3] it is necessary to *maximize* the probability of a *correct* decision, $P(C)$. Now

$$P(C) = \int_{-\infty}^{\infty} \ldots \int_{-\infty}^{\infty} P(C|R)p(r_1, r_2, \ldots, r_n)\mathrm{d}r_1 \, \mathrm{d}r_2 \ldots$$
$$\ldots \mathrm{d}r_n \tag{3.2.99}$$

where $P(C|R)$ is the conditional probability of a correct decision *given* the received vector R, and $p(r_1, r_2, \ldots, r_n)$ is the value of the joint probability density function of the n random variables, corresponding to the n components of R, at the given value of R. The vector R may here have *any* of its possible values and is not confined to the *particular* value received. Equation 3.2.99 may be written more simply as

$$P(C) = \int_{-\infty}^{\infty} P(C|R)p(R)\mathrm{d}R \tag{3.2.100}$$

Since $p(R)$ is nonnegative, it can be seen that $P(C)$ is maximized by maximizing $P(C|R)$ for *every possible* value of the vector R. For any *given* received vector R, $P(C|R)$ is maximized by selecting as the detected value of S_i, the vector S_j for which

$$P(S_j|R) > P(S_i|R), \quad i = 0, 1, \ldots, l-1, \quad i \neq j \quad (3.2.101)$$

$P(S_i|R)$ is the conditional probability of S_i, given R, and so is the *a posteriori* probability of S_i.

Thus the detection process that *minimizes the probability of error*, selects from the l vectors $\{S_i\}$ the vector which *maximizes* $P(S_i|R)$ and which therefore has the greatest *a posteriori* probability of being correct. R is here the *given* received vector. Clearly, the detected element value is j, where j is the value of i that maximizes $P(S_i|R)$. If the maximum *a posteriori* probability is shared by two or more of the l vectors $\{S_i\}$, the detector may select *any one* of these vectors as the detected vector S_j.

From Equation 3.1.9,

$$P(S_i|R) = \frac{p(R|S_i)P(S_i)}{p(R)} \quad (3.2.102)$$

where the terms are defined as follows. $P(S_i)$ is the *a priori* probability of S_i.

$$p(R) = p(r_1, r_2, \ldots, r_n)$$

is the value of the joint probability density function of the random variables with sample values $r_1, r_2, \ldots, r_n$, when the random variables have the given values $r_1, r_2, \ldots, r_n$.

$$p(R|S_i) = p(r_1, r_2, \ldots, r_n|s_{i1}, s_{i2}, \ldots, s_{in})$$

is the value of the conditional joint probability density function of the random variables with sample values $r_1, r_2, \ldots, r_n$, when the random variables have the given values $r_1, r_2, \ldots, r_n$, and given the values $s_{i1}, s_{i2}, \ldots, s_{in}$.

From Equations 3.2.101 and 3.2.102, the detection process that minimizes the probability of error, selects from the l vectors $\{S_i\}$, the vector S_j for which

$$P(S_j)p(R|S_j) > P(S_i)p(R|S_i), \quad i = 0, 1, \ldots, l-1, \quad i \neq j \quad (3.2.103)$$

$$\text{or} \quad \frac{p(R|S_j)}{p(R|S_i)} > \frac{P(S_i)}{P(S_j)}, \quad i = 0, 1, \ldots, l-1, \quad i \neq j \quad (3.2.104)$$

This is *Bayes' likelihood ratio test* for a set of l hypotheses, where the cost of no error is always zero and the cost of an error is always 1. It is therefore the *ideal observer detection system* for an l-level signal element. When $l = 2$, this becomes the ideal observer detection system for a binary signal-element, as described in Section 3.2.3.

The detection process given by Equation 3.2.104 minimizes the probability of error in the detection of the element value i, for *any* joint probability density function of the n noise samples $\{w_j\}$ and not just for the joint Gaussian probability density assumed here.

In the important case where the l vectors $\{S_i\}$ are *equally likely* to be received,

$$P(S_i) = \frac{1}{l} \qquad (3.2.105)$$

and the detection process, that minimizes the probability of error, selects from the l vectors $\{S_i\}$, the vector S_j for which

$$\frac{p(R|S_j)}{p(R|S_i)} > 1, \quad i = 0, 1, \ldots, l-1, \quad i \neq j \qquad (3.2.106)$$

or $\qquad p(R|S_j) > p(R|S_i), \quad i = 0, 1, \ldots, l-1, \quad i \neq j \qquad (3.2.107)$

Thus the detected element value is j, where j is the value of i that *maximizes* $p(R|S_i)$.

Since the noise components $\{w_j\}$ in the given received vector R are sample values of statistically independent Gaussian random variables with zero mean and variance σ^2, it follows from Equation 3.2.71 that r_j is a sample value of a Gaussian random variable with a mean value s_{ij} and a variance σ^2. Furthermore, the different $\{r_j\}$ are sample values of statistically independent Gaussian random variables. Thus the conditional probability density function of the random variable with sample value r_j, given S_i, is

$$p(r_j|S_i) = \frac{1}{\sqrt{(2\pi\sigma^2)}} \exp\left(-\frac{(r_j - s_{ij})^2}{2\sigma^2}\right) \qquad (3.2.108)$$

where $j = 1, 2, \ldots, n$, and so

$$p(R|S_i) = p(r_1, r_2, \ldots, r_n|S_i)$$
$$= p(r_1|S_i)p(r_2|S_i) \ldots p(r_n|S_i)$$
$$= \prod_{j=1}^{n} \frac{1}{\sqrt{(2\pi\sigma^2)}} \exp\left(-\frac{(r_j - s_{ij})^2}{2\sigma^2}\right)$$

$$= \frac{1}{(2\pi\sigma^2)^{\frac{1}{2}n}} \exp\left(-\frac{1}{2\sigma^2}\sum_{j=1}^{n}(r_j - s_{ij})^2\right)$$

$$= \frac{1}{(2\pi\sigma^2)^{\frac{1}{2}n}} \exp\left(-\frac{1}{2\sigma^2}|R - S_i|^2\right) \qquad (3.2.109)$$

where $|R - S_i|$ is the *distance* between the vectors R and S_i.

It can be seen from Equation 3.2.109 that when $p(R|S_i)$ is *maximum*, $|R - S_i|$ is *minimum*, so that the detected element value j satisfies

$$|R - S_j| < |R - S_i|, \quad i = 0, 1, \ldots, l-1, \quad i \neq j \qquad (3.2.110)$$

and the distance between R and S_j *is less* than the distance between R and any of the other $\{S_i\}$. Thus the detection process, that *minimizes the probability of error*,[3] selects the element value j for which S_j is *closest* to R.

3.2.9   Optimum detection process for m l-level signals

Suppose now that the n-component signal vector S_i in Equation 3.2.98 instead of being an individual l-level signal-element becomes the sum of m l-level signal elements, the ith of which is the n-component row vector $s_i Y_i$. s_i is here a scalar quantity that has one of l possible values and carries the element value of the ith signal-element. Y_i is the n-component row vector,

$$Y_i = y_{i1} \; y_{i2} \cdots y_{in} \qquad (3.2.111)$$

that characterizes or identifies the ith signal-element. $m \leqslant n$ and the m vectors $\{Y_i\}$ are assumed to be linearly independent, so that the resultant signal vector

$$\sum_{i=1}^{m} s_i Y_i$$

has l^m possible values. The linear independence of the m $\{Y_i\}$ is a *sufficient* (but not in fact necessary) condition for the resultant signal vector to have a different value for every different combination of the values of the $\{s_i\}$, and therefore for the unique detectability of the m $\{s_i\}$ from the resultant signal vector.

Let Y be the $m \times n$ matrix of rank m whose ith row is given by the n-component row vector Y_i, and let S be the m-component row vector

$$S = s_1 \ s_2 \dots s_m \qquad (3.2.112)$$

Then

$$\sum_{i=1}^{m} s_i Y_i = SY \qquad (3.2.113)$$

From Equation 3.2.98, the n received sample values are given by the n-component row vector

$$R = SY + W \qquad (3.2.114)$$

Equation 3.2.114 is obtained from Equation 3.2.98 by replacing the signal vector S_i with SY, and the only real difference between these two vectors is that SY has l^m possible values, whereas S_i has only l possible values. It follows that the optimum detection process for SY is basically the same as that described for S_i in Section 3.2.8. The detected vector SY uniquely determines the corresponding vector S, whose m components $\{s_i\}$ carry the element values of the m signal-elements and therefore the information being transmitted.

If each component s_i of S is equally likely to have any one of its l possible values and if the m $\{s_i\}$ are statistically independent, then the vector S is equally likely to have any one of its l^m possible values, and so is the vector SY. Under these conditions the detection process that minimizes the probability of error in the detection of S from the received vector R, selects from the l^m possible vectors $\{SY\}$ the vector at the minimum distance from R and takes the corresponding vector S as the detected vector. In other words, it selects as the detected vector S the one of its l^m possible values that minimizes $|R - SY|$. This detection process minimizes the probability of the incorrect detection of *one or more* of the $\{s_i\}$, from the received vector R. It may be shown that, at very high signal/noise ratios, the detection process also minimizes the probability of error in the detection of any *individual* s_i (see Section 5.1).

Strictly speaking, a *different* symbol should be used for each of the l^m different possible values of the vector S, or at least for the value of S *actually received* and for a *possible* value of S, which includes the value actually received and $l^m - 1$ *other* values. This has not been done for two reasons. Firstly, so long as it is borne in mind that, for any *given* received vector R, the vector S in Equation 3.2.114 is the *particular* value of S received, whereas the *set* of vectors, designated by $\{SY\}$ or $\{S\}$, is the collection of *all possible* values of SY or S, respectively, there is in fact no confusion between the two meanings ascribed to the symbol S. The introduction of additional symbols would therefore introduce an unnecessary complication into the notation used. Secondly, the use of the symbol S with the two meanings emphasizes the important fact that the different possible values ascribed to S in the detection process are the *actual* possible values of this vector and, of course, include the particular value received. In the estimation processes described in Section 3.3, the receiver does not have any prior knowledge of the signal vector S, so that the possible values ascribed to S in an estimation process are not necessarily or even usually the actual possible values of this vector. In this case, the possible value ascribed to S is designated S_a, to emphasize the fact that its value is *not* usually one of the actual possible values of S.

The detection process is implemented as follows. It generates in turn each of the l^m possible vectors $\{SY\}$ and for each of these it measures the distance $|R - SY|$ from SY to the received vector R. Each time the measured distance is smaller than the smallest distance previously measured, the new distance together with the associated vector S are stored, replacing the corresponding quantities previously stored. Whenever the measured distance is equal to or greater than the stored distance, the new distance together with the associated vector S are ignored. Thus at the end of the process, the stored vector S is the required detected vector.

The l^m possible values of the signal vector SY may be represented as points in an n-dimensional vector space. An error occurs in the detection of S when the received signal vector SY is *not* the one of the l^m possible values of SY at the minimum distance from the received vector R. The exact evaluation of the probability of error in the detection of S is difficult, but an approximate upper bound to the error probability may readily be obtained. Let the minimum distance between any two of the possible vectors $\{SY\}$ be $2d$, and let $S_b Y$ and $S_c Y$ be two of the $\{SY\}$ at this minimum distance. The probability of an error in the detection of S is generally less than the probability that the signal vector $S_b Y$, received together with the noise vector W, is incorrectly detected as the signal vector $S_c Y$. The received n-component vector is here

$$R = S_b Y + W \qquad (3.2.115)$$

and the decision boundary between $S_b Y$ and $S_c Y$ is the $(n-1)$-dimensional hyperplane that perpendicularly bisects the line joining $S_b Y$ and $S_c Y$. Thus $S_b Y$ is at a distance d from the decision boundary and the probability that $S_b Y$ is incorrectly detected as $S_c Y$ is $Q(d/\sigma)$, as in Equation 3.2.96. It may now be shown that the probability of error in the detection of S from R, in Equation 3.2.114, is generally less than $Q(d/\sigma)$. At high signal/noise ratios, the error probability is in fact $Q(d_0/\sigma)$, where d_0 is no more than 10 or 20% (that is, 1 or 2 dB) above d, the discrepancy between d and d_0 decreasing as the signal/noise ratio increases. The approximate upper bound $Q(d/\sigma)$ to the probability of error in the detection of S from R may be justified heuristically as follows.

If, in the detection of S from R, the received signal vector SY is at a distance $2d$ from one of the other possible vectors $\{SY\}$, the probability of error in the detection of S from R must exceed $Q(d/\sigma)$, since there are many other possible vectors $\{SY\}$ at various distances of more than $2d$ from the received vector SY and *each* of these is separated from the received vector SY by a decision boundary that is an $(n-1)$-dimensional hyperplane perpendicularly bisecting the line joining it to the received vector SY. However, at high signal/noise ratios with error rates of the order of 1 in 10^5 or 1 in 10^6, an increase of only 10% in the distance from the received vector SY to the decision boundary that separates it from another of the possible vectors $\{SY\}$, reduces the probability that SY will be incorrectly detected as the other possible vector SY, by a factor of about 10 times. Thus the probability of error in the detection of S is determined largely by the distance from SY to the *nearest* decision boundary.

If the received vector SY is at the minimum distance $2d$ from i of the other possible vectors $\{SY\}$, then the presence of these i vectors (in the absence of the other possible $\{SY\}$) may raise the probability of error in the detection of S to nearly $iQ(d/\sigma)$. Clearly, if *every* possible vector SY has several other possible vectors $\{SY\}$ at the minimum distance $2d$, then the average probability of error in the detection of S exceeds $Q(d/\sigma)$. On the other hand, there may be several of the possible vectors $\{SY\}$, or even the large majority of these, that have no other of the possible vectors $\{SY\}$ at or near the minimum distance $2d$, and the probability of error in the detection of any of these is negligible compared with $Q(d/\sigma)$. Under these conditions, the average probability of error may be only a small fraction of $Q(d/\sigma)$. With practical signals there is normally considerable symmetry in the possible values of SY. The two effects just mentioned now tend to offset each other to a significant degree, with the result that, at high signal/noise ratios, the average probability of error in the detection of SY is $Q(d_0/\sigma)$, where d_0 is usually no more than 1 or 2 dB greater than d, and rarely less than d.

An error in the detection of S involves an error in the detection of *one or more* of the m $\{s_i\}$ and therefore an error in the detection of one or more of the m signal-elements that comprise SY. Clearly, the average probability of error in the detection of an individual received signal-element is less than the probability of error in the detection of S. Thus $Q(d/\sigma)$ may be taken as an approximate upper bound to the probability of error in the detection of an individual received signal-element, and $Q(d/\sigma)$ gives an estimate of the signal/noise ratio for a given element-error rate, the estimate becoming more accurate as the signal/noise ratio increases.

3.3 LINEAR ESTIMATION OF UNKNOWN PARAMETERS

3.3.1 *Maximum-likelihood estimation of an unknown parameter*

Consider the received waveform

$$r(t) = sy(t) + w(t) \tag{3.3.1}$$

where $sy(t)$ is the signal waveform and $w(t)$ the noise waveform. The information in the transmitted signal is carried in the value of the real scalar s.

The received waveform $r(t)$ is sampled n times at the time instants $\{jT\}$, for $j = 1, 2, \ldots, n$, to give the n components of the row vector

$$R = r_1\ r_2 \ldots r_n \tag{3.3.2}$$

where $r_j = r(jT)$. The corresponding n sample values of $y(t)$ are given by the n components of the row vector

$$Y = y_1\ y_2 \ldots y_n \tag{3.3.3}$$

where $y_j = y(jT)$, and the n sample values of $w(t)$ are given by the n components of the row vector

$$W = w_1\ w_2 \ldots w_n \tag{3.3.4}$$

where $w_j = w(jT)$. As before, the $\{w_j\}$ are statistically independent Gaussian random variables with zero mean and variance σ^2.

Clearly, the n sample values of the signal waveform $sy(t)$ are given by the n components of the row vector

$$sY = sy_1 \ sy_2 \dots sy_n \tag{3.3.5}$$

and the received n-component vector is

$$R = sY + W \tag{3.3.6}$$

It is assumed that the mean-square value (or expected energy) of the parameter s is given by

$$E[s^2] = \overline{s^2} \tag{3.3.7}$$

where $\overline{s^2}$ is a constant. Also, the $\{w_j\}$ are statistically independent of s.

The receiver is assumed to have prior knowledge of Y, but no prior knowledge of s or σ^2. Given the received vector R, it is required to make the *best linear estimate* of s.

Let s now be the *actual* value of the parameter being estimated, and let s_a be any *possible* value of s as far as the receiver is concerned, before it has estimated s from R.

In any one transmission of the signal waveform, s and s_a are *sample values* of the corresponding random variables. No particular assumptions are made here about s. It may have any one of two or more discrete values, or any one of a continuous range (infinite number) of values. Thus the parameter s will normally (or often) have different values in different transmissions. The possible values of s and the prior knowledge the receiver has of these values determine the probability density function $p(s_a)$ of the random variable with sample value s_a. The less prior knowledge the receiver has of s (that is, of the random variable with sample value s), the wider the range of values of s_a with a nonnegligible probability density $p(s_a)$. When the receiver has *no* prior knowledge of s, as is assumed here, s_a is equally likely to have *any* real value, so that $p(s_a)$ is constant over *all* real values of s_a. Thus the receiver must accept both say $s_a = 10^{-10}$ and $s_a = 10^{10}$ as being equally likely estimates of s. Furthermore, the possible values of s now have no influence on $p(s_a)$.

The best linear estimate of s from R is here taken to be the value of s_a that maximizes $p(s_a | R)$, the value of the conditional probability density function of the random variable with sample value s_a, at the given value s_a and given R, which is the *a posteriori* probability density of s_a. Thus the best estimate of s is taken to be the *maximum a posteriori* estimate. The estimate of s may, of course, be optimized according to other criteria, so that this is not necessarily the most suitable estimate in every application.

From Bayes' theorem, as given by Equation 3.1.8,

$$p(s_a|R) = \frac{p(s_a)}{p(R)}p(R|s_a) \tag{3.3.8}$$

Since $p(R)$ does not depend on s_a and $p(s_a)$ is constant for all real values of s_a, the value of s_a that maximizes $p(s_a|R)$ also maximizes $p(R|s_a)$, and vice versa. Thus the receiver must choose s_a to maximize $p(R|s_a)$ which is the likelihood function of s_a. This gives the *maximum-likelihood estimate* of s. Clearly, when the receiver has no prior knowledge of s, the maximum-likelihood estimate of s is also the maximum *a posteriori* estimate and is therefore the required estimate.

It can be seen from Equation 3.3.8 that when the receiver has *some* prior knowledge of s, which means that $p(s_a)$ is not a constant, the maximum *a posteriori* estimate of s is no longer the same as the maximum-likelihood estimate and is furthermore a *better* estimate since it makes use of the available prior knowledge of s.

The receiver may, of course, be designed for use with a noise signal having *any* joint probability density for the n samples $\{w_j\}$ and not just for the joint Gaussian probability density assumed here. If $s = s_a$ in the received signal given by Equation 3.3.6, assuming for the moment that s may take on any value, then

$$r_j = s_a y_j + w_j \tag{3.3.9}$$

for $j = 1, 2, \ldots, n$, so that, for the given received sample r_j and the given value of s, r_j is a sample value of a Gaussian random variable with mean $s_a y_j$ and variance σ^2, the n $\{r_j\}$ being sample values of statistically independent Gaussian random variables. Thus

$$p(R|s_a) = p(r_1|s_a y_1)p(r_2|s_a y_2) \ldots p(r_n|s_a y_n)$$

$$= \prod_{j=1}^{n} \frac{1}{\sqrt{(2\pi\sigma^2)}} \exp\left[-\frac{(r_j - s_a y_j)^2}{2\sigma^2}\right]$$

$$= \frac{1}{(2\pi\sigma^2)^{\frac{1}{2}n}} \exp\left[-\frac{1}{2\sigma^2} \sum_{j=1}^{n} (r_j - s_a y_j)^2 \right]$$

$$= \frac{1}{(2\pi\sigma^2)^{\frac{1}{2}n}} \exp\left[-\frac{1}{2\sigma^2} |R - s_a Y|^2 \right] \tag{3.3.10}$$

where $|R - s_a Y|$ is the *length* of the vector $R - s_a Y$ or the *distance* between the vectors R and $s_a Y$.

It follows from Equation 3.3.10 that when the receiver has no prior knowledge of s, it must choose s_a to *minimize* $|R - s_a Y|$. Thus the best estimate s_a of s is that which minimizes the distance between R and $s_a Y$ in the n-dimensional vector space. Let x be the value of s_a that minimizes $|R - s_a Y|$.

It is clear that if the receiver has no prior knowledge of Y, then, as far as the receiver is concerned, Y is the n-component vector Y_a which is equally likely to have any length and any direction in the n-dimensional vector space, so that $s_a Y_a$ is equally likely to lie at any point in the vector space. Since the minimum value of $|R - s_a Y_a|$ is now zero, the best estimate of s is given by s_a in the equation

$$s_a Y_a = R \tag{3.3.11}$$

However, as Y_a is equally likely to be any real n-component vector, the receiver cannot from the value of R in Equation 3.3.11 obtain any useful estimate of s.

Assume now that the receiver has prior knowledge of Y. This means that, as far as the receiver is concerned, $s_a Y$ is equally likely to lie at any point in the one-dimensional subspace spanned by Y, that is, on the line traced out by $s_a Y$ as s_a varies over all real values. Furthermore, since Y is known, any point $s_a Y$, in the one-dimensional subspace spanned by Y, uniquely determines the corresponding value of s_a.

If x is the value of s_a that minimizes $|R - s_a Y|$, xY is the point in the one-dimensional subspace spanned by Y, at the *minimum distance* from R. Thus, by the projection theorem, xY is the *orthogonal projection* of R on to the one-dimensional subspace. This is illustrated in Figure 3.5.

The vector $R - xY$ is orthogonal to the vector Y, so that the inner product of $R - xY$ and Y is

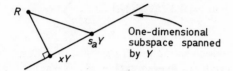

Fig. 3.5   Maximum-likelihood estimation of s from R

$$(R - xY)Y^T = 0 \tag{3.3.12}$$

Thus
$$xYY^T = RY^T \tag{3.3.13}$$

or
$$x = RY^T(YY^T)^{-1} \tag{3.3.14}$$

so that
$$x = |Y|^{-2}RY^T \tag{3.3.15}$$

where
$$|Y|^2 = YY^T \tag{3.3.16}$$

RY^T is the 1×1 matrix whose component is the inner product of R and Y. RY^T is determined by a *linear* filter that is *matched* to Y. The filter multiplies each component of the received vector R by the corresponding component of Y, and adds the n products to give RY^T. $|Y|^2$ is the squared length of the vector Y and is assumed to be known at the receiver. Hence, as can be seen from Equation 3.3.15, the matched filter enables the estimate x of the parameter s to be derived from R.

In Equation 3.3.12, $(R - xY)Y^T$ is a 1×1 *matrix* whereas the inner product of $R - xY$ and Y is a *scalar* quantity, so that, strictly speaking, the inner product of $R - xY$ and Y is equal to the *component* of the 1×1 matrix $(R - xY)Y^T$, and *not* to the matrix itself. It is with this understanding that the inner product of two vectors is written in terms of the corresponding matrices. Again, in Equation 3.3.16, $|Y|^2$ is a *scalar* quantity whereas YY^T is a 1×1 *matrix*. $|Y|^2$ is of course equal to the *component* of the matrix YY^T, and not to the matrix YY^T itself. In any equation relating 1×1 matrices and scalars, the 1×1 matrices will be taken as the corresponding component values.

From Equations 3.3.6 and 3.3.14,

$$
\begin{aligned}
x &= (sY + W)Y^T(YY^T)^{-1} \\
&= sYY^T(YY^T)^{-1} + WY^T(YY^T)^{-1} \\
&= s + u
\end{aligned}
\tag{3.3.17}
$$

where

$$u = W Y^T (Y Y^T)^{-1} \tag{3.3.18}$$

Since $Y^T(YY^T)^{-1}$ is a linear transformation, u is a sample value of a Gaussian random variable with zero mean. Thus, from Equation 3.3.17, the expected or average value of x (taken over all its *possible* values) is

$$E[x] = E[s+u] = E[s] \tag{3.3.19}$$

x, s and u here being treated as random variables. This means that x is an *unbiased* estimate of s, so that the expected or average value of the error in x is zero. Clearly, if s is a constant, $E[x] = s$.

The estimate of s given by x in Equation 3.3.14 or 3.3.15 is the value of s_a that maximizes $p(R|s_a)$, so that x is the *maximum-likelihood estimate* of s. Under the assumed conditions, x is also the value of s_a that maximizes $p(s_a|R)$, the *a posteriori* probability density of s_a, so that x is furthermore the *maximum a posteriori estimate* of s. Since x is derived from the received vector R by means of a *linear* transformation, x is a *linear* estimate of s. In the sense that x is the value of s_a, corresponding to both the maximum-likelihood and maximum *a posteriori* estimates of s, x is the *best linear estimate* of s from R, obtainable when the receiver knows Y but has no prior knowledge of s or σ^2.

3.3.2 Minimum mean-square error estimation of an unknown parameter

Consider again the received vector R given by Equation 3.3.6, under the various conditions previously assumed. It is now required to obtain a linear estimate x of the unknown parameter s, such that the mean-square error in x, $E[(x-s)^2]$, is minimized. x and s here are random variables, and the corresponding vectors R and W are random vectors. We are here concerned with the 'expectation' or average taken over all possible sample values of the random variable $(x-s)^2$.

To obtain a *linear* estimate of s from R, it is assumed that the n-component vector R is fed to the n input-terminals of an $n \times 1$ linear network Z^T, whose output terminal holds the required estimate x. The mathematical symbol Z^T represents an $n \times 1$ matrix or an n-component column vector, whose jth component is z_j, Z^T being the transpose of the $1 \times n$ matrix Z. Thus

$$x = RZ^T$$
$$= sYZ^T + WZ^T \qquad (3.3.20)$$

where sYZ^T is the signal component of x and WZ^T is the noise component.

Clearly, the estimation process must work correctly for all possible values of the unknown parameter s. Furthermore, the linear network Z^T is fixed (or time invariant) and is therefore independent of (or unaffected by) the particular value of s in any transmission. We are not therefore concerned here with learning or adaptive processes.

Since the n components of W are statistically independent Gaussian random variables with zero mean and variance σ^2, WZ^T is a Gaussian random variable with zero mean and variance

$$E[(WZ^T)^2] = E\left[\left(\sum_{j=1}^{n} w_j z_j\right)^2\right] = E\left[\sum_{j=1}^{n} (w_j z_j)^2\right]$$
$$= \sum_{j=1}^{n} \sigma^2 z_j^2$$
$$= \sigma^2 |Z|^2 \qquad (3.3.21)$$

where $|Z|$ is the length of the vector Z.

A necessary (but not sufficient) condition for the minimum mean-square error in the estimate x is that the signal/noise *ratio* in Equation 3.3.20 be maximized. This can be seen from the fact that if the signal/noise ratio is *not* maximized, then the network can always be adjusted to *reduce* the noise variance at the output of the network Z^T without changing the output signal component. Since the output noise component has zero mean and, for a given magnitude, is equally likely to be either positive or negative, the reduction in the noise variance is always accompanied by a reduction of the mean-square error in the resultant output signal x, regardless of the particular value of the signal component in x.

Let J be the vector of unit length that has the same direction as Z in the n-dimensional vector space, so that J is collinear with Z. Hence

$$J = |Z|^{-1}Z = Z(ZZ^T)^{-\frac{1}{2}} \qquad (3.3.22)$$

where, of course, $|Z|^{-1}$ is a scalar quantity and $(ZZ^T)^{-\frac{1}{2}}$ is a 1×1 matrix. The signal/noise ratio in Equation 3.3.20 is now given by

$$\frac{E[(sYZ^T)^2]}{E[(WZ^T)^2]} = \frac{E[(sYJ^T)^2]}{\sigma^2} \qquad (3.3.23)$$

from Equation 3.3.21. Thus the signal/noise ratio is maximum when $E[(sYJ^T)^2]$ is maximum. But the orthogonal projection of the signal vector sY on to the unit vector J, or more precisely, on to the one-dimensional subspace spanned by J in the n-dimensional vector space, is the n-component vector $(sYJ^T)J$, the square of whose length is $(sYJ^T)^2$. This can be seen from Figure 3.6, where 0 is the origin of the n-dimensional vector space and θ is the angle between the two vectors.

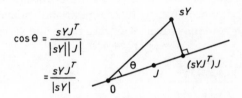

$$\cos\theta = \frac{sYJ^T}{|sY||J|}$$

$$= \frac{sYJ^T}{|sY|}$$

Fig. 3.6 Orthogonal projection of s Y on to the one-dimensional subspace spanned by J

Clearly, the length of the vector given by the orthogonal projection of sY on to the one-dimensional subspace spanned by J, is maximum when $\theta = 0$ and J is collinear with Y, so that

$$J = |Y|^{-1}Y = Y(YY^T)^{-\frac{1}{2}} \qquad (3.3.24)$$

Thus, for the maximum signal/noise ratio in x,

$$Z = qY \qquad (3.3.25)$$

where q is any real scalar. From Equation 3.3.20,

$$x = qRY^T$$
$$= qsYY^T + qWY^T \qquad (3.3.26)$$

The value of q must now be selected to minimize the mean-square error in x.

It is clear from Equations 3.3.15 and 3.3.26 that the linear transformation of the received vector R, giving the minimum mean-square error in the estimate of s, is a scalar multiple of that giving the maximum-likelihood estimate. Each estimate is obtained by means of a *matched filter*, so that in each case the signal/noise ratio is maximized.[1-4]

The mean-square error in x is

$$E[(x-s)^2] = E[(s(qYY^T-1)+qWY^T)^2]$$

$$= E[s^2(q|Y|^2-1)^2+q^2(WY^T)^2+2sq(q|Y|^2-1)WY^T]$$

$$= E[s^2](q|Y|^2-1)^2+E[q^2(WY^T)^2] \text{ (since } E[sWY^T] = 0)$$

$$= \overline{s^2}(q^2|Y|^4-2q|Y|^2+1)+q^2\sigma^2|Y|^2$$

(from Equations 3.3.7 and 3.3.21)

$$= (\overline{s^2}|Y|^2+\sigma^2)|Y|^2q^2-2\overline{s^2}|Y|^2q+\overline{s^2}$$

$$= \left[(\overline{s^2}|Y|^2+\sigma^2)^{\frac{1}{2}}|Y|q-\frac{\overline{s^2}|Y|}{(\overline{s^2}|Y|^2+\sigma^2)^{\frac{1}{2}}}\right]^2$$

$$+\overline{s^2}\left(1-\frac{\overline{s^2}|Y|^2}{\overline{s^2}|Y|^2+\sigma^2}\right) \tag{3.3.27}$$

The mean-square error is minimum when

$$(\overline{s^2}|Y|^2+\sigma^2)^{\frac{1}{2}}|Y|q-\frac{\overline{s^2}|Y|}{(\overline{s^2}|Y|^2+\sigma^2)^{\frac{1}{2}}} = 0 \tag{3.3.28}$$

that is, when

$$q = \frac{\overline{s^2}}{\overline{s^2}|Y|^2+\sigma^2} \tag{3.3.29}$$

and the minimum value of the mean-square error, obtained when q satisfies Equation 3.3.29, is

$$\overline{s^2}\left(1-\frac{\overline{s^2}|Y|^2}{\overline{s^2}|Y|^2+\sigma^2}\right) = \frac{\overline{s^2}\sigma^2}{\overline{s^2}|Y|^2+\sigma^2} \tag{3.3.30}$$

as can be seen from Equations 3.3.27 and 3.3.28.

The linear network Z^T that minimizes the mean-square error in x is now

$$Z^T = qY^T = \left(|Y|^2+\frac{\sigma^2}{\overline{s^2}}\right)^{-1}Y^T \tag{3.3.31}$$

and, from Equation 3.3.20,

$$x = \left(|Y|^2 + \frac{\sigma^2}{\overline{s^2}}\right)^{-1} R Y^T$$

$$= \left(1 + \frac{\sigma^2}{\overline{s^2}|Y|^2}\right)^{-1} s + \left(|Y|^2 + \frac{\sigma^2}{\overline{s^2}}\right)^{-1} W Y^T \qquad (3.3.32)$$

bearing in mind that $Y Y^T = |Y|^2$. Clearly, x is a *biased* estimate of s.

From Equation 3.3.29,

$$0 \leqslant q \leqslant |Y|^{-2} \qquad (3.3.33)$$

and the higher the signal/noise ratio $\overline{s^2}|Y|^2/\sigma^2$, the nearer q approaches to $|Y|^{-2}$.

At high signal/noise ratios, $\overline{s^2}|Y|^2 \gg \sigma^2$, and now for the minimum mean-square error,

$$q \simeq |Y|^{-2} = (Y Y^T)^{-1} \qquad (3.3.34)$$

Thus, from Equation 3.3.25,

$$Z^T = q Y^T \simeq Y^T (Y Y^T)^{-1} \qquad (3.3.35)$$

and the minimum mean-square error estimate of s is

$$x = R Z^T \simeq R Y^T (Y Y^T)^{-1} \qquad (3.3.36)$$

From Equation 3.3.14, x is also the *maximum-likelihood estimate* of s. Clearly, x is now an unbiased estimate of s, as can be seen by setting $\overline{s^2}|Y|^2 \gg \sigma^2$ in Equation 3.3.32.

Since q is a positive real scalar with a value less than $|Y|^{-2}$, it can be seen from Equations 3.3.15 and 3.3.26 that the minimum mean-square error estimate of s always has the same sign as the maximum-likelihood estimate, but, in the presence of noise, it is always smaller in magnitude. q decreases to $\overline{s^2}/\sigma^2$ as the signal/noise ratio $\overline{s^2}|Y|^2/\sigma^2$ decreases to zero, as can be seen from Equation 3.3.29. When both $\overline{s^2}|Y|^2/\sigma^2$ and $\overline{s^2}/\sigma^2$ decrease to zero, q decreases to zero.

It is clear from Equations 3.3.26 and 3.3.29 that in order to make a linear estimate of x having the minimum mean-square error, a prior knowledge is needed of the signal/noise ratio $\overline{s^2}|Y|^2/\sigma^2$. When the receiver has no prior knowledge of this ratio, which is often the case in practice, the best that can be done is to minimize the mean-square error, given an *unbiased estimate* of s.

The estimate of s, obtained from R by a linear network Z^T (where the mathematical symbol Z^T represents an $n \times 1$ matrix, as before), is given by Equation 3.3.20. The noise component WZ^T is here a Gaussian random variable with zero mean and variance $\sigma^2|Z|^2$, as can be seen from Equation 3.3.21. The condition for x to be an unbiased estimate of s is that

$$E[x] = E[sYZ^T + WZ^T] = E[sYZ^T] = E[s] \quad (3.3.37)$$

which must also hold when s is a constant. Thus

$$YZ^T = 1 \quad (3.3.38)$$

The mean-square error in the estimate x is now

$$\begin{aligned} E[(x-s)^2] &= E[(sYZ^T + WZ^T - s)^2] \\ &= E[(WZ^T)^2] \\ &= \sigma^2|Z|^2 \end{aligned} \quad (3.3.39)$$

from Equation 3.3.21.

Thus, for the linear estimate x to have the minimum mean-square error, *given* that it is an unbiased estimate, the vector Z must be selected to minimize $|Z|$, given Equation 3.3.38. For these two conditions to be satisfied it is clear that Z must be collinear with Y, so that

$$Z = qY \quad (3.3.40)$$

where q is a real scalar. From Equation 3.3.38, $qYY^T = 1$, so that

$$q = (YY^T)^{-1} \quad (3.3.41)$$

and, from Equation 3.3.40,

$$x = RZ^T = RY^T(YY^T)^{-1} \quad (3.3.42)$$

as in Equation 3.3.14. The mean-square error in x is now

$$\sigma^2|(YY^T)^{-1}Y|^2 = \frac{\sigma^2}{|Y|^2} \quad (3.3.43)$$

as can be seen from Equation 3.3.39.

It follows that the linear estimate of s that minimizes the mean-square error, subject to it being an unbiased estimate, is also the maximum-likelihood estimate of s. At high signal/noise ratios, this estimate becomes the true minimum mean-square error estimate of s.

3.3.3 Maximum-likelihood estimation of a set of m unknown parameters

Suppose now that the n-component signal vector sY in Equation 3.3.6 is replaced by the sum of m signal vectors, the ith of which is the n-component row vector s_iY_i. s_i is here a scalar quantity whose value it is required to estimate at the receiver, for each i. Y_i is the n-component row vector

$$Y_i = y_{i1} \ y_{i2} \ldots y_{in} \tag{3.3.44}$$

that characterizes or identifies the ith signal-element. $m \leqslant n$, and the m vectors $\{Y_i\}$ are assumed to be linearly independent, so that the resultant n-component signal vector

$$\sum_{i=1}^{m} s_i Y_i$$

uniquely determines the m $\{s_i\}$. If the $\{Y_i\}$ are not linearly independent, it is not possible to obtain a linear estimate of the $\{s_i\}$ from $\sum_{i=1}^{m} s_i Y_i$, since there is now an infinite set of possible values of the $\{s_i\}$ for any given resultant signal vector. The linear independence of the $\{Y_i\}$ is therefore a *necessary and sufficient* condition for the linear estimation of the $\{s_i\}$ from $\sum_{i=1}^{m} s_i Y_i$.

Let Y be the $m \times n$ matrix of rank m whose ith row is given by the n-component row vector Y_i, and let S be the m-component row vector

$$S = s_1 \ s_2 \ldots s_m \tag{3.3.45}$$

Then

$$\sum_{i=1}^{m} s_i Y_i = SY \tag{3.3.46}$$

and S is uniquely determined by SY.

From Equation 3.3.6, the n received sample values are now given by the n-component row vector

$$R = SY + W \tag{3.3.47}$$

where SY has replaced sY. As before, the n components of the noise vector W are statistically independent Gaussian random variables with zero mean and variance σ^2, and W is statistically independent of SY.

The receiver is assumed to have prior knowledge of the m vectors $\{Y_i\}$ and hence of the matrix Y, but no prior knowledge of S or σ^2. Given the received vector R, it is required to make the best linear estimate of S.

Let S now be the *actual* value of the vector being estimated, and let S_a be any *possible* value of S as far as the receiver is concerned, before it has estimated S from R. In any one transmission of the signal waveform, S and S_a are *sample values* of the corresponding random vectors, and S normally has different values in different transmissions.

Since the receiver knows Y but has no prior knowledge of S, S_a is equally likely to be any real m-component vector and $S_a Y$ is equally likely to lie at any point in the m-dimensional subspace spanned by the m $\{Y_i\}$, the receiver having prior knowledge of this subspace. Let

$$S_a Y = F \qquad (3.3.48)$$

where the jth component of F is f_j. It can be seen from Equation 3.3.47 that if $SY = F$, then, for $j = 1, 2, \ldots, n$,

$$r_j = f_j + w_j \qquad (3.3.49)$$

so that, for the given received sample r_j and the given value of f_j, r_j is a sample value of a Gaussian random variable with mean f_j and variance σ^2, the n $\{r_j\}$ being sample values of statistically independent Gaussian random variables.

The best available estimate of S from R is here taken to be the value of S_a that maximizes $p(S_a|R)$, the *a posteriori* probability density of S_a. This is a process of *maximum a posteriori estimation*.

From Bayes' theorem, as given by Equation 3.1.8,

$$p(S_a|R) = \frac{p(S_a)}{p(R)} p(R|S_a) \qquad (3.3.50)$$

Since $p(R)$ does not depend on S_a, and $p(S_a)$ is constant for all vectors $\{S_a\}$, the value of S_a that maximizes $p(S_a|R)$ also maximizes $p(R|S_a)$, and vice versa. Thus the receiver must choose S_a to maximize $p(R|S_a)$. This is a process of *maximum-likelihood estima-*

tion. Clearly, the maximum-likelihood estimate of *any* parameter is the maximum *a posteriori* estimate when the receiver has no prior knowledge of the parameter, this relationship applying for any joint probability density function and not just for the joint Gaussian probability density assumed here. Now,

$$
\begin{aligned}
p(R|S_a) &= p(R|F) \\
&= p(r_1|f_1) \; p(r_2|f_2) \dots p(r_n|f_n) \\
&= \prod_{j=1}^{n} \frac{1}{\sqrt{(2\pi\sigma^2)}} \exp\left[-\frac{(r_j-f_j)^2}{2\sigma^2} \right] \\
&= \frac{1}{(2\pi\sigma^2)^{\frac{1}{2}n}} \exp\left[-\frac{1}{2\sigma^2} \sum_{j=1}^{n} (r_j-f_j)^2 \right] \\
&= \frac{1}{(2\pi\sigma^2)^{\frac{1}{2}n}} \exp\left[-\frac{1}{2\sigma^2} |R-F|^2 \right]
\end{aligned}
\tag{3.3.51}
$$

Thus the receiver must choose S_a to *minimize* $|R-F|$, the *distance* between R and F, so that, from Equation 3.3.48, the receiver must choose S_a to minimize $|R-S_aY|$.

Let the *n*-component vector XY be the orthogonal projection of R on to the *m*-dimensional subspace spanned by the $\{Y_i\}$, where X is the appropriate *m*-component row vector. The square of the distance between R and S_aY is

$$
\begin{aligned}
&(R-S_aY)(R-S_aY)^T \\
&= (R-XY+XY-S_aY)(R-XY+XY-S_aY)^T \\
&= (R-XY)(R-XY)^T + (R-XY)(XY-S_aY)^T \\
&\quad + (XY-S_aY)(R-XY)^T + (XY-S_aY)(XY-S_aY)^T \\
&= (R-XY)(R-XY)^T + (XY-S_aY)(XY-S_aY)^T
\end{aligned}
\tag{3.3.52}
$$

since $R-XY$ is orthogonal to $XY-S_aY$.

Now $(R-XY)(R-XY)^T$ is not dependent on S_a, and $(XY-S_aY)(XY-S_aY)^T$ is nonnegative, so that $(R-S_aY)(R-S_aY)^T$ is minimum when $S_aY = XY$ or $S_a = X$. Thus the point at the *minimum distance* from R, in the *m*-dimensional subspace spanned by the $\{Y_i\}$, is the *orthogonal projection* of R on to this subspace. This is the well-known projection theorem.

It follows now that $p(R|S_a)$ and therefore $p(S_a|R)$ is maximum when $S_a = X$.

Since $R - XY$ is orthogonal to the m-dimensional subspace spanned by the $\{Y_i\}$, it is orthogonal to each of the $\{Y_i\}$, so that

$$(R - XY)Y_i^T = 0 \qquad (3.3.53)$$

for $i = 1, 2, \ldots, m$, and

$$(R - XY)Y^T = 0 \qquad (3.3.54)$$

or
$$XYY^T = RY^T \qquad (3.3.55)$$

or
$$X = RY^T(YY^T)^{-1} \qquad (3.3.56)$$

Since Y is an $m \times n$ matrix of rank m, YY^T is nonsingular. YY^T is also real, symmetric and positive definite. $Y^T(YY^T)^{-1}$ is a real $n \times m$ matrix of rank m. In the particular case where $m = n$, $R = XY$ so that $X = RY^{-1}$, Y now being an $n \times n$ nonsingular matrix.

If the received vector R is fed to the n input terminals of the $n \times m$ linear network $Y^T(YY^T)^{-1}$, the signals at the m output terminals are the components $\{x_i\}$ of the vector X, where X is the best estimate the detector can make of S, under the assumed conditions.

From Equations 3.3.47 and 3.3.56,

$$\begin{aligned} X &= (SY + W)Y^T(YY^T)^{-1} \\ &= SYY^T(YY^T)^{-1} + WY^T(YY^T)^{-1} \\ &= S + U \end{aligned} \qquad (3.3.57)$$

where
$$U = WY^T(YY^T)^{-1} \qquad (3.3.58)$$

and the m-component vector U is the noise vector at the output of the network $Y^T(YY^T)^{-1}$. The ith component of U is u_i.

Each component of the noise vector U is a sample value of a Gaussian random variable with zero mean and a variance which is *not* normally equal to σ^2 and which may differ from one component to another.

Equation 3.3.57 shows that the network $Y^T(YY^T)^{-1}$ performs a process of *exact linear equalization* on the received vector R, such that the ith component of X is

$$x_i = s_i + u_i \qquad (3.3.59)$$

for $i = 1, 2, \ldots, m$, and all intersymbol interference is eliminated at the output of the network. Thus, from Equation 3.3.57, the expected or average value of the vector X (taken over all its *possible* values) is

$$E[X] = E[S+U] = E[S] \qquad (3.3.60)$$

X, S and U here being treated as random vectors. This means that X is an *unbiased* estimate of S, and the expected or average value of the error in X is zero. Clearly, if S is a constant vector, $E[X] = S$.

The linear estimate of S given by X in Equation 3.3.56 is the value of S_a that maximizes $p(R|S_a)$, so that X is the *maximum-likelihood estimate* of S. Under the assumed conditions, X is also the value of S_a that maximizes $p(S_a|R)$, the *a posteriori* probability density of S_a, so that X is furthermore the *maximum a posteriori estimate* of S. In this sense, X is the *best linear estimate* of S from R, obtainable when the receiver knows Y but has no prior knowledge of S or σ^2.

It can be seen from Equation 3.3.50 that when the receiver has some prior knowledge of S, which means that $p(S_a)$ is not constant, the maximum *a posteriori* estimate of S is no longer the same as the maximum-likelihood estimate and is also a *better* estimate since it makes use of the prior knowledge of S.

3.3.4 Example

A three-component received vector is given by

$$R = s_1 Y_1 + s_2 Y_2 + W \qquad (3.3.61)$$

where s_1 and s_2 are real scalars, and

$$R = r_1 \ r_2 \ r_3 \qquad (3.3.62)$$

$$Y_1 = 1 \ \ 1 \ \ 0 \qquad (3.3.63)$$

$$Y_2 = 0 \ \ 1 \ \ 1 \qquad (3.3.64)$$

$$W = w_1 \ w_2 \ w_3 \qquad (3.3.65)$$

The components of W are statistically independent Gaussian random variables with zero mean and variance σ^2. The receiver has prior knowledge of Y_1 and Y_2 but no prior knowledge of s_1, s_2 or σ^2.

Determine the maximum-likelihood estimate of the two-component vector (s_1, s_2).

Let

$$S = s_1 \ s_2 \tag{3.3.66}$$

and

$$Y = \begin{bmatrix} 1 & 1 & 0 \\ 0 & 1 & 1 \end{bmatrix} \tag{3.3.67}$$

so that

$$R = SY + W \tag{3.3.68}$$

Clearly, it is required to obtain the maximum-likelihood estimate of S from R.

Consider the three-dimensional vector space containing R. The maximum-likelihood estimate of S is the two-component vector $X = (x_1, x_2)$, where XY is the orthogonal projection of R on to the two-dimensional subspace spanned by Y_1 and Y_2. This is illustrated in Figure 3.7.

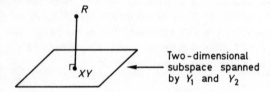

Fig. 3.7   *Maximum-likelihood estimation of S from R*

Since $R - XY$ is orthogonal to the subspace,

$$(R - XY)Y^T = 0 \tag{3.3.69}$$

or

$$X = RY^T(YY^T)^{-1} \tag{3.3.70}$$

But

$$YY^T = \begin{bmatrix} 2 & 1 \\ 1 & 2 \end{bmatrix} \tag{3.3.71}$$

so that

$$(YY^T)^{-1} = \begin{bmatrix} \frac{2}{3} & -\frac{1}{3} \\ -\frac{1}{3} & \frac{2}{3} \end{bmatrix} \tag{3.3.72}$$

and

$$Y^T(YY^T)^{-1} = \begin{bmatrix} 1 & 0 \\ 1 & 1 \\ 0 & 1 \end{bmatrix} \begin{bmatrix} \frac{2}{3} & -\frac{1}{3} \\ -\frac{1}{3} & \frac{2}{3} \end{bmatrix} = \begin{bmatrix} \frac{2}{3} & -\frac{1}{3} \\ \frac{1}{3} & \frac{1}{3} \\ -\frac{1}{3} & \frac{2}{3} \end{bmatrix} \tag{3.3.73}$$

Thus

$$X = (r_1, r_2, r_3) \begin{bmatrix} \frac{2}{3} & -\frac{1}{3} \\ \frac{1}{3} & \frac{1}{3} \\ -\frac{1}{3} & \frac{2}{3} \end{bmatrix}$$

$$= (\tfrac{2}{3}r_1 + \tfrac{1}{3}r_2 - \tfrac{1}{3}r_3, -\tfrac{1}{3}r_1 + \tfrac{1}{3}r_2 + \tfrac{2}{3}r_3) \quad (3.3.74)$$

or

$$x_1 = \tfrac{2}{3}r_1 + \tfrac{1}{3}r_2 - \tfrac{1}{3}r_3 \quad (3.3.75)$$

$$x_2 = -\tfrac{1}{3}r_1 + \tfrac{1}{3}r_2 + \tfrac{2}{3}r_3 \quad (3.3.76)$$

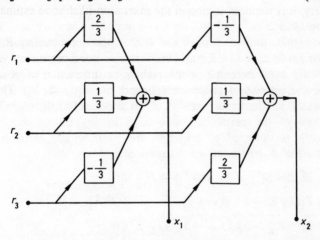

Fig. 3.8 Linear network for the maximum-likelihood estimation of S from R

The network $Y^T(YY^T)^{-1}$ is shown diagrammatically in Figure 3.8, where each square is a multiplier that multiplies the input signal by the quantity marked inside the square.

3.3.5 Minimum mean-square error estimation of a set of m unknown parameters

Consider again the received *n*-component vector R, given by Equation 3.3.47 and described in Section 3.3.3.

It is now required to obtain from R a linear estimate X of the *m*-component vector S, such that the mean-square error in X, $E[|X-S|^2]$, is minimized. As before, the receiver has prior knowledge of Y, but for the moment no assumptions will be made concerning the prior knowledge of S or σ^2. X, S, R and W here are random vectors.

To obtain a *linear* estimate of S from R, it is assumed that the n-component vector R is fed to the n input terminals of an $n \times m$ linear network Z^T, whose m output terminals hold the required estimate X. The mathematical symbol Z^T represents an $n \times m$ matrix. Thus

$$X = RZ^T$$
$$= SYZ^T + WZ^T \tag{3.3.77}$$

where both SYZ^T and WZ^T are m-component row vectors. SYZ^T is the signal component of X and WZ^T is the noise component.

The random vector S may be discrete or continuous, and S will normally have different sample values in different transmissions. Thus the estimation process must work correctly for all possible values of S. The linear network Z^T is fixed and is therefore independent of the particular value of S in any transmission. The network is designed to minimize the mean-square error in the estimate of S, where the mean-square error is

$$E[|X-S|^2] = E[|SYZ^T + WZ^T - S|^2]$$
$$= E[(SYZ^T - S + WZ^T)(SYZ^T - S + WZ^T)^T]$$
$$= E[(SYZ^T - S)(SYZ^T - S)^T + (WZ^T)(WZ^T)^T$$
$$\qquad\qquad + 2(SYZ^T - S)(WZ^T)^T]$$
$$= E[|SYZ^T - S|^2] + E[|WZ^T|^2] \tag{3.3.78}$$

since W is statistically independent of S and $E[WZ^T] = 0$, so that

$$E[2(SYZ^T - S)(WZ^T)^T] = 0$$

$|X-S|$ is, of course, the *distance* between X and S in the m-dimensional Euclidean vector space containing X and S. $E[|SYZ^T - S|^2]$ is the mean-square error due to signal distortion (attenuation and intersymbol interference) alone, and $E[|WZ^T|^2]$ is the mean-square error due to noise alone.

Let Z_i be the n-component row vector given by the ith row of Z, whose jth component is z_{ij}, and let

$$WZ^T = U \tag{3.3.79}$$

where U is an m-component row vector that is not necessarily related in any way to the noise vector U in Equation 3.3.57. The ith component of U is u_i. Now

$$E[|WZ^T|^2] = E[|U|^2] = \sum_{i=1}^{m} E[u_i^2] \qquad (3.3.80)$$

and

$$E[u_i^2] = E[(WZ_i^T)^2] = \sigma^2 |Z_i|^2 \qquad (3.3.81)$$

from Equation 3.3.21, since the $\{w_j\}$ are statistically independent Gaussian random variables with zero mean and variance σ^2. Thus

$$E[|WZ^T|^2] = \sigma^2 \sum_{i=1}^{m} |Z_i|^2 \qquad (3.3.82)$$

Suppose first that Z_j, one of the vectors $\{Z_i\}$ forming the rows of Z, does not lie in the m-dimensional subspace spanned by the $\{Y_i\}$. If Z_j is replaced by its orthogonal projection on to the subspace (which again will be known as Z_j), there is no change in the inner product of Z_j and SY, since of course SY lies in the subspace. Thus there is no change in $E[|SYZ^T - S|^2]$. However the length of Z_j is now smaller, so that $E[|WZ^T|^2]$ is reduced, as can be seen from Equation 3.3.82. Thus $E[|X - S|^2]$ is reduced (Equation 3.3.78).

Clearly, for the minimum mean-square error in X it is necessary (but not sufficient) that the m $\{Z_i\}$ lie in the m-dimensional subspace spanned by the $\{Y_i\}$. This condition is ensured by letting

$$Z = QY \qquad (3.3.83)$$

where Q is an $m \times m$ real matrix. The problem is therefore to find the $m \times m$ matrix Q that minimizes the mean-square error in X.

Let the $m \times m$ matrix C be the correlation matrix for the m $\{s_i\}$, so that

$$C = E[S^T S] \qquad (3.3.84)$$

where the component in the ith row and jth column of C is

$$c_{ij} = E[s_i s_j] \qquad (3.3.85)$$

It is assumed that C is nonsingular, which implies that no s_i is entirely determined by the other $\{s_i\}$, each s_i having a component that is independent of the others. C is now a real symmetric positive-definite matrix, which means that all its eigenvalues (characteristic roots) are real and positive.

The following are two important relationships that are used in the subsequent analysis. Firstly,

$$E[W^T W] = \sigma^2 I \tag{3.3.86}$$

where I is an $n \times n$ identity matrix. Secondly, if A and B are any two n-component row vectors with real components, the inner product of A and B is

$$AB^T = BA^T = \text{tr}(A^T B) = \text{tr}(B^T A) \tag{3.3.87}$$

where $\text{tr}(A^T B)$ is the trace of the $n \times n$ matrix $A^T B$ and is the sum of the components along the main diagonal of that matrix. The trace of a square matrix is a linear function of that matrix.

It follows from Equations 3.3.78 and 3.3.83 that the mean-square error in the estimate of S is

$$
\begin{aligned}
E[|X - S|^2] &= E[|SYY^T Q^T - S|^2 + |WY^T Q^T|^2] \\
&= E[(SYY^T Q^T - S)(SYY^T Q^T - S)^T + (WY^T Q^T)(WY^T Q^T)^T] \\
&= E[SYY^T Q^T QYY^T S^T - 2SQYY^T S^T + SS^T + WY^T Q^T QYW^T] \\
&= E[\text{tr}(QYY^T S^T SYY^T Q^T - 2S^T SYY^T Q^T + S^T S + \\
&\qquad\qquad QYW^T WY^T Q^T)] \\
&= \text{tr}\left[Q(YY^T CYY^T + \sigma^2 YY^T)Q^T - 2CYY^T Q^T + C\right] \tag{3.3.88}
\end{aligned}
$$

The $m \times m$ matrix $(YY^T CYY^T + \sigma^2 YY^T)$ is real, symmetric and positive-definite, since C, YY^T and $YY^T CYY^T$ are real, symmetric and positive-definite. It follows that there is an $m \times m$ real nonsingular matrix G such that

$$GG^T = YY^T CYY^T + \sigma^2 YY^T \tag{3.3.89}$$

Now, from Equation 3.3.88,

$$
\begin{aligned}
E[|X - S|^2] &= \text{tr}\left[QGG^T Q^T - 2CYY^T Q^T + C\right] \\
&= \text{tr}\left[(QG - CYY^T (G^T)^{-1})(QG - CYY^T (G^T)^{-1})^T + \right. \\
&\qquad\qquad \left. C - CYY^T (G^T)^{-1} G^{-1} YY^T C^T\right] \\
&= \text{tr}\left[(QG - CYY^T (G^T)^{-1})(QG - CYY^T (G^T)^{-1})^T\right] + \\
&\qquad\qquad \text{tr}\left[C(I - YY^T (G^T)^{-1} G^{-1} YY^T C)\right] \tag{3.3.90}
\end{aligned}
$$

where I is an $m \times m$ identity matrix.

In Equation 3.3.90, $\text{tr}\left[C(I - YY^T (G^T)^{-1} G^{-1} YY^T C)\right]$ is independent of Q and $\text{tr}\left[(QG - CYY^T (G^T)^{-1})(QG - CYY^T (G^T)^{-1})^T\right]$ is nonnegative, so that $E[|X - S|^2]$ is minimum when

$$\text{tr}\left[(QG - CYY^T (G^T)^{-1})(QG - CYY^T (G^T)^{-1})^T\right] = 0 \tag{3.3.91}$$

that is, when

$$QG - CYY^T(G^T)^{-1} = 0 \tag{3.3.92}$$

or
$$
\begin{aligned}
Q &= CYY^T(G^T)^{-1}G^{-1} \\
&= CYY^T(YY^TCYY^T + \sigma^2 YY^T)^{-1} \\
&= (YY^T + \sigma^2 C^{-1})^{-1} \tag{3.3.93}
\end{aligned}
$$

Since YY^T and C are real symmetric matrices, so also is Q.

From Equations 3.3.90 and 3.3.91, the minimum value of $E[|X - S|^2]$ is

$$\mathrm{tr}\left[C(I - YY^T(G^T)^{-1}G^{-1}YY^TC)\right]$$

$$= \mathrm{tr}\left[C(I - YY^T(YY^TCYY^T + \sigma^2 YY^T)^{-1}YY^TC)\right]$$

$$= \mathrm{tr}\left[C(I - (I + \sigma^2(YY^TC)^{-1})^{-1})\right] \tag{3.3.94}$$

where I is an $m \times m$ identity matrix. In the absence of noise, $\sigma^2 = 0$, and it is clear from Equation 3.3.94 that the minimum mean-square error in X is now zero, as would be expected.

From Equations 3.3.83 and 3.3.93, the linear network Z^T that minimizes $E[|X - S|^2]$ is given by the $n \times m$ matrix

$$Z^T = Y^TQ^T = Y^T(YY^T + \sigma^2 C^{-1})^{-1} \tag{3.3.95}$$

and the minimum mean-square error estimate of S is

$$X = RY^T(YY^T + \sigma^2 C^{-1})^{-1}$$

$$= SYY^T(YY^T + \sigma^2 C^{-1})^{-1} + WY^T(YY^T + \sigma^2 C^{-1})^{-1} \tag{3.3.96}$$

from Equation 3.3.77. Clearly, X is a *biased* estimate of S.

It can be seen from Equation 3.3.95 that, in order to implement the linear network Z^T, the receiver requires prior knowledge of the matrices Y and C, as well as of the noise level σ^2.

At high signal/noise ratios, $\sigma^2 C^{-1} \to 0$, so that

$$Z^T \simeq Y^T(YY^T)^{-1} \tag{3.3.97}$$

and
$$X \simeq RY^T(YY^T)^{-1} = S + WY^T(YY^T)^{-1} \tag{3.3.98}$$

X is here the maximum-likelihood estimate of S, and X is obtained from R by a process of *exact linear equalization*. This means that all intersymbol interference is eliminated at the output of the network and the expected value of X is

$$E[X] \simeq E[S + WZ^T] = E[S] \tag{3.3.99}$$

so that X is an unbiased estimate of S. The result is to be expected since it is now more important to minimize the intersymbol interference than the noise.

Suppose now that the receiver has no prior knowledge of S or σ^2, and hence of the signal/noise ratio. Under these conditions the receiver has no prior knowledge of the $\{s_i\}$, so that it cannot assume, before the estimation process, that either $|s_i| \gg |s_j|$, $|s_i| \ll |s_j|$ or $|s_i| \simeq |s_j|$, for any $i \neq j$, in the actual received vector S. Clearly, if the estimate of s_i contains a component dependent on s_j, which implies the presence of intersymbol interference, the receiver cannot neglect the possibility that $|s_j| \gg |s_i|$, when an interference component equal to only a small fraction of s_j could cause a serious error in the estimate of s_i. Thus the receiver cannot tolerate *any* intersymbol interference in its estimates of the $\{s_i\}$. Alternatively, if the receiver has no prior knowledge of the $\{s_i\}$, it must assume that each s_i in the given vector S is a sample value of a random variable with a constant probability density over *all* real values and therefore with infinite (or at least extremely large) variance. Obviously, under these conditions, *no* intersymbol interference is acceptable. In fact, if the receiver has *no* prior knowledge of S, the best it can do is to minimize the mean-square error, subject to the *exact linear equalization* of the received signal. But it can be seen from the preceding analysis that this is essentially what is done by the linear network that minimizes the mean-square error at high signal/noise ratios, since this both minimizes the mean-square error *and* accurately equalizes the signal. Thus, for the minimum mean-square error in X, *given* the exact linear equalization of the signal, X satisfies Equation 3.3.98 and so is the *maximum-likelihood* estimate of S. Clearly, this vector X is the best linear estimate of S both in the sense that it minimizes the mean-square error, assuming exact linear equalization, and also in the sense that it is the maximum-likelihood estimate.

The equivalence of the maximum-likelihood estimate of S and the estimate that minimizes the mean-square error, given the exact linear equalization of the signal, can be derived more rigorously as follows. A linear estimate of S is given by RZ^T, where Z is an appropriate $n \times m$ matrix. The network Z^T that achieves the exact linear equalization of the received signal SY must satisfy the equation $YZ^T = I$, where I is an $m \times m$ identity matrix. Thus it must set to *zero* the term $E[|SYZ^T - S|^2]$ in the expression for the mean-square error in RZ^T, given by Equation 3.3.78, where of course $X = RZ^T$. Furthermore, from this equation, the network Z^T that minimizes the mean-square error in RZ^T, given the exact linear equalization of the received signal, must minimize $E[|WZ^T|^2]$, given that $YZ^T = I$. But, from the analysis leading to Equation 3.3.83, it is clear that now $Z = QY$, where Q is an $m \times m$ matrix. It therefore follows that $Z^T = Y^T(YY^T)^{-1}$, so that RZ^T is also the maximum-likelihood estimate of S. Finally, it can be seen from Equation 3.3.82, that the mean-square error in this vector RZ^T, is the sum of the squares of the components of the $n \times m$ matrix $Y^T(YY^T)^{-1}$, multiplied by the noise variance σ^2.

It is evident now that the maximum-likelihood estimate of S is the estimate X that minimizes the mean-square error, given the exact linear equalization of the received signal. But the exact linear equalization of the received signal means that each component of X is equal to the corresponding component of S, added to a noise component, so that X is an *unbiased* estimate of S. Thus the maximum-likelihood estimate of S is the unbiased estimate X that minimizes the mean-square error in X. This assumes, of course, that the components of the received noise vector W are statistically independent Gaussian random variables with zero mean and fixed variance.

In general, when the signal/noise ratio is not high and when the *given* received signal-vector S is a sample value of a random vector, such that S normally has different values in different transmissions, the minimum mean-square error estimate of S is to be preferred to the maximum-likelihood estimate. This is because it is now more important to minimize the errors in the different estimates than it is to avoid a bias in these estimates. An example of a minimum mean-square error estimate is considered in Chapter 4. It often happens, however, that the vector S is essentially unchanged over a number of successive transmissions, and it is required to obtain the best estimate of S from the individual estimates obtained in the corresponding transmissions, without the use of unduly complex equipment. The resultant estimate of S is therefore taken as the *average* of the individual estimates or more often as the average of the appropriately weighted values of these estimates (the individual estimates being multiplied by suitable scalar quantities). The resultant estimate has the same bias as the individual estimates but an appreciably reduced noise component. When the individual estimates minimize the mean-square error, they are biased and so is the resultant estimate of S, which is not however a minimum mean-square error estimate. Furthermore, as the number of the individual estimates increases, the mean-square error in the resultant estimate does not decrease towards zero but rather towards the value determined by the bias in the estimate, which remains unchanged. The resultant estimate of S could, of course, be converted to the corresponding minimum mean-square error estimate by means of the appropriate linear transformation, and the mean-square error in this estimate would decrease towards zero as the number of the individual estimates increases. This would, however, unduly complicate the system. On the other hand, when the individual estimates are maximum-likelihood estimates, they are unbiased and so is the resultant estimate. Now, as the number of individual estimates increases, the mean-square error in the resultant estimate decreases towards zero. Thus the resultant estimate usually has a lower mean-square error than it does in the case where the individual estimates minimize the mean-square error. In this application, therefore, the maximum-likelihood estimate is to be preferred to the minimum mean-square error estimate. Examples of maximum-likelihood estimates are considered in Chapters 5 and 6.

Suppose next that the mean-square value (or expected energy) of the ith component of S is

$$E[s_i^2] = s^2 \qquad (3.3.100)$$

for $i = 1, 2, \ldots, m$, where s is a constant. Assume also that the components of S are uncorrelated and have zero mean, so that

$$E[s_i s_j] = 0 \qquad (3.3.101)$$

when $i \neq j$. Thus

$$C = s^2 I \qquad (3.3.102)$$

where I is an $m \times m$ identity matrix. These conditions often hold, at least approximately, in practice.

From Equation 3.3.95, the linear network Z^T is now given by

$$Z^T = Y^T \left(YY^T + \frac{\sigma^2}{s^2} I \right)^{-1} \qquad (3.3.103)$$

and the minimum mean-square error estimate of S is

$$X = RY^T \left(YY^T + \frac{\sigma^2}{s^2} I \right)^{-1}$$

$$= SYY^T\left(YY^T + \frac{\sigma^2}{s^2}I\right)^{-1} + WY^T\left(YY^T + \frac{\sigma^2}{s^2}I\right)^{-1} \qquad (3.3.104)$$

It is clear from Equation 3.3.103 that, in order to implement the $n \times m$ network Z^T, the receiver requires prior knowledge of the matrix Y and of the signal/noise ratio s^2/σ^2.

In the particular case where S has only one component, Z^T and X become as in Equations 3.3.31 and 3.3.32, respectively.

It can be seen from Equation 3.3.103 that at high signal/noise ratios, when $\sigma^2/s^2 \ll 1$ (assuming for convenience that each $|Y_i|$ is in the neighbourhood of unity), Z^T is given by Equation 3.3.97, so that Z^T performs a process of exact equalization and eliminates all intersymbol interference.

At very low signal/noise ratios, when $\sigma^2/s^2 \gg 1$, the matrix YY^T in $\left(YY^T + \dfrac{\sigma^2}{s^2}I\right)^{-1}$ becomes masked by the matrix $\dfrac{\sigma^2}{s^2}I$, so that the linear network Z^T becomes more nearly like the network $\dfrac{s^2}{\sigma^2}Y^T$ which is *matched* to SY. Again, this is to be expected, because it is now more important to minimize the noise than the intersymbol interference, and this is achieved by the matched network which maximizes the output signal/noise ratio.

The structure and function both of linear networks that are matched to the received signal and of linear networks that eliminate intersymbol interference will now be studied in some detail.

3.3.6 Matched and inverse networks

It can be seen that the $n \times m$ matrix

$$Z^T = Y^T(YY^T)^{-1} \qquad (3.3.105)$$

is such that

$$YZ^T = I \qquad (3.3.106)$$

where I is an $m \times m$ identity matrix. This means, of course, that the m row vectors $\{Y_i\}$, given by the rows of Y, and the m row vectors $\{Z_i\}$, given by the rows of Z, satisfy the equations

$$Y_i Z_i^T = 1 \qquad (3.3.107)$$

and

$$Y_i Z_j^T = 0 \qquad (3.3.108)$$

where $i = 1, 2, \ldots, m, j = 1, 2, \ldots, m$, and $i \neq j$. Thus Z^T is a right inverse of Y. Furthermore, both the $\{Y_i\}$ and the $\{Z_i\}$ are sets of m linearly independent vectors that span the *same* m-dimensional subspace of the n-dimensional vector space. Clearly, Z^T is the *particular* right inverse of Y for which this holds.

It follows that the $n \times m$ network $Y^T(YY^T)^{-1}$ performs a process of matrix inversion on the received vector $SY + W$, replacing SY by S and W by $WY^T(YY^T)^{-1}$, to obtain an unbiased estimate of S. Furthermore, since the rows of Y and the columns of $Y^T(YY^T)^{-1}$ span the same subspace, the linear transformation $Y^T(YY^T)^{-1}$ involves more than just a simple matrix inversion, as will now be shown.

The $n \times m$ network $Y^T(YY^T)^{-1}$ may be split into two separate parts, the $n \times m$ network Y^T and the $m \times m$ network $(YY^T)^{-1}$, as shown in Figure 3.9. As before, $R = SY + W$.

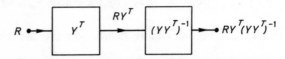

Fig. 3.9 *Two main constituents of the network* $Y^T(YY^T)^{-1}$

The network Y^T may now itself be divided into m separate networks, each of which is a linear filter *matched* to the corresponding vector Y_i, as shown in Figure 3.10.

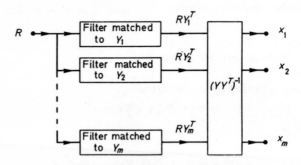

Fig. 3.10 *Network* $Y^T(YY^T)^{-1}$ *in expanded form*

The filter matched to Y_i forms the inner product of R and Y_i. Thus it multiplies each component of R by the corresponding component of Y_i and adds the products, to give RY_i^T.

The m matched filters together perform the linear transformation

Y^T, and form the $n \times m$ network Y^T in Figure 3.9. The m-component output signal vector from Y^T is RY^T, whose ith component is RY_i^T, as can be seen from Figure 3.10.

The network Y^T is *matched* to the signal SY and performs a process of *matched filter estimation* on the received vector R. The output signal RY^T from the network Y^T uniquely determines the vector XY, which if fed to the input of Y^T would give this output signal. Thus

$$RY^T = XYY^T \qquad (3.3.109)$$

and

$$(R - XY)Y^T = 0 \qquad (3.3.110)$$

as in Equation 3.3.54. It follows that XY is the orthogonal projection of R on to the m-dimensional subspace spanned by the $\{Y_i\}$, so that the matched network Y^T performs a process of *orthogonal projection*.

The network $(YY^T)^{-1}$ transforms the vector RY^T to the vector

$$RY^T(YY^T)^{-1} = XYY^T(YY^T)^{-1} = X \qquad (3.3.111)$$

so that $(YY^T)^{-1}$ is an *inverse* network, which reverses the transformation by means of which X has been converted to XYY^T.

Clearly, Y^T is equivalent to a *matched filter* and $(YY^T)^{-1}$ to an *inverse filter*.

The wanted component in the output signal RY_i^T from the ith matched filter in Figure 3.10 is $s_i Y_i Y_i^T$. The matched filter maximizes the ratio of the energy level of this signal to the average energy level of the noise component WY_i^T in RY_i^T. However, RY_i^T also contains $m - 1$ components $\{s_j Y_j Y_i^T\}$ due to the other components of the vector S, so that there may be considerable intersymbol interference in RY_i^T from the other components of S. The inverse network $(YY^T)^{-1}$ processes the $\{RY_i^T\}$ to eliminate all intersymbol interference, and suitably adjusts the levels of the resultant signals to give the $\{x_i\}$ at its output terminals.

The matched network Y^T, in Figure 3.9, increases the signal/noise ratio essentially because it forms the orthogonal projection of R on to the appropriate subspace in the n-dimensional vector space. The precise mechanism by which the signal/noise ratio is increased is as follows.

The *smallest* subspace of the n-dimensional vector space which contains SY for *all* values of S is the m-dimensional subspace spanned by the $\{Y_i\}$. If R is projected orthogonally on to this subspace, the projection XY comprises the original signal-vector SY added to a noise vector V, as shown in Figure 3.11.

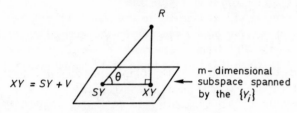

Fig. 3.11   *Projected vector XY*

Clearly

$$W = R - SY \qquad (3.3.112)$$

and

$$V = XY - SY \qquad (3.3.113)$$

V is the orthogonal projection of W on to the m-dimensional subspace spanned by the $\{Y_i\}$. If θ is the angle between W and V, the length of V is

$$|V| = |W \cos \theta| \qquad (3.3.114)$$

so that

$$|V| \leqslant |W| \qquad (3.3.115)$$

Now

$$|W| = \left(\sum_{j=1}^{n} w_j^2 \right)^{\frac{1}{2}} \qquad (3.3.116)$$

so that the mean-square value of $|W|$ is

$$E[|W|^2] = E\left[\sum_{j=1}^{n} w_j^2 \right] = \sum_{j=1}^{n} E[w_j^2] = \sum_{j=1}^{n} \sigma^2 = n\sigma^2 \qquad (3.3.117)$$

It follows that the average or expected noise energy in R is $n\sigma^2$.

The energy of a *given* signal-vector SY is

$$e = {}'|SY|^2 \qquad (3.3.118)$$

so that the signal/noise ratio in R is $e/n\sigma^2$, this being the ratio of the signal energy to the *average* noise energy.

Since V is the orthogonal projection of W on to the m-dimensional subspace spanned by the $\{Y_i\}$, the orthogonal projection of V on to any direction in the m-dimensional subspace is the *same* as the projection of the corresponding vector W. Let v_i be the orthogonal projection of V on to the ith of m orthogonal directions in the m-dimensional subspace. Then

$$|V| = \left(\sum_{i=1}^{m} v_i^2 \right)^{\frac{1}{2}} \qquad (3.3.119)$$

and the mean-square value of $|V|$ is

$$E[|V|^2] = E\left[\sum_{i=1}^{m} v_i^2 \right] = \sum_{i=1}^{m} E[v_i^2] = \sum_{i=1}^{m} \sigma^2 = m\sigma^2 \quad (3.3.120)$$

This uses the fact that the noise component along *any* direction in the vector space has zero mean and variance σ^2.

Clearly, the average noise energy in XY is $m\sigma^2$, whereas the signal energy is $e = |SY|^2$ as before. Thus the signal/noise ratio in XY is $e/m\sigma^2$.

It follows that when the received vector R in the n-dimensional vector space is projected orthogonally on to the m-dimensional subspace containing the signal-vector SY, the signal/noise ratio in the projected vector is n/m times that in the vector R. Clearly, the smaller the ratio m/n, the greater the improvement in signal/noise ratio, when R is projected on to the subspace.

To summarize, the vector X, which is an unbiased linear estimate of S, is obtained by the orthogonal projection of the received vector R on to the m-dimensional subspace spanned by the $\{Y_i\}$, followed by a process of exact linear equalization that eliminates the inter-symbol interference in the components of X and appropriately adjusts their levels. The whole process is achieved by feeding R to the n input terminals of the $n \times m$ network $Y^T(YY^T)^{-1}$, whose m output terminals then hold the vector X. X is the best linear estimate that can be made of S, when the receiver has no prior knowledge of S or σ^2.

3.3.7 Example

A three-component received vector is given by

$$R = s_1 Y_1 + s_2 Y_2 + W \qquad (3.3.121)$$

where s_1 and s_2 are real scalars, and

$$R = r_1 \ \ r_2 \ \ r_3 \qquad (3.3.122)$$

$$Y_1 = 1 \ \ 1 \ \ 0 \qquad (3.3.123)$$

$$Y_2 = 0 \ \ 1 \ \ 1 \qquad (3.3.124)$$

$$W = w_1 \ w_2 \ w_3 \tag{3.3.125}$$

The components of W are statistically independent Gaussian random variables with zero mean and variance σ^2. The receiver has prior knowledge of Y_1 and Y_2 but no prior knowledge of s_1, s_2 or σ^2.

Determine the matched and inverse networks required to obtain the maximum-likelihood estimate of the two-component vector (s_1, s_2). Then evaluate the advantage gained in tolerance to noise through the use of the matched and inverse networks.

Let

$$S = s_1 \ s_2 \tag{3.3.126}$$

and

$$Y = \begin{bmatrix} 1 & 1 & 0 \\ 0 & 1 & 1 \end{bmatrix} \tag{3.3.127}$$

so that

$$R = SY + W \tag{3.3.128}$$

and

$$r_1 = s_1 + w_1 \tag{3.3.129}$$

$$r_2 = s_1 + s_2 + w_2 \tag{3.3.130}$$

$$r_3 = s_2 + w_3 \tag{3.3.131}$$

Consider the three-dimensional vector space containing R. The maximum-likelihood estimate of S is the two-component vector $X = (x_1, x_2)$, where XY is the orthogonal projection of R on to the two-dimensional subspace spanned by Y_1 and Y_2.

It has previously been shown that

$$X = RY^T(YY^T)^{-1} \tag{3.3.132}$$

where

$$Y^T(YY^T)^{-1} = \begin{bmatrix} \frac{2}{3} & -\frac{1}{3} \\ \frac{1}{3} & \frac{1}{3} \\ -\frac{1}{3} & \frac{2}{3} \end{bmatrix} \tag{3.3.133}$$

Furthermore,

$$Y^T = \begin{bmatrix} 1 & 0 \\ 1 & 1 \\ 0 & 1 \end{bmatrix} \tag{3.3.134}$$

is the *matched network*

and

$$(YY^T)^{-1} = \begin{bmatrix} \frac{2}{3} & -\frac{1}{3} \\ -\frac{1}{3} & \frac{2}{3} \end{bmatrix} \tag{3.3.135}$$

is the *inverse network*.

The linear network $Y^T(YY^T)^{-1}$ in *expanded form* becomes as

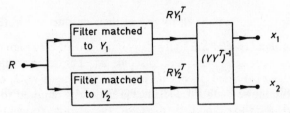

Fig. 3.12 Network $Y^T(YY^T)^{-1}$ *in expanded form*

shown in Figure 3.12. The output signals from the matched filters in Figure 3.12 are

$$RY_1{}^T = r_1 + r_2 = 2s_1 + s_2 + w_1 + w_2 \qquad (3.3.136)$$

and

$$RY_2{}^T = r_2 + r_3 = 2s_2 + s_1 + w_2 + w_3 \qquad (3.3.137)$$

The output signal from the first matched filter contains an inter-symbol-interference component s_2 from the received signal-element $s_2 Y_2$, and the output signal from the second matched filter contains an intersymbol interference component s_1 from the received signal-element $s_1 Y_1$. It is evident from the magnitude of the intersymbol interference that RY_1^T and RY_2^T could not be used satisfactorily for the *estimation* of the corresponding parameters s_1 and s_2. Furthermore, since the receiver has no prior knowledge of s_1 or s_2, it cannot assume, before the estimation process, that either $|s_1| \gg |s_2|, |s_2| \gg |s_1|$ or $|s_1| \simeq |s_2|$. The receiver must clearly take account of these different possibilities. Thus the receiver cannot in fact tolerate *any* intersymbol interference in its estimates of s_1 and s_2.

Each matched filter performs a process of matched-filter estimation on the corresponding received element $s_i Y_i$, such that in the output signal from the matched filter, the ratio of the *wanted* signal energy (for the given s_i) to the expected noise energy is maximum and given by

$$\frac{(2s_i)^2}{E[(w_i + w_{i+1})^2]} = \frac{4s_i^2}{E[w_i^2 + w_{i+1}^2 + 2w_i w_{i+1}]} = \frac{2s_i^2}{\sigma^2} \qquad (3.3.138)$$

In Equation 3.3.138, $E[w_i w_{i+1}] = 0$, $E[w_i^2] = \sigma^2$, and $i = 1$ or 2. The energy of the intersymbol interference component has not been considered here.

The two matched filters together perform the linear transformation

Y^T on the received vector R, to give at their outputs

$$RY^T = (RY_1^T, RY_2^T)$$
$$= (2s_1+s_2+w_1+w_2, 2s_2+s_1+w_2+w_3) \quad (3.3.139)$$

The output signal from the inverse network $(YY^T)^{-1}$, when RY^T is fed to its input, is

$$RY^T(YY^T)^{-1}$$
$$= (2s_1+s_2+w_1+w_2, 2s_2+s_1+w_2+w_3) \begin{bmatrix} \frac{2}{3} & -\frac{1}{3} \\ -\frac{1}{3} & \frac{2}{3} \end{bmatrix}$$
$$= (\tfrac{2}{3}(2s_1+s_2+w_1+w_2)-\tfrac{1}{3}(2s_2+s_1+w_2+w_3),$$
$$\qquad -\tfrac{1}{3}(2s_1+s_2+w_1+w_2)+\tfrac{2}{3}(2s_2+s_1+w_2+w_3))$$
$$= (s_1+\tfrac{2}{3}w_1+\tfrac{1}{3}w_2-\tfrac{1}{3}w_3, s_2-\tfrac{1}{3}w_1+\tfrac{1}{3}w_2+\tfrac{2}{3}w_3) \quad (3.3.140)$$

and, from Equation 3.3.132, this is the vector $X = (x_1, x_2)$. Thus

$$x_1 = s_1+\tfrac{2}{3}w_1+\tfrac{1}{3}w_2-\tfrac{1}{3}w_3 \quad (3.3.141)$$

$$x_2 = s_2-\tfrac{1}{3}w_1+\tfrac{1}{3}w_2+\tfrac{2}{3}w_3 \quad (3.3.142)$$

Clearly, there is no intersymbol interference in either x_1 or x_2, and X is an unbiased estimate of S.

The signal/noise ratio in x_1 is

$$\frac{s_1^2}{E[(\tfrac{2}{3}w_1+\tfrac{1}{3}w_2-\tfrac{1}{3}w_3)^2]} = \frac{s_1^2}{E[\tfrac{4}{9}w_1^2+\tfrac{1}{9}w_2^2+\tfrac{1}{9}w_3^2]}$$
$$= \frac{s_1^2}{\tfrac{4}{9}\sigma^2+\tfrac{1}{9}\sigma^2+\tfrac{1}{9}\sigma^2} = 1.5\frac{s_1^2}{\sigma^2} \quad (3.3.143)$$

since $E[w_i w_j] = 0$ for $i \neq j$, and $E[w_i^2] = \sigma^2$.

Similarly, the signal/noise ratio in x_2 is

$$\frac{s_2^2}{E[(-\tfrac{1}{3}w_1+\tfrac{1}{3}w_2+\tfrac{2}{3}w_3)^2]} = 1.5\frac{s_2^2}{\sigma^2} \quad (3.3.144)$$

It is clear from Equations 3.3.129 and 3.3.131 that s_1 and s_2 could be estimated from the values of r_1 and r_3, respectively, since r_1 is independent of s_2 and r_3 is independent of s_1.

The signal/noise ratio in r_1 is

$$\frac{s_1^2}{E[w_1^2]} = \frac{s_1^2}{\sigma^2} \quad (3.3.145)$$

and the signal/noise ratio in r_3 is

$$\frac{s_2^2}{E[w_3^2]} = \frac{s_2^2}{\sigma^2} \qquad (3.3.146)$$

Comparing Equations 3.3.143 and 3.3.144 with Equations 3.3.145 and 3.3.146, it can be seen that the linear network $Y^T(YY^T)^{-1}$ increases the signal/noise ratio by 1.5 times, and so achieves an advantage of 1.8 dB in tolerance to the Gaussian noise. Comparing Equations 3.3.143 and 3.3.144 with Equation 3.3.138, it can be seen that the removal of the intersymbol interference in the two components of RY^T reduces the wanted-signal/noise ratio by 1.2 dB, bearing in mind that the energy of the intersymbol interference in each component of RY^T is neglected here.

3.4 DETECTION OF A SIGNAL WITH UNKNOWN PARAMETERS IN NOISE

3.4.1 Composite hypothesis testing

It has been shown that in the application of Bayes' likelihood ratio test to the detection of the value of i from the received n-component vector R in Equation 3.2.3, where

$$R = S_i + W \qquad (3.4.1)$$

and $i = 0$ or 1, the decision rule that minimizes the average cost is given by Equations 3.2.27 and 3.2.28. This decision rule is a test to decide between one hypothesis and another. It is a *simple hypothesis test*.[1]

When S_0 and S_1, instead of having fixed values, may each have a range of values given by the points in the regions B_0 and B_1, respectively, of the n-dimensional vector space, the decision rule becomes a test to decide between one composite hypothesis and another. It is a *composite hypothesis test*.[1] The received n-component vector now becomes

$$R = F + W \qquad (3.4.2)$$

where the n-component row vector F replaces S_i, and F may now have any one of possibly an infinite set of values, a subset of which comprise the region B_0 and the complementary subset of which (that is, the remainder of the possible values) comprise the region B_1.

Clearly, B_0 and B_1 are disjoint. The two regions B_0 and B_1 together do not normally occupy the whole of the n-dimensional vector space, but usually some subspace which may or may not be bounded (extend to infinity along its orthogonal axes). The decision rule becomes a test to decide between the hypothesis H_0 that $F \in B_0$ and the hypothesis H_1 that $F \in B_1$. The prior probabilities of the two events $F \in B_0$ and $F \in B_1$ are

$$P(F \in B_0) = q_0 \qquad (3.4.3)$$

and
$$P(F \in B_1) = q_1 \qquad (3.4.4)$$

where
$$q_0 + q_1 = 1 \qquad (3.4.5)$$

The problem of deciding whether to accept H_0 or H_1 is that of deciding whether the signal vector F in Equation 3.4.2 lies in the region B_0 or B_1 of the n-dimensional vector space.

As before, the n-dimensional vector space containing R is divided into two complementary *decision* regions which together fill the *whole* of the vector space. If R lies in the decision region D_0, the hypothesis H_0 is accepted, and if R lies in the decision region D_1, the hypothesis H_1 is accepted. The probability that H_1 is accepted, for a *given F* that satisfies H_0 and so lies in the region B_0 of the vector space, is

$$\alpha(F) = \int_{D_1} p(R|F, F \in B_0) dR \qquad (3.4.6)$$

and the probability that H_0 is accepted, for a *given F* that satisfies H_1 and so lies in the region B_1 of the vector space, is

$$\beta(F) = \int_{D_0} p(R|F, F \in B_1) dR \qquad (3.4.7)$$

$p(R|F, F \in B_i)$ is the value of the conditional probability density function of the random vector with sample value R, when the random vector has this sample value, given the *particular* value of the vector F, which lies in the region B_i where $i = 0$ or 1.

Thus the probability of accepting H_1 when H_0 is true, is

$$\alpha' = \int_{B_0} \alpha(F) p(F|F \in B_0) dF$$

$$= \int_{B_0} \left[\int_{D_1} p(R|F, F \in B_0) dR \right] p(F|F \in B_0) dF$$

$$= \int_{D_1}\left[\int_{B_0} p(R|F, F \in B_0)p(F|F \in B_0)\mathrm{d}F\right]\mathrm{d}R \qquad (3.4.8)$$

where $p(F|F \in B_0)$ is the value of the conditional probability density function of the random vector with sample value F, when the vector has the given value F and given that F lies in the region B_0. The probability of accepting H_0 when H_1 is true, is

$$\beta' = \int_{B_1} \beta(F)p(F|F \in B_1)\mathrm{d}F$$

$$= \int_{B_1}\left[\int_{D_0} p(R|F, F \in B_1)\mathrm{d}R\right]p(F|F \in B_1)\mathrm{d}F$$

$$= \int_{D_0}\left[\int_{B_1} p(R|F, F \in B_1)p(F|F \in B_1)\mathrm{d}F\right]\mathrm{d}R \qquad (3.4.9)$$

From Equation 3.2.19, the average cost in the testing of the two hypotheses becomes

$$G = q_0C_{00} + q_1C_{11} + q_0C_\alpha\alpha' + q_1C_\beta\beta' \qquad (3.4.10)$$

where C_{00}, C_{11}, C_α and C_β are as previously defined.

Proceeding exactly as for the simple hypothesis test, it can be shown[1] that the *minimum average cost* is obtained by applying the following decision rule to the received vector R.

Let

$$L(R) = \frac{\displaystyle\int_{B_1} p(R|F, F \in B_1)p(F|F \in B_1)\mathrm{d}F}{\displaystyle\int_{B_0} p(R|F, F \in B_0)p(F|F \in B_0)\mathrm{d}F} \qquad (3.4.11)$$

and

$$K = \frac{q_0C_\alpha}{q_1C_\beta} \qquad (3.4.12)$$

If $L(R) < K$, accept H_0 and detect i as 0. If $L(R) > K$, accept H_1 and detect i as 1.

This decision rule is known as *Bayes' generalized likelihood ratio test*.[1]

The values of R for which $L(R) < K$ lie in the decision region D_0 of the n-dimensional vector space. The values of R for which $L(R) > K$ lie in the decision region D_1. The *decision boundary* between the two decision regions D_0 and D_1 is given by

$$L(R) = K \qquad (3.4.13)$$

$L(R)$ is the *generalized likelihood ratio* and K is the *threshold*.

When R lies in the decision region D_0, the receiver decides that F lies in the region B_0, and when R lies in the decision region D_1, the receiver decides that F lies in the region B_1.

3.4.2　Optimum detection with limited prior knowledge of the received signal

Consider now the n-component received vector

$$R = \sum_{i=1}^{m} s_i Y_i + W = SY + W \qquad (3.4.14)$$

which is the sum of m signal-elements $\{s_i Y_i\}$ and a noise vector W, where $m \leqslant n$. The $\{s_i\}$ are scalar quantities that carry the element values, and the n-component row vectors $\{Y_i\}$ characterize or identify the corresponding signal-elements, each of which is, of course, an n-component vector. The $\{Y_i\}$ are linearly independent and so span an m-dimensional subspace of the n-dimensional vector space. Y is the $m \times n$ matrix of rank m whose ith row is Y_i, and S is the m-component row vector whose ith component is s_i. The components of W are statistically independent Gaussian random variables with zero mean and variance σ^2, as before. Each signal element $s_i Y_i$ is binary coded, such that

$$s_i = \pm k \qquad (3.4.15)$$

where k is a positive constant. The m $\{s_i\}$ are statistically independent and equally likely to have either binary value. The information in the received signal is contained entirely in the binary values of the $\{s_i\}$, the $\{s_i\}$ being statistically independent of the $\{w_i\}$.

It is assumed that the receiver has prior knowledge of the m vectors $\{Y_i\}$ and therefore of the matrix Y, but it has no prior knowledge of k or σ^2, other than that k is positive. The absence of any prior knowledge of k is taken to be complete, in the sense that the receiver does not know that k has the same or even related values for the different $\{s_i\}$. Thus, as far as the receiver is concerned, s_i may take on any positive or negative value which is in no way related to the value of s_j, for any $j \neq i$. Let s_i now be the *actual* value of this parameter in Equation 3.4.15, and let s_{ia} be any *possible* value of s_i, as far as the receiver is concerned, before it has detected or estimated s_i. Thus

$$s_{ia} = \pm k_i \qquad (3.4.16)$$

for each i, where k_i has *any* positive real value and is in no way related to k_j, for any $j \neq i$. The $\{s_{ia}\}$ are statistically independent and equally likely to have either binary value.

Clearly, the binary value of each signal element is determined by the *sign* of the corresponding s_i, and not by its magnitude. Thus, to determine the binary values of the m signal-elements, the receiver must simply detect the *signs* of the m $\{s_i\}$. The problem now is to determine the decision regions and hence the decision boundaries in the vector space, that *minimize the probability of error* in the detection of each element value from R, under the assumed conditions.

Let S be the *actual* value of the received m-component signal vector, and let S_a be any *possible* value of S, as far as the receiver is concerned. Thus S_a is equally likely to be any real m-component vector, and $S_a Y$ is equally likely to lie at *any* point in the m-dimensional subspace spanned by the m $\{Y_i\}$, within the n-dimensional vector space.

Consider the detection of the binary value of the ith signal element $s_i Y_i$, that is, the detection of the sign of s_i from the received vector R. The receiver must here decide, with the minimum probability of error, whether $s_i > 0$ or $s_i < 0$.

The m-dimensional subspace spanned by the $\{Y_i\}$ can be divided into two regions B_0 and B_1, where B_0 contains all vectors $\{S_a Y\}$ for which $s_{ia} < 0$, and B_1 contains all vectors $\{S_a Y\}$ for which $s_{ia} > 0$. B_0 and B_1 are disjoint and together fill the whole of the m-dimensional subspace. Clearly, the values of $S_a Y$ for which $s_{ia} < 0$ are separated from the values of $S_a Y$ for which $s_{ia} > 0$, by the values of $S_a Y$ for which $s_{ia} = 0$. These values of $S_a Y$ satisfy

$$S_a Y = \sum_{\substack{j=1 \\ j \neq i}}^{m} s_{ja} Y_j \qquad (3.4.17)$$

for all real values of the $m-1$ $\{s_{ja}\}$ for which $j \neq i$. Thus they form the $(m-1)$-dimensional subspace spanned by the $m-1$ $\{Y_j\}$ for which $j \neq i$. This subspace forms the *boundary* between the regions B_0 and B_1 of the m-dimensional subspace. If the receiver decides that SY lies in the region B_0, it detects s_i as having the binary value

$-k_i$. Similarly, if the receiver decides that SY lies in the region B_1, it detects s_i as having the binary value k_i.

Let F be the n-component row vector

$$F = S_a Y \qquad (3.4.18)$$

so that, if $S = S_a$, assuming for the moment that S may take on any value, then

$$R = F + W \qquad (3.4.19)$$

as in Equation 3.4.2.

Since F is equally likely to lie at any point in the m-dimensional subspace spanned by the m $\{Y_i\}$,

$$\left.\begin{array}{l} p(F|F \in B_0) = c \\ p(F|F \in B_1) = c \end{array}\right\} \qquad (3.4.20)$$

for all permissible values of F. The scalar quantity c is a vanishingly small positive constant. Thus, from Equation 3.4.11,

$$L(R) = \frac{\displaystyle\int_{B_1} p(R|F, F \in B_1)\mathrm{d}F}{\displaystyle\int_{B_0} p(R|F, F \in B_0)\mathrm{d}F} = \frac{\displaystyle\int_{B_1} p(R|F)\mathrm{d}F}{\displaystyle\int_{B_0} p(R|F)\mathrm{d}F} \qquad (3.4.21)$$

and from Equation 3.3.51,

$$p(R|F) = \frac{1}{(2\pi\sigma^2)^{\frac{1}{2}n}} \exp\left(-\frac{1}{2\sigma^2}|R - F|^2\right) \qquad (3.4.22)$$

Let XY be the orthogonal projection of R on to the m-dimensional subspace spanned by the m vectors $\{Y_i\}$. Then from Equation 3.4.22, bearing in mind that $R - XY$ is orthogonal to $XY - F$,

$$p(R|F) = \frac{1}{(2\pi\sigma^2)^{\frac{1}{2}n}} \exp\left[-\frac{1}{2\sigma^2}(|R - XY|^2 + |XY - F|^2)\right]$$

$$= \frac{1}{(2\pi\sigma^2)^{\frac{1}{2}n}} \exp\left(-\frac{1}{2\sigma^2}|R - XY|^2\right) \exp\left(-\frac{1}{2\sigma^2}|F - XY|^2\right) \qquad (3.4.23)$$

Inserting this value of $p(R|F)$ in Equation 3.4.21,

$$L(R) = \frac{\displaystyle\int_{B_1} \exp\left(-\frac{1}{2\sigma^2}|F - XY|^2\right)\mathrm{d}F}{\displaystyle\int_{B_0} \exp\left(-\frac{1}{2\sigma^2}|F - XY|^2\right)\mathrm{d}F} \qquad (3.4.24)$$

According to the decision rule, the value of $L(R)$ corresponding to the received vector R is compared with the threshold K in Equation 3.4.12. Since it is required to minimize the *probability of error* in the detection of the ith element, which is equally likely to have either binary value, $q_0 = q_1 = \frac{1}{2}$ and $C_\alpha = C_\beta = 1$, so that $K = 1$. Thus, depending upon whether $L(R) < 1$ or $L(R) > 1$, the receiver accepts the hypothesis H_0 that SY lies in B_0 or the hypothesis H_1 that SY lies in B_1, respectively. Hence, depending upon whether $L(R) < 1$ or $L(R) > 1$, s_i is detected as $-k_i$ or k_i, respectively.

Suppose now that $x_i < 0$, so that XY lies in the region B_0 of the m-dimensional subspace spanned by the $\{Y_i\}$, as can be seen from the definition of B_0. XY is shown in Figure 3.13.

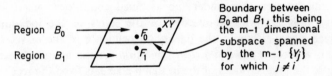

Fig. 3.13   m-dimensional subspace spanned by the $\{Y_i\}$

Consider the signal vector F_0 in B_0 and the corresponding signal vector F_1 in B_1 which is such that the boundary between B_0 and B_1 (the $(m-1)$-dimensional subspace spanned by the $m-1$ vectors $\{Y_j\}$ for which $j \neq i$) perpendicularly bisects the line joining F_0 and F_1. F_0 and F_1 are shown in Figure 3.13. For *any* vector F_0 in B_0, F_0 is *closer* to XY than is F_1, so that for every F_0 and the corresponding F_1,

$$|F_0 - XY| < |F_1 - XY| \tag{3.4.25}$$

or

$$\exp\left(-\frac{1}{2\sigma^2}|F_0 - XY|^2\right) > \exp\left(-\frac{1}{2\sigma^2}|F_1 - XY|^2\right) \tag{3.4.26}$$

It follows that whenever $x_i < 0$, XY lies in B_0 and

$$\int_{B_0} \exp\left(-\frac{1}{2\sigma^2}|F - XY|^2\right)dF > \int_{B_1} \exp\left(-\frac{1}{2\sigma^2}|F - XY|^2\right)dF \tag{3.4.27}$$

so that, from Equation 3.4.24, $L(R) < 1$. Similarly, it can be shown that whenever $x_i > 0$, XY lies in B_1 and $L(R) > 1$.

Clearly, for the minimum probability of error, s_i is detected as

$-k_i$ when $L(R) < 1$, and s_i is detected as k_i when $L(R) > 1$. But it has just been shown that when $x_i < 0$, XY lies in the region B_0 and $L(R) < 1$; when $x_i > 0$, XY lies in the region B_1 and $L(R) > 1$. Thus, for the minimum probability of error under the assumed conditions, s_i is detected as $-k_i$ when $x_i < 0$, and s_i is detected as k_i when $x_i > 0$.

X is the maximum-likelihood estimate of S and is given by Equation 3.3.56. Thus, if the received n-component vector R is fed to the n input terminals of the $n \times m$ network $Y^T(YY^T)^{-1}$, the m output terminals of this network hold the m-component vector X. This is an unbiased estimate of S, as can be seen from Equation 3.3.57. The network performs a process of exact linear equalization on the received signal.

It can be seen that under the assumed conditions, where the receiver has no prior knowledge of the levels of the individual received signal-elements, the linear independence of the m vectors $\{Y_i\}$ is a *necessary and sufficient* condition for the unique detectability of the binary values of the m signal-elements from the received signal vector SY.

To summarize, the probability of error in a detection process is minimized if R is projected orthogonally on to the m-dimensional subspace spanned by the $\{Y_i\}$ to give the vector XY, and if, for each i, s_i is detected as k_i or $-k_i$ depending upon whether $x_i > 0$ or $x_i < 0$, respectively. This is the optimum detection process when the detector has prior knowledge of Y but no prior knowledge of σ^2 or of the levels of the individual received signal-elements. The detection process is implemented by feeding the received vector R to the n input terminals of the network $Y^T(YY^T)^{-1}$, to give the vector X at the m output terminals. Each x_i is now compared with a decision threshold of zero, to give the detected binary value of the corresponding s_i. Thus a process of exact linear equalization is first used to obtain the maximum-likelihood *estimate* of S, and S is then *detected* from this estimate.

3.4.3 *Probability of error*

SY has 2^m possible values and lies in the subspace spanned by the $\{Y_i\}$. The detection process just described can be considered to divide the m-dimensional subspace into 2^m decision regions, each corresponding to a different one of the 2^m possible values of SY. The process then determines which of these regions contains the vector XY, and the possible value of S corresponding to this region is the detected value of S. This method of analysis enables the probability of error to be evaluated and brings out some important properties of the detection process.

The decision regions are defined by m decision boundaries. The ith boundary is the locus of all points traced out by

$$\sum_{j=1}^{m} a_j Y_j$$

where $a_i = 0$ and the $\{a_j\}$, for $j \neq i$, may vary independently over all real values. The boundary is therefore traced out by AL_i for all real values of the $(m-1)$-component row vector A, where L_i is the $(m-1) \times n$ matrix obtained by deleting the ith row from Y. The boundary is clearly the $(m-1)$-dimensional subspace spanned by the $m-1$ vectors $\{Y_j\}$ for which $j \neq i$. It divides the m-dimensional subspace spanned by all $\{Y_j\}$ into two regions. If the vector XY lies in one of these, x_i is positive, and if XY lies in the other, x_i is negative. The former gives a detected value of k_i for s_i and the latter a detected value of $-k_i$. k_i may, of course, have any positive value.

From Equation 3.4.14, the received n-component vector is

$$R = SY + W \qquad (3.4.28)$$

where S is an m-component row vector and Y is an $m \times n$ matrix of rank m.

Assume for the moment that $m = 2$, so that the rows of Y span a two-dimensional subspace. The decision boundaries and decision regions in this subspace are shown in Figure 3.14.

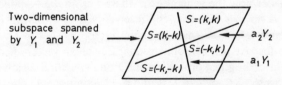

Fig. 3.14   Decision regions in a two-dimensional subspace

The lines traced out by $a_1 Y_1$ and $a_2 Y_2$, for all real values of a_1 and a_2, intersect at the origin where $a_1 = a_2 = 0$. It can be seen that if the orthogonal projection of R on to the subspace is $XY = x_1 Y_1 + x_2 Y_2$ and if this lies on the line traced out by $a_2 Y_2$, then $x_2 = a_2$ and $x_1 = 0$. For all $\{XY\}$ above this line, $x_1 > 0$ so that s_1 is detected as k_1, and for all $\{XY\}$ below this line, $x_1 < 0$ so that s_1 is detected as $-k_1$. Thus the line $a_2 Y_2$ is the decision boundary that separates the decision regions corresponding to the two possible values of s_1. Again, if XY lies on the line traced out by $a_1 Y_1$, then $x_1 = a_1$ and $x_2 = 0$. For all $\{XY\}$ to the right of this line, $x_2 > 0$ so that s_2 is detected as k_2, and for all $\{XY\}$ to the left of this line, $x_2 < 0$ so that s_2 is detected as $-k_2$. Thus the line $a_1 Y_1$ is the decision boundary that separates the decision regions corresponding to the two possible values of s_2.

It can now be seen from Figure 3.14 that the *two* decision boundaries, given by the lines $a_1 Y_1$ and $a_2 Y_2$, separate the two-dimensional subspace spanned by Y_1 and Y_2 into *four* decision regions. Furthermore, it can be seen that each of the four decision regions contains a different one of the four possible vectors $\{SY\}$, the vector S having in each case the value of the detected vector S appropriate to that decision region. Thus the received signal vector SY is detected as the one of the four possible vectors $\{SY\}$ that lies in the *same* decision region as XY, and the possible value of S in the detected vector SY gives the detected vector S. The fact that the receiver has no prior knowledge of k merely means that the four possible values of SY are not here their *actual* values but rather the four possible values of $\pm k_1 Y_1 \pm k_2 Y_2$, where k_1 and k_2 have arbitrary positive values. Since the binary value of each signal element is carried by the *sign* of the corresponding s_i and not by its magnitude, the receiver has sufficient prior knowledge of the received signal in order to achieve satisfactory detection, just so long as the signal/noise ratio is sufficiently high.

Consider now the general case where S is an m-component vector and Y is an $m \times n$ matrix, the received signal satisfying Equations 3.4.28 and 3.4.15, so that the signal vector SY lies in an m-dimensional subspace of the n-dimensional Euclidean vector space.

Let the orthogonal projection of SY on to the ith decision boundary be the vector KL_i, where K is an $(m-1)$-component row vector and L_i is the $(m-1) \times n$ matrix obtained by deleting the ith row from Y. Since the vector $SY - KL_i$ is orthogonal to the $(m-1)$-dimensional subspace comprising the ith decision boundary, it is also orthogonal to any Y_j, where $j \neq i$. Thus

$$(SY - KL_i)L_i^T = 0 \qquad (3.4.29)$$

or
$$KL_iL_i^T = SYL_i^T \qquad (3.4.30)$$

or
$$K = SYL_i^T(L_iL_i^T)^{-1} \qquad (3.4.31)$$

Let d_i be the distance from SY to KL_i. Then

$$d_i^2 = (SY - KL_i)(SY - KL_i)^T$$

$$= (SY - KL_i)Y^TS^T - (SY - KL_i)L_i^TK^T$$

$$= (SY - KL_i)Y^TS^T \qquad \text{(from Equation 3.4.29)}$$

$$= [SY - SYL_i^T(L_iL_i^T)^{-1}L_i]Y^TS^T$$

$$\qquad\qquad\qquad\qquad \text{(from Equation 3.4.31)}$$

$$= SY[I - L_i^T(L_iL_i^T)^{-1}L_i]Y^TS^T \qquad (3.4.32)$$

where I is an $n \times n$ identity matrix.

Let S_i be the $(m-1)$-component row vector obtained by deleting the ith component s_i from S. Then

$$d_i^2 = (s_iY_i + S_iL_i)[I - L_i^T(L_iL_i^T)^{-1}L_i](s_iY_i + S_iL_i)^T \qquad (3.4.33)$$

which reduces to

$$d_i = k[Y_i(I - L_i^T(L_iL_i^T)^{-1}L_i)Y_i^T]^{\frac{1}{2}} \qquad (3.4.34)$$

d_i is a function of i but is independent of the binary values of the $\{s_i\}$.

An error occurs in the detection of the binary value of s_i, which is the binary value of the ith signal element, when XY and SY lie on opposite sides of the ith decision boundary. Thus the probability of an error in the detection of the ith signal element is the probability that the orthogonal projection of the noise vector W on to the perpendicular from SY on to the ith decision boundary, is in the direction towards the boundary and exceeds the perpendicular distance from SY to the boundary. But it has been shown that the value of the orthogonal projection of W on to *any* direction in the n-dimensional vector space is a Gaussian random variable with zero mean and variance σ^2. Since d_i is the perpendicular distance from SY to the ith decision boundary, the probability of an error in the detection of the ith signal element is

$$p_i = \int_{d_i}^{\infty} \frac{1}{\sqrt{(2\pi\sigma^2)}} \exp\left(-\frac{w^2}{2\sigma^2}\right) dw$$

$$= \int_{d_i/\sigma}^{\infty} \frac{1}{\sqrt{2\pi}} \exp(-\tfrac{1}{2}w^2) dw$$

$$= Q\left(\frac{d_i}{\sigma}\right) \tag{3.4.35}$$

where d_i is given by Equation 3.4.34.

In the general case where d_i is a function of i, different signal-elements in a group of m have different error probabilities. A simple upper bound to the average element error probability is given by the value of p_i for the smallest d_i. Let d_i have its minimum value when $i = j$. Then the average element-error probability is less than or equal to

$$p_j = Q\left(\frac{d_j}{\sigma}\right) \tag{3.4.36}$$

A more direct but perhaps less illuminating derivation of the element error probability gives the result in a simpler form, as follows.

From Equation 3.3.59, the ith component of X is

$$x_i = s_i + u_i \tag{3.4.37}$$

$s_i = \pm k$ and u_i is the ith component of the m-component row vector

$$U = WZ^T \tag{3.4.38}$$

where $$Z^T = Y^T(YY^T)^{-1} \tag{3.4.39}$$

An error occurs in the detection of the binary value of s_i when x_i has the opposite sign to s_i, that is, when u_i has a magnitude greater than k and the opposite sign to s_i. But, from Equation 3.3.81, u_i is a Gaussian random variable with zero mean and variance $\sigma^2|Z_i|^2$, where Z_i is the n-component row vector given by the ith row of the $m \times n$ matrix Z. Thus the probability of error in the detection of the binary value of the ith signal element is

$$p_i = \int_k^{\infty} \frac{1}{\sqrt{(2\pi\sigma^2|Z_i|^2)}} \exp\left(-\frac{u^2}{2\sigma^2|Z_i|^2}\right) du$$

$$= \int_{k/\sigma|Z_i|}^{\infty} \frac{1}{\sqrt{2\pi}} \exp(-\tfrac{1}{2}u^2)du$$

$$= Q\left(\frac{k}{\sigma|Z_i|}\right) \tag{3.4.40}$$

where $|Z_i|$ is the length of the vector that forms the ith column of $Y^T(YY^T)^{-1}$.

A simple upper bound to the average element error probability is given by the value of p_i for the largest $|Z_i|$. Let $|Z_i|$ have its maximum value when $i = j$. Then the average element error probability is less than or equal to

$$p_j = Q\left(\frac{k}{\sigma|Z_j|}\right) \tag{3.4.41}$$

3.4.4 Example

A three-component received vector is given by

$$R = s_1 Y_1 + s_2 Y_2 + W \tag{3.4.42}$$

where

$$Y_1 = 1 \quad 1 \quad 0 \tag{3.4.43}$$

and

$$Y_2 = 0 \quad 1 \quad 1 \tag{3.4.44}$$

$s_1 = \pm k$ and $s_2 = \pm k$, where k is a positive constant. s_1 and s_2 are statistically independent and equally likely to have either binary value. The three components of W are statistically independent Gaussian random variables with zero mean and variance σ^2.

Obtain an expression for the probability of error in the detection of the binary values of s_1 and s_2, firstly for the case where the receiver has prior knowledge of Y_1 and Y_2 but no prior knowledge of k or σ^2 other than that k is positive, and secondly for the case where the receiver has prior knowledge of Y_1, Y_2 and k. It is assumed that the appropriate optimum detection process is used in each case.

Let

$$S = s_1 \quad s_2 \tag{3.4.45}$$

and

$$Y = \begin{bmatrix} 1 & 1 & 0 \\ 0 & 1 & 1 \end{bmatrix} \tag{3.4.46}$$

so that $$R = SY + W \tag{3.4.47}$$

The four possible values of the received signal-vector SY are

$$T_1 = kY_1 + kY_2 \tag{3.4.48}$$

$$T_2 = -kY_1 + kY_2 \tag{3.4.49}$$

$$T_3 = kY_1 - kY_2 \tag{3.4.50}$$

$$T_4 = -kY_1 - kY_2 \tag{3.4.51}$$

The positions of T_1, T_2, T_3 and T_4 in the two-dimensional subspace spanned by Y_1 and Y_2 are shown in Figure 3.15.

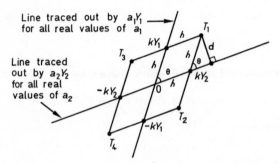

Fig. 3.15   Signal vectors and decision boundaries in the two-dimensional subspace

When the receiver has *no* prior knowledge of k or σ^2, other than that k is positive, it cannot assume that k has the same value for both s_1 and s_2 or even that its values for s_1 and s_2 are in any way related. Under these conditions, the detection process that minimizes the probability of error, forms the orthogonal projection XY of R on to the two-dimensional subspace spanned by Y_1 and Y_2, where X is the two-component row vector (x_1, x_2), and detects the binary values of s_1 and s_2 from the signs of x_1 and x_2, respectively.

X is the maximum-likelihood estimate of S and is given by

$$X = RY^T(YY^T)^{-1} \tag{3.4.52}$$

as can be seen from Equation 3.3.70 and Figure 3.7. Thus, from

Equation 3.3.73, when R is fed to the three input terminals of the 3×2 network

$$Y^T(YY^T)^{-1} = \begin{bmatrix} \frac{2}{3} & -\frac{1}{3} \\ \frac{1}{3} & \frac{1}{3} \\ -\frac{1}{3} & \frac{2}{3} \end{bmatrix} \tag{3.4.53}$$

X is obtained at the two output terminals, as shown in Figure 3.8.

The probability of error in the detection of the binary value of s_1 or s_2 may now be evaluated from either Equation 3.4.35 or 3.4.40, and each of these gives the error probability

$$p_i = Q\left(\frac{\sqrt{3}k}{\sqrt{2}\sigma}\right) \tag{3.4.54}$$

for both s_1 and s_2.

A more instructive derivation of the probability of error in the detection of s_1 or s_2 is as follows.

The length of each of the four vectors $\pm kY_1$ and $\pm kY_2$, in Figure 3.15, is

$$h = (k^2 + k^2)^{\frac{1}{2}} = \sqrt{2}k \tag{3.4.55}$$

and the angle θ between the vectors kY_1 and kY_2 is given by

$$\cos\theta = \frac{(kY_1)(kY_2)^T}{|kY_1| \, |kY_2|} = \frac{k^2 Y_1 Y_2^T}{h^2} = \frac{1}{2} \tag{3.4.56}$$

so that $\theta = 60°$.

It is now evident from the symmetry in Figure 3.15 that the distance from each of the four vectors $\{T_i\}$ to each of the two decision boundaries is

$$d = h\sin\theta = \sqrt{\tfrac{3}{2}}k \tag{3.4.57}$$

An error occurs in the detection of the binary value of the ith signal-element when XY and the received signal-vector SY lie on opposite sides of the ith decision boundary. This occurs when the orthogonal projection of the noise vector W on to the perpendicular from SY on to the decision boundary is in the direction towards the boundary and exceeds the perpendicular distance d from SY to the boundary. Since the value of the projection of W is a Gaussian random variable with zero mean and variance σ^2, it follows that the probability of error in the detection of either signal element is

$$Q\left(\frac{d}{\sigma}\right) = Q\left(\frac{\sqrt{3}k}{\sqrt{2}\sigma}\right) \tag{3.4.58}$$

as in Equation 3.4.54.

When the receiver has prior knowledge of both Y and k, so that it knows the four possible values of SY, the detection process that minimizes the probability of error selects the one of the four possible vectors $\{SY\}$ that is nearest to the received vector R and takes the corresponding vector S as the detected vector.

The one of the four possible vectors $\{SY\}$ that is nearest to R is also the vector SY that is nearest to XY, the orthogonal projection of R on to the two-dimensional subspace. Thus, although the optimum detection process in fact operates directly on the received vector R without forming the vector XY, the operation of the detection process can be described more simply here if it is assumed that XY is first formed from R, and S is then detected from XY. This does not affect any of the important properties of the detection process and enables a simple comparison to be made with the other detection process, previously described.

The four possible values of the received signal-vector SY are the vectors $\{T_i\}$ given by Equations 3.4.48 to 3.4.51 and shown in Figure 3.16. SY is equally likely to be any of the four $\{T_i\}$. The

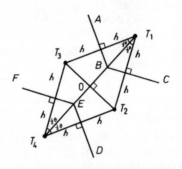

Fig. 3.16 *Signal vectors and decision boundaries in the two-dimensional subspace*

operation of the optimum detection process is equivalent to forming the orthogonal projection XY of the received vector R on to the two-dimensional subspace spanned by Y_1 and Y_2, and then selecting the vector T_i that lies in the same decision region as XY. The corresponding vector S is the detected vector. The decision boundaries in the subspace are the perpendicular bisectors of the lines joining the different pairs of the four vectors $\{T_i\}$. Thus the decision boundaries are the lines ABC, CBED, ABEF, DEF, and each of the corresponding four decision regions contains one of the four vectors $\{T_i\}$, this being the detected value of XY if it lies in that region.

The distance between T_1 and T_2, T_2 and T_4, T_4 and T_3, T_3 and T_1, is $2h$ in each case. The distance between T_1 and T_4 is

$$4h \cos \tfrac{1}{2}\theta = 2\sqrt{3}h \qquad (3.4.59)$$

since $\theta = 60°$, and the distance between T_2 and T_3 *is*

$$4h \sin \tfrac{1}{2}\theta = 2h \qquad (3.4.60)$$

Since the decision boundary separating any two of the four vectors $\{T_i\}$ is the perpendicular bisector of the line joining these vectors, it is clear that the distance to the *nearest* decision boundary is h, for each of the four $\{T_i\}$.

At high signal/noise ratios with additive Gaussian noise, even a very small increase in the distance to a decision boundary produces a considerable reduction in the corresponding error probability. Thus the error probability is effectively determined by the *nearest decision boundary*, the remaining boundaries having in comparison a negligible effect on the error probability. Furthermore, if only every second or third T_i has a decision boundary at a distance equal to the minimum between any T_i and its associated decision boundaries, or alternatively, if some of the $\{T_i\}$ have two or three decision boundaries at this minimum distance, then in either case the error probability changes by no more than two or three times, which at high signal/noise ratios represents a change of only a fraction of 1 dB in the tolerance to the Gaussian noise, and this can normally be neglected. The distance to the *nearest* decision boundary, for all $\{T_i\}$ and their associated boundaries, is thus a reasonably reliable measure of the tolerance to additive white Gaussian noise, so long as the signal/noise ratio remains high.

From Equation 3.4.55, the distance to the nearest decision boundary is $\sqrt{2}k$, so that the probability of error in the detection of either signal element is approximately

$$\int_{\sqrt{2}k}^{\infty} \frac{1}{\sqrt{2\pi}\sigma^2} \exp\left(-\frac{w^2}{2\sigma^2}\right) dw = Q\left(\frac{\sqrt{2}k}{\sigma}\right) \qquad (3.4.61)$$

From Equations 3.4.58 and 3.4.61, it can be seen that, for a given probability of error, the detection process just described tolerates a noise variance that is 4/3 times greater than that tolerated by the previous detection process, so that the second detection process has an advantage of 1.25 dB in tolerance to additive white Gaussian noise.

It can be shown that as the number of signal elements, m, increases, so does the advantage in tolerance to noise of the better of the two detection processes. When m is between 10 and 20, advantages of the order of 10 or more dB may be obtained by the better detection process.

3.4.5   Assessment of detection processes

In the detection process described in Section 3.4.2, a process of exact linear equalization is first applied to the received vector R to give the maximum-likelihood estimate X of the vector S, and S is then detected from X. This detection process can be applied to the general case where each of the m signal elements may have any one of l different values, where $l \geqslant 2$. To achieve the best available tolerance to the Gaussian noise with the given process, where each s_i is equally likely to have any of its possible values and is statistically independent of the other $\{s_i\}$, the detected value of each s_i is taken as the one of its l possible values that is closest to the corresponding x_i. Thus each x_i is compared with $l-1$ decision thresholds, that are placed half way between adjacent possible values of the corresponding s_i, and the detected value of s_i is its possible value between the same decision thresholds as x_i. The possible values of s_i here are, of course, their *actual* values.

Clearly the receiver now requires prior knowledge of the levels of the individual received signal-elements, so that it must know the l^m possible vectors $\{SY\}$. However, since the detection process is a simple development of that described in Section 3.4.2, it has become suboptimum in the sense that it does not achieve the best available tolerance to noise for the given prior knowledge of the received signal used at the receiver. It does, however, have one important advantage over the optimum detection process, which will now be described.

Let the optimum detection process, that detects S as the one of its possible values for which SY is closest to R, be the System 1, and let the suboptimum detection process just described be the System 2.

The weakness of System 1 is that l^m sequential operations must be performed in a detection process, and l^m can become very large even for modest values of l and m. Furthermore, any useful reduction in the number of sequential operations, by the use of parallel working, would be prohibitively expensive.

System 2, on the other hand, involves an $n \times m$ matrix transformation on the received vector R, followed by the comparison of the m output signals with the appropriate decision thresholds. If the components of R are stored in analogue form, the matrix transformation may be implemented directly in analogue form by nm attenuators, with inverters where necessary and possibly some amplifiers. On the other hand, if the components of R are stored in digital form (that is, as binary-coded numbers), the matrix transformation may be implemented either by the appropriate sequence of nm multiplications, performed by a single digital multiplier, or alternatively by any one of several different iterative processes, the description of which is however beyond the scope of the present analysis. The most important of these are studied in Chapters 5 and 6. Whichever arrangement is used for the implementation of System 2, it does not involve more complex equipment than System 1 nor does it require an excessive number of sequential operations. For instance, System 2 (or at least any one of several simple modifications of this system) could be implemented, without the use of unduly complex equipment, to handle values of l, m and n in the ranges 2–8, 2–16 and 2–32, respectively, in an application where successive groups of m l-level signal elements are transmitted at an information rate of up to 10 000 bits/s. This would not be possible with System 1. Clearly, System 2 can be used at very much higher transmission rates than can System 1.

Systems 1 and 2 have further interesting properties, which are as follows. System 1 involves the direct detection of S from the received vector R, whereas System 2 involves first a process of linear equalization giving the maximum-likelihood estimate of S and then the detection of S from this estimate. Systems 1 and 2 are in fact the two basic techniques that may be applied to the detection of any received group of digital signals. Any other technique can be considered as some combination or development of these two. Furthermore, whereas System 1 is the best available detection process, System 2 is probably the least effective detection process that would be of any real practical value. Any useful detection process would therefore have a tolerance to noise somewhere between that of System 1 and that of System 2, Systems 1 and 2 representing the two extreme cases.

The tolerance to noise of System 2 can be increased some way towards that of System 1 through the use of the appropriate non-linear equalizer in place of the linear equalizer $Y^T(YY^T)^{-1}$, and this does not significantly increase either the equipment complexity or the number of sequential operations in a detection process (as is shown in Chapter 4). Alternatively, one of the simpler techniques used to implement System 2 may be modified by suitable changes in some of the operations and by the application of suitable non-linearities, to raise the tolerance to noise close to that of System 1 (as is shown in Chapter 5). However it now turns out that the resulting system no longer involves a simple process of equalization followed by detection, but is really an alternative implementation of System 1, in which the iterative process first selects a very small subset of the l^m possible vectors $\{SY\}$, as a subset normally containing the vector SY at the minimum distance from R, and then operates on this subset in the same manner as does System 1. It thus achieves nearly the same tolerance to noise as does System 1 but without the need for an excessive number of sequential operations. Clearly, System 2 provides the key to the practical implementation of System 1!

A detection process for a very large or infinite sequence of signal elements, such as is obtained in a synchronous serial data-transmission system, can obviously not operate on the *whole* sequence of elements, so that at any one time it can operate on only a *portion* of the received signal. However, when the vectors $\{Y_i\}$ are formed by the effects of signal distortion in transmission, acting on the serial stream of transmitted signal elements, the received signal has a simple structure leading to a correspondingly simple detection process. The optimum detection process, equivalent to System 1, now becomes the optimum detection process described in Chapter 5. The suboptimum detection process, equivalent to System 2, becomes the well known linear equalizer followed by the appropriate decision thresholds, as described in Chapter 4.

3.5 COLOURED GAUSSIAN NOISE

It has been assumed throughout that the Gaussian noise components $\{w_i\}$ are statistically independent, with zero mean and a fixed variance σ^2. Suppose now that the received n-component vector is

$$R_v = SY_v + V \tag{3.5.1}$$

where SY_v and V are n-component signal and noise vectors, respectively. S is an m-component row vector, whose components are the values to be detected or estimated, and Y_v is an $m \times n$ matrix of rank m. The components of V are jointly Gaussian random variables with zero mean and covariance matrix

$$N = E[V^T V] \tag{3.5.2}$$

Assume that the $n \times n$ matrix N is nonsingular, so that N is a real symmetric positive-definite matrix.

It may be shown[1] that there exists an $n \times n$ real nonsingular upper-triangular matrix M, such that

$$MM^T = N^{-1} \tag{3.5.3}$$

All the components of M below the main diagonal are zero, and all the components along the main diagonal are nonzero. Clearly,

$$N = (M^T)^{-1} M^{-1} \tag{3.5.4}$$

If now the vector R_v is fed to the n input terminals of the $n \times n$ linear network given by σM, the n-component vector held at the n output terminals of this network is

$$R = \sigma R_v M$$

$$= \sigma S Y_v M + \sigma V M \tag{3.5.5}$$

where $\sigma S Y_v M$ and $\sigma V M$ are the n-component signal and noise vectors, respectively.

Since σM is a linear network and $E[V] = 0$, $E[\sigma V M] = 0$.

Thus the covariance matrix of the noise vector $\sigma V M$ is

$$E[\sigma^2 (VM)^T VM] = E[\sigma^2 M^T V^T VM]$$

$$= \sigma^2 M^T N M = \sigma^2 M^T (M^T)^{-1} M^{-1} M = \sigma^2 I \tag{3.5.6}$$

where I is an $n \times n$ identity matrix.

Clearly the n Gaussian noise components at the output of the network are uncorrelated and therefore statistically independent, with zero mean and variance σ^2. Thus, from Equation 3.5.5,

$$R = SY + W \qquad (3.5.7)$$

where

$$Y = \sigma Y_v M \qquad (3.5.8)$$

and

$$W = \sigma V M \qquad (3.5.9)$$

SY and W are the n-component signal and noise vectors, respectively, at the output of the network σM. Y is an $m \times n$ matrix of rank m, and W is an n-component row vector whose components are statistically independent Gaussian random variables with zero mean and variance σ^2. Thus the vector R at the output of the network σM is of the general type assumed throughout Sections 3.1 to 3.4.

The network σM is a *noise whitening network* or prewhitener. Clearly, *any* received vector R_v, containing coloured Gaussian noise with a covariance matrix that is nonsingular, may be converted by a noise whitening network into the corresponding vector R, containing statistically independent Gaussian noise components with fixed variance. Furthermore, any of the detection or estimation processes previously described may be applied to the vector R at the output of the noise whitening network. It will now be shown that if this vector R is used in any detection or estimation process for S, that is optimum according to some particular criterion (minimum mean-square error estimation, maximum-likelihood estimation, or minimum error-probability detection), then the resultant process achieves the optimum detection or estimation of S from R_v, according to the given criterion.[2]

Consider three different arrangements for detecting or estimating S, as shown in Figure 3.17. σM is here the $n \times n$ linear noise whitening network given by the real nonsingular upper-triangular matrix σM. In System 2 the noise whitening network is used to obtain the vector R from R_v, and in System 3 the inverse of the noise whitening network is used to obtain R_v from R. $\sigma^{-1} M^{-1}$, the inverse of σM, clearly exists. The detection or estimation processes are all optimum according to the same given criterion.

Since System 1 is, by definition, the optimum detection or estimation process for S from R_v, it is clear that System 2 cannot perform *better* than System 1. Furthermore, System 3 must give the *same* performance as System 1, since in each case S is detected or estimated optimally from R_v. But System 3 cannot perform *better* than

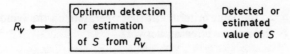

System 1: Detection or estimation of S from R_v

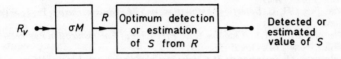

System 2: Detection or estimation of S from R_v

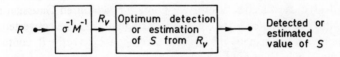

System 3: Detection or estimation of S from R

Fig. 3.17 Reversibility of noise-whitening network

System 2, since System 2 is, by definition, the optimum detection or estimation process for S from R. Thus System 2 must give the *same* performance as System 1, which means that the detection or estimation of S from R_v in System 2 is optimum according to the same criterion as that in System 1. Clearly, Systems 1 and 2 perform the same resultant operation on R_v.

If the matrix M is singular, M^{-1} does not exist and the linear transformation σM is no longer *reversible*. System 3 cannot now be implemented as shown. Thus the matrices M and N must be non-singular for the preceding results to hold. This is, however, generally the case.

It can now be seen that all the detection and estimation processes studied here may be applied to a received vector containing coloured Gaussian noise (correlated noise components), simply by feeding the vector through a noise-whitening network and then operating, in the manner described, on the output vector from this network. The resultant detection or estimation process, operating on the received vector at the input to the noise whitening network, is optimum according to the same criterion as that operating on the vector at the output of the noise whitening network.

REFERENCES

1. Hancock, J. C. and Wintz, P. A. *Signal Detection Theory*, McGraw-Hill, New York (1966)
2. Van Trees, H. L. *Detection, Estimation, and Modulation Theory*, Pt. 1, Wiley, New York (1968)
3. Wozencraft, J. M. and Jacobs, I. M. *Principles of Communication Engineering*, Wiley, New York (1965)
4. Schwartz, M. *Information Transmission, Modulation and Noise*, McGraw-Hill Kogakusha, Tokyo (1970)

4 Linear and nonlinear transversal equalizers

4.1 INTRODUCTION

We have now considered the most likely types of distortion that may be introduced into a baseband signal, when this is transmitted over a linear time-invariant channel. We have also studied the principles of both detection and estimation of some parameter of a received signal, from one or more sample values of this signal. We are therefore now in a position to consider the various possible detection processes that may be applied to a received digital data signal which has been subjected to linear distortion. In this and the next chapter we shall consider only *serial* signals.

Detection processes for distorted digital signals may be classified into two separate groups. In the first of these, the received sampled digital signal is fed through an *equalizer* that corrects the distortion introduced by the channel and restores the received signal into a copy of the transmitted signal, neglecting for the moment the effects of noise. The resultant received signal is then detected in the conventional manner, as normally applied to a serial digital signal in the absence of intersymbol interference. In other words, the equalizer acts as the *inverse* of the channel, so that the channel and equalizer together introduce no signal distortion, and each signal element is detected as it arrives, independently of the others, by comparing the corresponding sample value with the appropriate threshold level (or levels).[30-135] In the second group of detection processes, the decision process itself is modified to take account of the signal distortion that has been introduced by the channel, and often no attempt is made to remove or even reduce the signal distortion prior to the actual decision process. Although no equalizer is now required, the decision process is sometimes considerably more complex than that when an equalizer is used.

In this chapter we shall be concerned with both linear and nonlinear equalizers for a linear baseband channel.[53-135] We shall study various design techniques for such equalizers and compare their relative performances. A synchronous serial binary data-transmission system is assumed throughout.

When the signal distortion varies with time, not only must the receiver detect the element values of the received data signal, but it must also continuously keep adjusting the equalizer so that this is held correctly adjusted in relation to the channel. If the sampled impulse-response of the channel varies significantly during the time required to process an individual received signal-element through the equalizer, this cannot be held correctly adjusted for the channel and correct equalization is no longer obtained. Thus there is clearly an upper limit to the rate of variation of the signal distortion that can be handled by an equalizer. Furthermore, the conventional method used to hold an equalizer correctly adjusted for the channel, which is considered in Section 4.5, places an additional and more severe restriction on the maximum acceptable rate of change in the channel response. It is therefore assumed here that the channel is either time invariant or else varies only slowly with time, which means that there is a truly negligible variation in the sampled impulse-response of the channel over a time interval equal to at least some ten times the duration of this response, and in some cases up to at least a hundred times, such that the channel can always be equalized accurately.

At transmission rates of up to 2400 bits/s over the switched telephone network, it is often reasonable to assume that the attenuation and delay characteristics of a given telephone circuit are unlikely to vary significantly over the duration of any one transmission,[141] and it is now sufficient for the equalizer to be adjusted correctly at the *start* of transmision and held in this setting for the rest of the transmission. Such an equalizer, which is adjusted automatically by means of the received signal, is known as an *automatic* equalizer. The *synchronizing* signal sent at the start of transmission normally contains a sequence of signal elements whose values have a fixed and known pattern. This is used by the receiver as a *training signal* to adjust the equalizer for the correct equalization of the channel.

At transmission rates above 2400 bits/s over the switched telephone network or above 100 bits/s over HF radio links, the sampled impulse-response of the channel may vary significantly with time,[141] and it is necessary now for the equalizer to be continuously readjusted. Such an equalizer is known as an *adaptive* equalizer.

As before, a training signal is sent at the start of transmission, for the initial adjustment of the equalizer. However, to avoid reducing the data transmission rate, once the transmission of data has started, an uninterrupted sequence of data elements is transmitted and no further training signals are sent. The received data signal *itself* is now used by the receiver to maintain the correct adjustment of the equalizer. Since much of the recent published literature on equalizers is concerned with these problems,[53-135] the available techniques for holding an equalizer correctly adjusted for a time-varying channel are not considered here in any detail, but the basic principles involved are surveyed briefly in Section 4.5.

4.2 BASIC ASSUMPTIONS

Consider the synchronous serial binary data-transmission system shown in Figure 4.1. The signal at the input to the transmitter filter

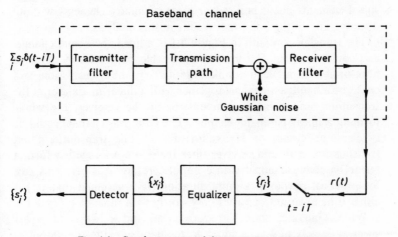

Fig. 4.1 *Synchronous serial data-transmission system*

is a sequence of regularly spaced impulses, the ith of which occurs at time $t = iT$ seconds and has the value (area)

$$s_i = \pm k \qquad (4.2.1)$$

where k is a positive constant. Each impulse $s_i\delta(t-iT)$ is a binary

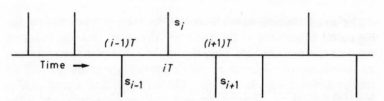

Fig. 4.2   A typical sequence of signal elements at the input to the baseband channel

polar signal-element, and a typical sequence of such elements is shown in Figure 4.2. In practice, of course, a rectangular or rounded waveform would be used, with the appropriate change in the transmitter filter. The $\{s_i\}$ are statistically independent and equally likely to have either binary value. Where this condition is not satisfied by the $\{s_i\}$, it can normally be achieved by 'scrambling' the transmitted sequence of element values and appropriately 'descrambling' the corresponding detected element values at the receiver, to obtain a copy of the original transmitted sequence.[1-4] Multilevel signal-elements could be used in place of binary elements without affecting any of the important results in this analysis.

The transmission path in Figure 4.1 is a linear baseband channel and could be a 600-Ω pair, a coaxial cable or a suitably designed fibre-optic communication link. Alternatively the transmission path could be a bandpass channel together with a linear modulator at the transmitter and a linear demodulator at the receiver, the whole forming a linear baseband channel. The bandpass channel could be a telephone circuit or HF radio link.[141] The transmitter filter, transmission path and receiver filter in Figure 4.1 together form a linear baseband channel whose impulse response is $y(t)$ and has, for practical purposes, a finite duration. It is assumed that $y(t)$ is either time invariant or varies only slowly with time.

White Gaussian noise with zero mean and a two-sided power spectral density of $\frac{1}{2}N_0$ is added to the data signal at the output of the transmission path, giving the Gaussian waveform $w(t)$ at the output of the receiver filter, that is, at the output of the baseband channel in Figure 4.1. Although many practical channels do not introduce significant levels of Gaussian noise, the relative tolerances of different data-transmission systems to additive white Gaussian noise is a good measure of their relative tolerances to most practical types of additive noise.[141]

The signal waveform at the output of the baseband channel in Figure 4.1 is

$$r(t) = \sum_i s_i y(t - iT) + w(t) \qquad (4.2.2)$$

and this is sampled once per signal element at the time instants $\{iT\}$, where i takes on all positive integer values. Various techniques are available for holding the sampling instants correctly synchronized to the received signal.[141]

It will for convenience be assumed that the transmitter and receiver filters have the same transfer function

$$H^{\frac{1}{2}}(f) = \begin{cases} T^{\frac{1}{2}} \cos \frac{1}{2}\pi fT, & -\dfrac{1}{T} < f < \dfrac{1}{T} \\ \\ 0, & \text{elsewhere} \end{cases} \qquad (4.2.3)$$

so that the resultant transfer function of the transmitter and receiver filters in cascade is the raised-cosine function

$$H(f) = \begin{cases} \frac{1}{2}T(1 + \cos \pi fT), & -\dfrac{1}{T} < f < \dfrac{1}{T} \\ \\ 0, & \text{elsewhere} \end{cases} \qquad (4.2.4)$$

Both $H^{\frac{1}{2}}(f)$ and $H(f)$ are real, nonnegative and even functions of f.

Alternatively, $H(f)$ could be any real, nonnegative and even function of f with a horizontal central portion and a sinusoidal roll-off about both $f = -1/2T$ and $f = 1/2T$, each roll-off having odd symmetry about the corresponding frequency $\pm 1/2T$. The transmitter and receiver filters again have equal transfer functions $H^{\frac{1}{2}}(f)$, but $H(f)$ now has a bandwidth somewhere between $1/2T$ and $1/T$ Hz (over positive frequencies).[141] It is assumed that in every case

$$\int_{-\infty}^{\infty} H(f)\mathrm{d}f = 1 \qquad (4.2.5)$$

Clearly, for any transmitter and receiver filters as just described, the receiver filter is *matched* to the signal at the output of the transmitter filter, so that when the transmission path introduces no signal distortion, the signal/noise ratio, at the appropriate sampling instants at the output of the receiver filter, is maximized.

Furthermore, it may be shown[141] that the resultant impulse-response $h(t)$ of the transmitter and receiver filters in cascade (which is also the impulse response $y(t)$ of the whole baseband channel in Figure 4.1 when the transmission path introduces no distortion, attenuation or delay) now satisfies

$$h(0) = 1 \qquad (4.2.6)$$

and
$$h(iT) = 0 \qquad (4.2.7)$$

for all nonzero integers $\{i\}$. The transmitter and receiver filters are not, of course, physically realizable, but can readily be made so by introducing a sufficient delay into their impulse response.

Thus, when the transmission path introduces no distortion, attenuation or delay, the received waveform at the output of the baseband channel in Figure 4.1 becomes

$$r(t) = \sum_i s_i h(t - iT) + w(t) \qquad (4.2.8)$$

and the sample value of this waveform at the time instant $t = iT$ is

$$r_i = s_i + w_i \qquad (4.2.9)$$

where $r_i = r(iT)$ and $w_i = w(iT)$.

The baseband channel in Figure 4.1 now introduces no inter-symbol interference or attenuation into the received sampled signal, and the best detection process is to detect each s_i from the corresponding r_i. When $r_i > 0$, s_i is detected as k, and when $r_i < 0$, s_i is detected as $-k$. Notice also that there is no delay in transmission, the transmitted binary signal-element $s_i \delta(t - iT)$, at the input to the baseband channel in Figure 4.1, being detected at the receiver from the sample r_i at the same time instant $t = iT$.

The Fourier transform of the ith signal-element, at the input to the transmitter filter in Figure 4.1, is $s_i \exp(-j2\pi f iT)$, where $j = \sqrt{-1}$, and the Fourier transform of the corresponding signal-element, at the input to the transmission path is $s_i \exp(-j2\pi f iT) H^{\frac{1}{2}}(f)$.

The energy spectral density of the latter signal element is

$$|s_i \exp(-j2\pi f iT) H^{\frac{1}{2}}(f)|^2 = s_i^2 H(f) \qquad (4.2.10)$$

since $H(f)$ is real and nonnegative. Thus the energy of the signal element at the input to the transmission path is

$$\int_{-\infty}^{\infty} s_i^2 H(f) \mathrm{d}f = s_i^2 \qquad (4.2.11)$$

from Equation 4.2.5.

If the impulse response of the transmitter filter in Figure 4.1 is $a(t)$, the waveform of the ith transmitted signal-element at the input to the transmission path is $s_i a(t-iT)$, and its energy is

$$\int_{-\infty}^{\infty} s_i^2 a^2(t-iT)dt = s_i^2 \qquad (4.2.12)$$

from Equation 4.2.11.

The autocorrelation function of this element waveform is

$$R_s(\tau) = \int_{-\infty}^{\infty} s_i a(t-iT) . s_i a(t-iT+\tau)dt \qquad (4.2.13)$$

and $R_s(\tau)$ is the inverse Fourier transform of the energy spectral density of the element waveform (Equation 4.2.10). Thus

$$R_s(\tau) = \int_{-\infty}^{\infty} s_i^2 H(f) \exp(j2\pi f\tau)df$$

$$= s_i^2 h(\tau) \qquad (4.2.14)$$

From Equation 4.2.7, $R_s(iT) = 0$ for nonzero integer values of i.

It follows that for any two signal-elements $s_i a(t-iT)$ and $s_l a(t-lT)$, at the input to the transmission path,

$$\int_{-\infty}^{\infty} s_i a(t-iT) . s_l a(t-lT)dt$$

$$= \int_{-\infty}^{\infty} s_i a(t-iT) . s_l a[t-iT+(i-l)T]dt$$

$$= \frac{s_l}{s_i} R_s[(i-l)T] = 0 \qquad (4.2.15)$$

for $i \neq l$, which means that the two element waveforms are *orthogonal*.

The energy of any sequence of n signal elements

$$\sum_{i=1}^{n} s_i a(t-iT)$$

at the input to the transmission path is

$$\int_{-\infty}^{\infty} \left[\sum_{i=1}^{n} s_i a(t-iT) \right]^2 dt = \int_{-\infty}^{\infty} \left[\sum_{i=1}^{n} s_i^2 a^2(t-iT) \right] dt$$

$$= \sum_{i=1}^{n} s_i^2 \qquad (4.2.16)$$

as can be seen from Equations 4.2.15 and 4.2.12. Thus the average transmitted energy per signal element and hence the average transmitted energy per bit of information is

$$\overline{s_i^2} = s_i^2 = k^2 \tag{4.2.17}$$

regardless of whether or not the $\{s_i\}$ are uncorrelated or have zero mean.

The power spectral density of the noise waveform $w(t)$, at the output of the receiver filter in Figure 4.1, is

$$\tfrac{1}{2}N_0|H^{\frac{1}{2}}(f)|^2 = \tfrac{1}{2}N_0H(f) \tag{4.2.18}$$

because $H(f)$ is real and nonnegative.

Since the additive white Gaussian noise, introduced at the output of the transmission path in Figure 4.1, has zero mean and since the receiver filter is linear, any sample w_i of the noise waveform $w(t)$, at the output of the receiver filter, is a Gaussian random variable with zero mean. The *variance* of any noise sample w_i is now its mean-square value, which is, of course, the average power of $w(t)$ and is

$$\int_{-\infty}^{\infty} \tfrac{1}{2}N_0H(f)\mathrm{d}f = \tfrac{1}{2}N_0 \tag{4.2.19}$$

from Equations 4.2.18 and 4.2.5.

From a well known theorem, the autocorrelation function of $w(t)$ is the inverse Fourier transform of its power spectral density and so is

$$R_n(\tau) = \int_{-\infty}^{\infty} \tfrac{1}{2}N_0H(f)\exp(\mathrm{j}2\pi f\tau)\mathrm{d}f$$

$$= \tfrac{1}{2}N_0h(\tau) \tag{4.2.20}$$

From Equation 4.2.7,

$$R_n(iT) = \tfrac{1}{2}N_0h(iT) = 0 \tag{4.2.21}$$

for any nonzero integer i. But the sampling instants for any two noise samples w_i and w_l are separated by a multiple of T seconds, so that, from Equation 4.2.21, the expected value of the product of w_i and w_l is equal to zero. Furthermore, the noise samples have zero mean, so that the expected value of the product of w_i and w_l is equal to the product of their respective mean values, and hence any two noise samples w_i and w_l are uncorrelated Gaussian random variables. It follows that the noise samples $\{w_i\}$ are *statistically independent* Gaussian random variables[140] with zero mean and variance

$$\sigma^2 = \tfrac{1}{2}N_0 \qquad (4.2.22)$$

The noise components are here designated $\{w_i\}$, as in Chapter 3, to emphasize the fact that they are statistically independent (or 'white').

Consider now the general case where the transmission path introduces signal distortion, so that $y(t) \neq h(t)$. If a unit impulse at time $t = 0$ is fed to the baseband channel in the absence of any other signals, the resultant waveform at the input to the sampler in Figure 4.1 is $y(t)$, which has for practical purposes a finite time duration that is less than $(g+1)T$ seconds but may or may not be greater than gT seconds, where g is the appropriate positive integer. The corresponding sample values at the sampler output form the *sampled impulse-response* of the baseband channel and are given by the $(g+1)$-component row vector

$$V = y_0 \; y_1 \ldots y_g \qquad (4.2.23)$$

where $y_i = y(iT)$. The delay in transmission, other than that involved in the time dispersion of the transmitted signal, is neglected here, so that $y_0 \neq 0$ and $y_i = 0$ for $i<0$ and $i>g$. Clearly, the baseband channel now disperses (or 'spreads out') each signal-element in time, the *shape* of each individual received signal-element being identical to that of any other element with the same value of s_i.

If the ith signal-element $s_i y(t-iT)$ is received alone and in the absence of noise, the corresponding sample values $\{r_l\}$, starting with r_i at time $t = iT$, are

$$s_i y_0 \; s_i y_1 \ldots s_i y_g \; 0 \; 0 \ldots \qquad (4.2.24)$$

all preceding $\{r_l\}$ being zero. It is assumed that the receiver has prior knowledge of both k and V, so that it knows the two *possible* sets of sample values corresponding to any individual received signal-element, although, *before* the detection of a particular received signal-element, it has no knowledge of *which* of its two possible sets of sample values are in fact received. Since there is a negligible variation in the sampled impulse-response of the channel during the processing of any individual received signal-element, the sampled impulse-response will be treated as though it were time invariant.

With the reception of an uninterrupted sequence of signal-elements in the presence of noise, the sample value of the received signal at the output of the baseband channel, at time $t = iT$, is

$$r_i = \sum_{l=0}^{g} s_{i-l} y_l + w_i \qquad (4.2.25)$$

If s_{i-h} is detected from r_i, where h is a nonnegative integer in the range 0 to g, the *wanted* signal-component in r_i is $s_{i-h} y_h$. However, r_i contains also an *intersymbol-interference* component

$$\sum_{\substack{l=0 \\ l \neq h}}^{g} s_{i-l} y_l$$

and a *noise* component w_i. Clearly, it may not be possible to detect s_{i-h} correctly from r_i, even in the absence of noise, unless the maximum value of the intersymbol-interference component has a magnitude less than $|ky_h|$. Under these conditions, the intersymbol interference on its own cannot prevent the correct detection of s_{i-h} from r_i, but it may seriously reduce the tolerance of the system to the additive noise. The evaluation of the precise effect of the intersymbol interference on the tolerance of the system to noise is in general quite difficult.[5–29]

The equalizer in Figure 4.1 operates on the received samples $\{r_i\}$ in such a way that the output sample value from the equalizer at time $t = (i+h)T$ is

$$x_{i+h} \simeq s_i + u_{i+h} \qquad (4.2.26)$$

where h is an appropriate nonnegative integer and u_{i+h} is the Gaussian noise component at the equalizer output, at time $t = (i+h)T$. Clearly, all intersymbol interference has been effectively eliminated in x_{i+h}.

In the detector, s_i is detected by comparing x_{i+h} with a decision threshold of zero. If $x_{i+h} > 0$, s_i is detected as k, and if $x_{i+h} < 0$, s_i is detected as $-k$. The detected value of s_i is designated s_i'.

A most powerful tool for the analysis of an equalizer that operates on a sampled signal is the z transform. This is described in Section 2.2.1. The particular importance of the z transform of a sampled signal is firstly that it is very simply related to the sample values of the signal, and secondly that the resultant z-transform of two sampled networks in cascade is the *product* of their respective z-transforms.

As shown in Section 2.2.1, the z transform of the sampled impulse-response of the baseband channel is

$$Y(z) = y_0 + y_1 z^{-1} + y_2 z^{-2} + \ldots + y_g z^{-g} \qquad (4.2.27)$$

where z^{-i} represents the time instant $t = iT$. The relationships between V, $Y(z)$ and t, for a typical baseband channel, are illustrated in Figure 4.3, which shows the effect of neglecting the time delay in

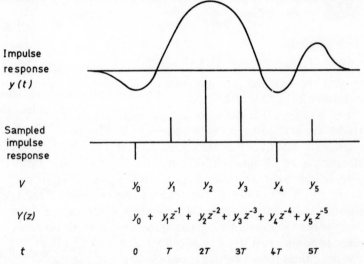

Fig. 4.3 *Relationship between $y(t)$, V and $Y(z)$ for a typical baseband channel*

transmission. The representation of a sequence of sample values in terms of the corresponding impulses is used here and in the following discussion as a convenient means of showing the sample values. Its deeper significance is considered in Section 2.2.1. The sample value at time $t = iT$ is equal to the *area* of the impulse at this time instant.

The z transform of the ith transmitted signal-element, at the input to the baseband channel in Figure 4.1, is $s_i z^{-i}$, and the z transform of the ith received signal-element at the output of the sampler is

$$s_i z^{-i} Y(z) = s_i y_0 z^{-i} + s_i y_1 z^{-i-1} + \ldots + s_i y_g z^{-i-g} \qquad (4.2.28)$$

The function of the equalizer is to convert $Y(z)$, the z transform of the baseband channel and sampler, to z^{-h}, which represents a simple delay of hT seconds with no signal distortion or attenuation, where h is an appropriate nonnegative integer. z^{-h} is the z transform of the baseband channel, sampler and equalizer. Thus, if the z transform of the equalizer is $C(z)$, the z transform of the ith received signal-element at the output of the equalizer is

$$s_i z^{-i} Y(z) C(z) \simeq s_i z^{-i-h} \qquad (4.2.29)$$

Equation 4.2.26 is now satisfied, so that s_i is detected from the sample value x_{i+h} at the output of the equalizer at time $t = (i+h)T$. Clearly

$$C(z) \simeq z^{-h} Y^{-1}(z) \qquad (4.2.30)$$

since, of course,

$$Y(z) Y^{-1}(z) = 1 \qquad (4.2.31)$$

so that the equalizer is equivalent to the *inverse* of the channel together with a *delay* of hT seconds.

Except where otherwise stated, a high signal/noise ratio is assumed. Under these conditions, as is indicated in Chapter 3, the best tolerance to additive Gaussian noise is achieved through the effective elimination of *all* intersymbol interference, which now tends to predominate over the noise in the received signal.[63] Again, except where specifically stated to the contrary, the equalizer is taken to equalize the channel sufficiently accurately to reduce the residual intersymbol interference to a negligible level, and the channel is taken to be such that it can be equalized accurately by a linear filter. The conditions for which this is possible are that $Y(z)$ has no roots (zeros) on the unit circle in the z plane, as is shown in Section 2.4.6.

4.3 LINEAR EQUALIZERS

4.3.1 Feedforward transversal filter

The most widely studied linear equalizer is the feedforward transversal filter.[30–107] It has been shown that the optimum linear receiver, using only linear processing of the signal prior to detection, includes such a transversal filter.[30–52] Assume now that the equalizer in Figure 4.1 is the linear feedforward transversal filter with m taps, shown in Figure 4.4.

The equalizer operates on sample values (or numbers) which may be in analogue or digital (binary coded) form. The signals (sample values) shown in Figure 4.4 are those present at the time instant

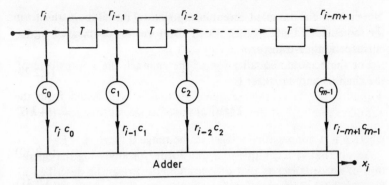

Fig. 4.4 Linear feedforward transversal equalizer for the baseband channel

$t = iT$. Each square marked T is here a storage element that holds the corresponding sample value r_j in either analogue or binary coded form. The storage elements are triggered at the time instants $\{jT\}$, for all positive integers $\{j\}$, and each time the transversal filter is triggered, the stored sample values $\{r_j\}$ are shifted one place to the right. Thus the extreme left-hand sample value at the input to the equalizer is that received at time $t = iT$, the next sample value is that received at time $t = (i-1)T$, and so on, each storage element introducing a delay of T seconds. Each circle in Figure 4.4 is a multiplier that multiplies its input signal by the quantity shown in the circle. The output signals from the multipliers, $\{r_{i-j}c_j\}$, are added together to give the output signal

$$x_i = \sum_{j=0}^{m-1} r_{i-j}c_j \qquad (4.3.1)$$

at time $t = iT$.

A practical linear feedforward transversal equalizer has between 8 and 64 taps, and is often implemented in a rather different form from that suggested by Figure 4.4. For instance, when the sample values are stored in digital form, only a single multiplier is used, the multiplier being shared between the different taps.

The transversal filter in Figure 4.4 has a sampled impulse-response given by the m-component row vector

$$C = c_0 \; c_1 \ldots c_{m-1} \qquad (4.3.2)$$

and the z transform of this sampled impulse-response is

$$C(z) = c_0 + c_1 z^{-1} + \ldots + c_{m-1} z^{-m+1} \qquad (4.3.3)$$

Notice that the sampled impulse-response of the filter is given by the sequence of its tap gains, which also form the coefficients in the corresponding z-transform.

For the accurate equalization of the channel, the z transform of the channel and equalizer is

$$Y(z)C(z) \simeq z^{-h} \qquad (4.3.4)$$

where h is a nonnegative integer in the range 0 to $m+g-1$.

An alternative technique to z transforms for analysing the operation of an equalizer is that of *convolution matrices*, as described in Section 2.2.2. The sampled impulse-response of two sampled networks in cascade is the *convolution* of their individual sampled impulse-responses, and is given by the product of one of the two sampled impulse-responses and the convolution matrix corresponding to the other.

Let Y be the $m \times (m+g)$ matrix whose ith row is given by the $(m+g)$-component row vector

$$Y_{i-1} = \overbrace{0 \ldots 0}^{i-1} \; y_0 \; y_1 \ldots y_g \; 0 \ldots 0 \qquad (4.3.5)$$

From Equation 4.3.4, the sampled impulse-response of the channel and equalizer is

$$\sum_{i=0}^{m-1} c_i Y_i = CY \simeq E_h \qquad (4.3.6)$$

where E_h is the $(m+g)$-component row vector given by

$$E_h = \overbrace{0 \ldots 0}^{h} \; 1 \; 0 \ldots 0 \qquad (4.3.7)$$

Y is the convolution matrix corresponding to the sampled impulse-response of the channel, and CY is the $(m+g)$-component row vector obtained by the convolution of the vectors V and C.

Various design techniques are available for evaluating the tap gains of the linear equalizer to satisfy Equation 4.3.6. With a finite feedforward transversal filter, Equation 4.3.6 will not in general be satisfied exactly, but instead

$$CY = E = e_0 \; e_1 \ldots e_{m+g-1} \qquad (4.3.8)$$

The *peak distortion* in the equalized signal is defined to be[63,69]

$$D_p = \frac{1}{|e_h|} \sum_{\substack{i=0 \\ i \neq h}}^{m+g-1} |e_i| \qquad (4.3.9)$$

the *mean-square distortion* is[63,69]

$$D_m = \frac{1}{e_h^2} \sum_{\substack{i=0 \\ i \neq h}}^{m+g-1} e_i^2 \qquad (4.3.10)$$

and the *mean-square error* (due to intersymbol interference) is[63,69]

$$k^2(e_h-1)^2 + k^2 \sum_{\substack{i=0 \\ i \neq h}}^{m+g-1} e_i^2 = k^2(E-E_h)(E-E_h)^T$$

$$= k^2|E-E_h|^2 \qquad (4.3.11)$$

where $(E-E_h)^T$ and $|E-E_h|$ are the transpose and length, respectively, of the $(m+g)$-component row vector $E-E_h$. In every case, the ideal (or desired) value of the sampled impulse-response of the equalized channel is taken to be E_h, so that $e_h \simeq 1$ in Equation 4.3.8. Again, only the effects of *intersymbol interference* are considered here, the effects of the noise being for the moment neglected. For a given equalizer it may be required to minimize any one of the three quantities just defined.

4.3.2 Zero forcing

In the early work on linear equalizers these were generally designed to minimize the peak distortion.[53-65] This is achieved, under certain conditions, by a technique known as 'zero forcing'.[56,57,63]

Assume that the component of V (Equation 4.2.11) of greatest magnitude is y_j and that the number of taps of the linear equalizer in Figure 4.4 is

$$m = 2l+1 \qquad (4.3.12)$$

where l is a positive integer. For reasons which will become clear later, the integer h in Equation 4.3.7 is now chosen to have the value

$$h = j+l \qquad (4.3.13)$$

In the technique of zero forcing, the tap gains of the linear equalizer are chosen to set e_h to 1 and $2l$ of the remaining $\{e_i\}$ in Equation 4.3.8 to zero. It has been shown[56,63] that when the peak distortion of the received signal

$$\frac{1}{|y_j|} \sum_{\substack{i=0 \\ i \neq j}}^{g} |y_i|$$

is less than unity, the peak distortion in the equalized signal, for the given value of h, is minimized when $e_i = 0$ for $h-l \leqslant i < h$ and $h < i \leqslant h+l$, with $e_h = 1$. From Equations 4.3.8, 4.3.12 and 4.3.13 we now have

$$CY = e_0 \ldots e_{j-1} \overbrace{0 \ldots 0}^{l} 1 \overbrace{0 \ldots 0}^{l} e_{j+m} \ldots e_{m+g-1} \qquad (4.3.14)$$

Let Z be the $m \times m$ matrix formed from Y by removing its first j and last $g-j$ columns. The first row of Z is now given by the m-component vector

$$Z_0 = y_j \; y_{j+1} \ldots y_g \; 0 \ldots 0 \qquad (4.3.15)$$

and each component along the main diagonal of Z has the value y_j. From Equation 4.3.14,

$$CZ = \overbrace{0 \ldots 0}^{l} 1 \overbrace{0 \ldots 0}^{l} \qquad (4.3.16)$$

When the peak distortion in the received signal is less than unity, the matrix Z is diagonally dominant and must therefore be nonsingular. Under these conditions, a unique vector C can always be found to satisfy Equation 4.3.16 and hence to minimize the peak distortion.[56]

For the exact equalization of the channel, $e_i = 0$ in Equation 4.3.14 for all values of i other than h. However, this requires the m tap gains to satisfy $m+g$ linear simultaneous equations, which is not normally possible.

Not only does the technique of zero forcing minimize the peak distortion in the equalized signal, when the peak distortion of the received signal is less than unity, but it also leads to some simple iterative processes for holding the equalizer correctly matched to a slowly time-varying channel.[53-65] The weakness of the technique is that it cannot be used in its simple form for equalizing severe time-varying signal distortion.

4.3.3 Separate equalization of two complementary factors of the channel response

Suppose that the z transform of the baseband channel in Figure 4.1 is

$$Y(z) = 1 + yz^{-1} \qquad (4.3.17)$$

which means, of course, that the sampled impulse-response of the channel is given by the two-component vector $(1,y)$.

The linear equalizer should now ideally have the z-transform

$$Y^{-1}(z) = (1 + yz^{-1})^{-1} \qquad (4.3.18)$$

Consider first the case where $|y| < 1$. Now $|yz^{-1}| < 1$, since $z^{-1} = \exp(-\sqrt{-1}\, 2\pi fT)$. Under these conditions, the binomial expansion can be applied to $(1 + yz^{-1})^{-1}$ to give

$$(1 + yz^{-1})^{-1} = 1 - yz^{-1} + y^2z^{-2} - y^3z^{-3} + \dots \qquad (4.3.19)$$

which is a convergent series[136] (see also Section 2.4.3).

The value of $(1 + yz^{-1})^{-1}$ can alternatively be obtained through the division of 1 by $(1 + yz^{-1})$, using a normal process of long division, as follows:[137-139]

$$
\begin{array}{r}
1 - yz^{-1} + y^2z^{-2} - \dots \\
1 + yz^{-1}\overline{)1} \\
\underline{1 + yz^{-1}} \\
-yz^{-1} \\
\underline{-yz^{-1} - y^2z^{-2}} \\
y^2z^{-2} \\
\underline{y^2z^{-2} + y^3z^{-3}} \\
-y^3z^{-3} \\
\dots\dots
\end{array}
$$

The linear transversal equalizer for a channel with z-transform $1 + yz^{-1}$ has the z-transform $(1 + yz^{-1})^{-1}$ and therefore has ideally an infinite number of taps. It is shown in Figure 4.5.

If $y \ll 1$, a good approximation to the ideal response for the equalized channel can be obtained with only the first few stages of the filter in Figure 4.5, since the gain introduced by the ith multiplier is $(-y)^i$ which decreases rapidly towards zero as i increases. Thus a practical equalizer for the given channel has a finite number of m

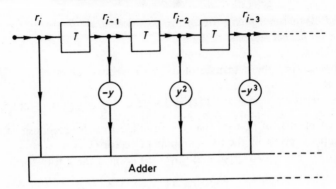

Fig. 4.5   Linear equalizer for a channel with z-transform $1+yz^{-1}$,
where $|y| < 1$

taps, as in Figure 4.4, and only approximately equalizes the channel.
Clearly, as y increases towards unity, a longer transversal equalizer
is needed, until when $y = 1$ equalization is no longer possible with a
finite transversal filter. The significance of this is considered in more
detail towards the end of this section.

The z transform of an m-tap linear transversal filter, that approximately equalizes the channel with z-transform $1+yz^{-1}$, where
$|y| < 1$, is obtained by taking the first m terms of $(1+yz^{-1})^{-1}$
and is

$$C(z) = 1 - yz^{-1} + y^2z^{-2} - \ldots + (-y)^{m-1}z^{-m+1} \quad (4.3.20)$$

Clearly, $(1+yz^{-1})C(z) \simeq 1$ and $C(z) \simeq (1+yz^{-1})^{-1}$.

Suppose now that $y = \frac{1}{2}$ and the transversal equalizer has six
taps. The z transform of the baseband channel is

$$Y(z) = 1 + \tfrac{1}{2}z^{-1} \quad (4.3.21)$$

and the z transform of the equalizer is

$$C(z) = 1 - \tfrac{1}{2}z^{-1} + \tfrac{1}{4}z^{-2} - \tfrac{1}{8}z^{-3} + \tfrac{1}{16}z^{-4} - \tfrac{1}{32}z^{-5} \quad (4.3.22)$$

the transversal equalizer being as shown in Figure 4.6.

The z transform of the channel and linear equalizer is now

$$
\begin{aligned}
&Y(z)C(z) \\
&= (1 + \tfrac{1}{2}z^{-1})(1 - \tfrac{1}{2}z^{-1} + \tfrac{1}{4}z^{-2} - \tfrac{1}{8}z^{-3} + \tfrac{1}{16}z^{-4} - \tfrac{1}{32}z^{-5}) \\
&= 1 - \tfrac{1}{64}z^{-6} \quad (4.3.23)
\end{aligned}
$$

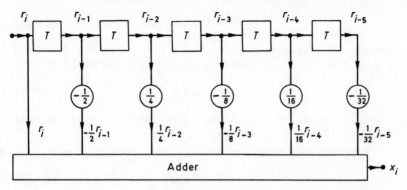

Fig. 4.6 *6-tap linear equalizer for a channel with z-transform* $1 + \frac{1}{2}z^{-1}$

so that

$$Y(z)C(z) \simeq 1 \qquad (4.3.24)$$

and

$$C(z) \simeq Y^{-1}(z) \qquad (4.3.25)$$

If the residual intersymbol interference at the output of the equalizer is not neglected, the sampled impulse-response of the channel and linear equalizer, represented in terms of the corresponding impulses, is $\delta(t) - \frac{1}{64}\delta(t - 6T)$, as shown in Figure 4.7. Clearly this is an arrangement of zero forcing.

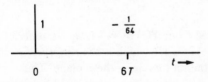

Fig. 4.7 *Sampled impulse-response of a channel, with z-transform* $1 + \frac{1}{2}z^{-1}$, *in cascade with the corresponding 6-tap transversal equalizer*

When the ith individual signal-element is received alone and in the absence of noise, the signal at the output of the equalizer is

$$s_i\delta(t - iT) - \tfrac{1}{64}s_i\delta(t - 6T - iT)$$

if represented in terms of impulses. Thus, when a continuous stream of data elements is received in the presence of Gaussian noise, the sample value of the signal at the output of the equalizer, at time $t = iT$, is

$$x_i = -\tfrac{1}{64}s_{i-6} + s_i + u_i \tag{4.3.26}$$

where u_i is a Gaussian random variable with zero mean and variance dependent on the transversal equalizer. From Equation 4.3.26,

$$x_i \simeq s_i + u_i \tag{4.3.27}$$

and s_i is detected from x_i, at time $t = iT$. When $x_i > 0$, s_i is detected as k, and when $x_i < 0$, s_i is detected as $-k$.

Suppose now that the z transform of the channel is given by

$$Y(z) = 1 + yz^{-1} \tag{4.3.28}$$

where $|y| > 1$. $(1 + yz^{-1})^{-1}$ does not here satisfy Equation 4.3.19, since the binomial expansion cannot now be applied to $(1 + yz^{-1})^{-1}$. This can be seen from the fact that the expansion of $(1 + yz^{-1})^{-1}$ given by Equation 4.3.19 has become a divergent series. The *same* divergent series is obtained if 1 is divided by $1 + yz^{-1}$, using a normal process of long division.

It is interesting to observe that the root of $1 + yz^{-1}$ is $-y$, so that when $|y| < 1$ the root lies *inside* the unit circle in the z-plane, and when $|y| > 1$ the root lies *outside* the unit circle.

To evaluate the m-tap linear transversal equalizer for the channel whose z transform is given by Equation 4.3.28, consider the channel with z-transform

$$M(z) = y + z^{-1} \tag{4.3.29}$$

which is obtained from $Y(z)$ by *reversing the order* of its coefficients. The root of $M(z)$ is $-1/y$ which is the *reciprocal* of the root of $Y(z)$ and lies inside the unit circle in the z plane. Thus

$$
\begin{aligned}
M^{-1}(z) &= (y + z^{-1})^{-1} = y^{-1}(1 + y^{-1}z^{-1})^{-1} \\
&= y^{-1}(1 - y^{-1}z^{-1} + y^{-2}z^{-2} - y^{-3}z^{-3} + \ldots) \\
&= y^{-1} - y^{-2}z^{-1} + y^{-3}z^{-2} - y^{-4}z^{-3} + \ldots
\end{aligned} \tag{4.3.30}
$$

since $|y^{-1}| < 1$, which is the condition that must be satisfied[136] for the binomial expansion of $(1 + y^{-1}z^{-1})^{-1}$. Equation 4.3.30 represents $M^{-1}(z)$ as a convergent series.

The value of $M^{-1}(z)$ can alternatively be obtained through the division of 1 by $M(z)$, using a normal process of long division, as follows.[137-139]

$$
y + z^{-1} \overline{\smash{\big)}\,
\begin{aligned}
& y^{-1} - y^{-2}z^{-1} + y^{-3}z^{-2} - \ldots \\
& 1 \\
& \underline{1 + y^{-1}z^{-1}} \\
& -y^{-1}z^{-1} \\
& \underline{-y^{-1}z^{-1} - y^{-2}z^{-2}} \\
& y^{-2}z^{-2} \\
& \underline{y^{-2}z^{-2} + y^{-3}z^{-3}} \\
& -y^{-3}z^{-3} \\
& \quad \ldots
\end{aligned}}
$$

The linear transversal equalizer for a channel with z-transform $M(z)$ has the z-transform $M^{-1}(z)$ and therefore has ideally an infinite number of taps. The z-transform of an m-tap linear transversal filter, that approximately equalizes the channel with z-transform $M(z)$, is obtained by taking the first m terms of $M^{-1}(z)$ and is

$$N(z) = y^{-1} - y^{-2}z^{-1} + y^{-3}z^{-2} - \ldots - (-y)^{-m}z^{-m+1} \quad (4.3.31)$$

Thus the z transform of the channel and linear equalizer is

$$M(z)N(z) \simeq 1 \quad (4.3.32)$$

Consider now the z-transform $C(z)$ which is obtained from $N(z)$ by *reversing the order* of its coefficients, so that

$$C(z) = -(-y)^{-m} - (-y)^{-m+1}z^{-1}$$
$$-(-y)^{-m+2}z^{-2} - \ldots + y^{-1}z^{-m+1} \quad (4.3.33)$$

Bearing in mind that $Y(z)$ is obtained from $M(z)$ by reversing the order of its coefficients, it can be seen from Equation 4.3.32 that

$$Y(z)C(z) \simeq z^{-m} \quad (4.3.34)$$

This follows from the fact that if the order of the coefficients of two polynomials in x are reversed, then so also is the order of the coefficients of their product, there being no other change in this product. Thus, if the sampled impulse-response of both the channel and the corresponding linear equalizer are reversed in time, then so also is the sampled impulse-response of the two in cascade, which implies that correct equalization of the channel is maintained, although there may be a change in the delay introduced by the equalizer. Clearly, when $|y| > 1$, $C(z)$ in Equation 4.3.33 is the

z transform of the required linear equalizer for the channel with z-transform $Y(z)$ in Equation 4.3.28.

The greater is the magnitude of y, the more rapidly do the magnitudes of the tap gains increase with their positions along the equalizer (measured from the input end), so that the smaller is the filter needed for a given accuracy of equalization.

Consider now the case where the z transform of the channel is

$$Y(z) = 1 + 2z^{-1} \qquad (4.3.35)$$

and the transversal equalizer has six taps. From Equation 4.3.33, the z transform of the equalizer is

$$C(z) = -(-2)^{-6} - (-2)^{-5}z^{-1} - (-2)^{-4}z^{-2} - \ldots + 2^{-1}z^{-5}$$

$$= -\tfrac{1}{64} + \tfrac{1}{32}z^{-1} - \tfrac{1}{16}z^{-2} + \ldots + \tfrac{1}{2}z^{-5} \qquad (4.3.36)$$

and the ith tap has a gain of $-(-\tfrac{1}{2})^{7-i}$.

Thus the z transform of the sampled impulse-response of the baseband channel and linear equalizer is

$$Y(z)C(z) = (1 + 2z^{-1})(-\tfrac{1}{64} + \tfrac{1}{32}z^{-1} - \ldots - \tfrac{1}{4}z^{-4} + \tfrac{1}{2}z^{-5})$$

$$= -\tfrac{1}{64} + z^{-6} \qquad (4.3.37)$$

so that

$$Y(z)C(z) \simeq z^{-6} \qquad (4.3.38)$$

Since, by definition, $Y(z)Y^{-1}(z) = 1$, it can be seen from Equation 4.3.38 that

$$C(z) \simeq z^{-6}Y^{-1}(z) \qquad (4.3.39)$$

which means that the transversal filter equalizes the channel and in addition introduces a delay of $6T$ seconds.

If the residual intersymbol interference at the output of the equalizer is not neglected, the sampled impulse-response of the channel and linear equalizer, represented in terms of the correspond-

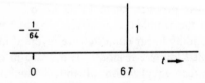

Fig. 4.8 *Sampled impulse-response of a channel, with z-transform $1 + 2z^{-1}$, in cascade with the corresponding 6-tap transversal equalizer*

ing impulses, is $-\frac{1}{64}\delta(t)+\delta(t-6T)$, as shown in Figure 4.8. Clearly this is an arrangement of zero forcing.

When the ith individual signal-element is received alone and in the absence of noise, the signal at the output of the equalizer is

$$-\frac{1}{64}s_i\delta(t-iT)+s_i\delta(t-6T-iT)$$

if represented in terms of impulses. Thus, when a continuous stream of data elements is received in the presence of Gaussian noise, the sample value of the signal at the output of the equalizer at time $t = (i+6)T$ is

$$x_{i+6} = s_i - \frac{1}{64}s_{i+6} + u_{i+6} \tag{4.3.40}$$

where u_{i+6} is a Gaussian random variable with zero mean and variance dependent upon the transversal equalizer. From Equation 4.3.40,

$$x_{i+6} \simeq s_i + u_{i+6} \tag{4.3.41}$$

and s_i is detected from the sign of x_{i+6}, at time $t = (i+6)T$.

It is interesting now to consider the relationship between the ideal and practical equalizers for a channel whose z transform is $Y(z) = 1+yz^{-1}$, where $|y|>1$. The ideal equalizer has the z-transform

$$\begin{aligned}
Y^{-1}(z) = (1+yz^{-1})^{-1} &= y^{-1}z(1+y^{-1}z)^{-1} \\
&= y^{-1}z(1-y^{-1}z+y^{-2}z^2-y^{-3}z^3+\ldots) \\
&= y^{-1}z-y^{-2}z^2+y^{-3}z^3-\ldots \tag{4.3.42}
\end{aligned}$$

which is a convergent series. The z transform given by Equation 4.3.42 is unfortunately not physically realizable, since z^i represents the time instant $-iT$ seconds or an *advance* in time of iT seconds. However, for a sufficiently large integer m,

$$Y^{-1}(z) \simeq y^{-1}z-y^{-2}z^2+y^{-3}z^3-\ldots-(-y)^{-m}z^m$$

$$\begin{aligned}
\text{and} \quad z^{-m}Y^{-1}(z) &\simeq y^{-1}z^{-m+1}-y^{-2}z^{-m+2} \\
&\quad +y^{-3}z^{-m+3}-\ldots-(-y)^{-m} \\
&= -(-y)^{-m}-(-y)^{-m+1}z^{-1}-(-y)^{-m+2}z^{-2} \\
&\quad -\ldots+y^{-1}z^{-m+1} \tag{4.3.43}
\end{aligned}$$

which is physically realizable and is in fact the z-transform $C(z)$ of the m-tap transversal equalizer given by Equation 4.3.33. This equalizer approximately equalizes the channel and at the same time introduces a delay of mT seconds into the equalized signal (Equation 4.3.34). The ideal equalizer with z-transform $Y^{-1}(z)$ of course introduces *no* delay into the equalized signal. The practical filter *must* introduce a delay in order that it may be physically realizable.

Consider next the case where the z transform of the channel is

$$Y(z) = (1 + \tfrac{1}{2}z^{-1})(1 + 2z^{-1}) = 1 + 2\tfrac{1}{2}z^{-1} + z^{-2} \quad (4.3.44)$$

A suitable equalizer is obtained by connecting in cascade the two six-tap transversal filters whose z transforms approximate to $(1 + \tfrac{1}{2}z^{-1})^{-1}$ and $z^{-6}(1 + 2z^{-1})^{-1}$, respectively, since the z transform of the channel and two filters is approximately

$$Y(z)(1 + \tfrac{1}{2}z^{-1})^{-1}z^{-6}(1 + 2z^{-1})^{-1} = z^{-6} \quad (4.3.45)$$

The equalizer is as shown in Figure 4.9. Each of the two transversal filters here equalizes or removes the corresponding *factor* of $Y(z)$.

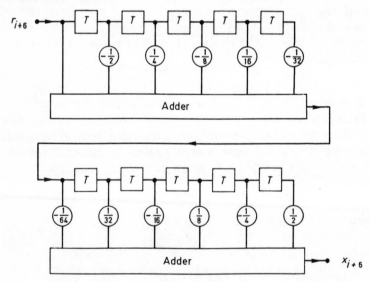

Fig. 4.9   Separate equalization of the two factors $1 + \tfrac{1}{2}z^{-1}$ and $1 + 2z^{-1}$ of the z-transform $1 + 2\tfrac{1}{2}z^{-1} + z^{-2}$

The equalizer in Figure 4.9 has the z-transform

$$(1-\tfrac{1}{2}z^{-1}+\tfrac{1}{4}z^{-2}-\tfrac{1}{8}z^{-3}+\tfrac{1}{16}z^{-4}-\tfrac{1}{32}z^{-5})(-\tfrac{1}{64}+\tfrac{1}{32}z^{-1}$$
$$-\tfrac{1}{16}z^{-2}+\tfrac{1}{8}z^{-3}-\tfrac{1}{4}z^{-4}+\tfrac{1}{2}z^{-5})$$
$$= -\tfrac{1}{64}(1-2\tfrac{1}{2}z^{-1}+5\tfrac{1}{4}z^{-2}-10\tfrac{5}{8}z^{-3}+21\tfrac{5}{16}z^{-4}-42\tfrac{21}{32}z^{-5}$$
$$+21\tfrac{5}{16}z^{-6}-10\tfrac{5}{8}z^{-7}+5\tfrac{1}{4}z^{-8}-2\tfrac{1}{2}z^{-9}+z^{-10}) \quad (4.3.46)$$

so that the equalizer is equivalent to a single transversal filter with 11 taps whose gains are given by the respective coefficients of the $\{z^{-i}\}$ on the right-hand side of Equation 4.3.46. The single transversal filter may clearly be used in place of the two filters in cascade, with some reduction in the equipment complexity but without affecting the performance of the system.

The z transform of the channel and linear equalizer is now

$$(1+2\tfrac{1}{2}z^{-1}+z^{-2})(-\tfrac{1}{64})(1-2\tfrac{1}{2}z^{-1}+5\tfrac{1}{4}z^{-2}-10\tfrac{5}{8}z^{-3}+21\tfrac{5}{16}z^{-4}$$
$$-42\tfrac{21}{32}z^{-5}+21\tfrac{5}{16}z^{-6}-10\tfrac{5}{8}z^{-7}+5\tfrac{1}{4}z^{-8}-2\tfrac{1}{2}z^{-9}+z^{-10})$$
$$= -\tfrac{1}{64}(1-64\tfrac{1}{64}z^{-6}+z^{-12})$$
$$\simeq -\tfrac{1}{64}+z^{-6}-\tfrac{1}{64}z^{-12} \simeq z^{-6} \quad (4.3.47)$$

If the residual intersymbol interference at the output of the equalizer is not neglected, the sampled impulse-response of the channel and linear equalizer, represented in terms of the corresponding impulses, is

$$-\tfrac{1}{64}\delta(t)+\delta(t-6T)-\tfrac{1}{64}\delta(t-12T)$$

as shown in Figure 4.10. As before, this is an arrangement of zero forcing.

When the ith individual signal-element is received alone and in the absence of noise, the signal at the output of the equalizer is

$$-\tfrac{1}{64}s_i\delta(t-iT)+s_i\delta(t-6T-iT)-\tfrac{1}{64}s_i\delta(t-12T-iT)$$

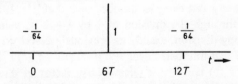

Fig. 4.10   Sampled impulse-response of a channel, with z-transform $1+2\tfrac{1}{2}z^{-1}+z^{-2}$, in cascade with the corresponding 11-tap transversal equalizer

if represented in terms of impulses. Thus, when a continuous stream of data elements is received in the presence of Gaussian noise, the sample value of the signal at the output of the equalizer, at time $t = (i+6)T$, is

$$x_{i+6} \simeq -\tfrac{1}{64}s_{i-6} + s_i - \tfrac{1}{64}s_{i+6} + u_{i+6} \qquad (4.3.48)$$

so that

$$x_{i+6} \simeq s_i + u_{i+6} \qquad (4.3.49)$$

and s_i is detected from the sign of x_{i+6}, at time $t = (i+6)T$. Clearly the residual intersymbol interference in x_{i+6} now originates from both the $(i-6)$th and $(i+6)$th received elements, but is still negligible.

The z transform of the equalizer for the given channel approximates to

$$z^{-h}(1+2\tfrac{1}{2}z^{-1}+z^{-2})^{-1} = z^{-h}(1+\tfrac{1}{2}z^{-1})^{-1}(1+2z^{-1})^{-1} \qquad (4.3.50)$$

where h is an appropriate nonnegative integer. It might be thought that this z transform could be evaluated directly through the division of 1 by $1+2\tfrac{1}{2}z^{-1}+z^{-2}$, using a normal process of long division. However this does not in fact give the required z-transform. The reason for the failure of the method is that the long division of 1 by $1+2\tfrac{1}{2}z^{-1}+z^{-2}$ gives a polynomial which is equal to the *product* of the polynomial obtained through the long division of 1 by $1+\tfrac{1}{2}z^{-1}$ and the polynomial obtained through the long division of 1 by $1+2z^{-1}$. The former polynomial approximates to $(1+\tfrac{1}{2}z^{-1})^{-1}$, but the latter polynomial forms a divergent series which does not therefore approximate to $z^{-h}(1+2z^{-1})^{-1}$, for *any* nonnegative integer h. Thus the long division of 1 by $1+2\tfrac{1}{2}z^{-1}+z^{-2}$ does not approximate to $z^{-h}(1+2\tfrac{1}{2}z^{-1}+z^{-2})^{-1}$. It is important to observe that $1+2\tfrac{1}{2}z^{-1}+z^{-2}$ has one root *inside* the unit circle in the z plane and one root *outside* the unit circle.

The principles and techniques that have just been considered may now be generalized quite simply so that they may be applied to any given channel.

Firstly, when y is a *complex* number such that $|y| < 1$, $(1+yz^{-1})^{-1}$ is obtained through the division of 1 by $1+yz^{-1}$, using a normal process of long division, exactly as previously described for the case where y is a *real* number. Secondly, if $Y(z) = Y_1(z)Y_2(z)\ldots Y_g(z)$ where each $Y_i(z)$ is a linear (elementary) factor of $Y(z)$, the polynomial obtained through the long division of 1 by $Y(z)$ is the *product* of the g polynomials obtained from the separate long division of 1 by the g factors $\{Y_i(z)\}$. Since $Y^{-1}(z) = Y_1^{-1}(z)Y_2^{-1}(z)\ldots Y_g^{-1}(z)$, it follows immediately that when *all* the roots of $Y(z)$ lie *inside* the unit circle in the z plane, the long division of 1 by $Y(z)$ gives $Y^{-1}(z)$. Suppose now that one or more roots of $Y(z)$ lie *outside* the unit circle, and let $Y_i(z)$ be one of the corresponding linear factors of $Y(z)$. The long division of 1 by $Y_i(z)$ gives a polynomial that forms a divergent series and is not therefore equal to $Y^{-1}(z)$. Since, as we have seen, the polynomial obtained from the long division of 1 by $Y(z)$ is the product of the polynomials obtained from the separate long division of 1 by the individual factors of $Y(z)$, it is clear that the long division of 1 by $Y(z)$ does *not* now give $Y^{-1}(z)$. Using these facts, the general design technique for a linear equalizer has been developed, as follows.

Suppose first that the z transform of the baseband channel in Figure 4.1 is given by

$$Y(z) = y_0 + y_1 z^{-1} + y_2 z^{-2} + \ldots + y_g z^{-g} \qquad (4.3.51)$$

where *all* the roots of $Y(z)$ lie *inside* the unit circle in the z plane.

To obtain the z-transform $C(z)$ of the m-tap linear equalizer for this channel, use a normal process of long division to divide 1 by $Y(z)$, as described for $Y(z) = 1 + yz^{-1}$ where $|y| < 1$, and take the first m terms of the resultant polynomial to give $C(z)$. Now

$$Y(z)C(z) \simeq 1 \qquad (4.3.52)$$

The equalizer is a minimum-delay network and applies zero forcing to the sampled impulse-response of the equalized channel. The main tap of the equalizer, that passes the wanted signal component through to the output, is here the first (or extreme left-hand) tap of the equalizer.

Suppose next that *all* the roots of $Y(z)$ lie *outside* the unit circle in the z plane. Reverse the order of the coefficients of $Y(z)$ to give

$$M(z) = y_g + y_{g-1} z^{-1} + y_{g-2} z^{-2} + \ldots + y_0 z^{-g} \qquad (4.3.53)$$

The roots of $M(z)$ are the reciprocals of the roots of $Y(z)$ and therefore lie *inside* the unit circle in the z plane. Use a normal process of long division to divide 1 by $M(z)$, and take the first m terms of the resultant polynomial to give the z-transform $N(z)$ of the equalizer for $M(z)$, such that

$$M(z)N(z) \simeq 1 \qquad (4.3.54)$$

Reverse the order of the coefficients of $N(z)$, to give $C(z)$, which is now the z transform of the m-tap linear equalizer for $Y(z)$, such that

$$Y(z)C(z) \simeq z^{-g-m+1} \qquad (4.3.55)$$

The equalizer is a maximum-delay network and applies zero forcing to the sampled impulse-response of the equalized channel. The main tap of the equalizer, that passes the wanted signal component through to the output, is here the last (or extreme right-hand) tap of the equalizer.

Suppose finally that

$$Y(z) = Y_1(z)Y_2(z) \qquad (4.3.56)$$

where $Y_1(z)$ and $Y_2(z)$ may each have several roots, with all the roots of $Y_1(z)$ inside the unit circle in the z plane and all the roots of $Y_2(z)$ outside the unit circle. Let

$$Y_1(z) = 1 + p_1 z^{-1} + p_2 z^{-2} + \ldots + p_{g-f} z^{-g+f} \qquad (4.3.57)$$

and

$$Y_2(z) = q_0 + q_1 z^{-1} + q_2 z^{-2} + \ldots + q_f z^{-f} \qquad (4.3.58)$$

where f is an integer in the range 0 to g. All coefficients here are real.

The z-transform $C_1(z)$ of the m_1-tap linear equalizer for $Y_1(z)$ is determined as follows. Use a normal process of long division to divide 1 by $Y_1(z)$, and take the first m_1 terms of the resultant polynomial to give $C_1(z)$. Thus

$$Y_1(z)C_1(z) \simeq 1 \qquad (4.3.59)$$

The z-transform $C_2(z)$ of the m_2-tap linear equalizer for $Y_2(z)$ is determined as follows. Let $M(z)$ be the polynomial determined from $Y_2(z)$ by reversing the order of its coefficients. Use a normal process of long division to divide 1 by $M(z)$. Take the first m_2 terms of the resultant polynomial and reverse the order of the coefficients, to give $C_2(z)$. Then

$$Y_2(z)C_2(z) \simeq z^{-f-m_2+1} \qquad (4.3.60)$$

The z transform of the linear equalizer for the baseband channel is now

$$C(z) = C_1(z)C_2(z) \qquad (4.3.61)$$

so that the z transform of the channel and equalizer is

$$Y(z)C(z) \simeq z^{-f-m_2+1} \qquad (4.3.62)$$

as can be seen from Equations 4.3.59 and 4.3.60. The equalizer here is neither a minimum nor a maximum delay network.

The equalization of either $Y_1(z)$ or $Y_2(z)$, in the manner just described, leads to a sampled impulse-response with a consecutive set of zeros between the wanted and unwanted components, so that $C(z)$ applies zero forcing to each of the two factors of $Y(z)$ and for practical purposes also to $Y(z)$ itself. It follows that zero forcing can in fact be applied to *any* linear baseband channel that can be equalized linearly. However, with a time-varying channel there is not necessarily a simple algorithm for holding the tap gains correctly adjusted for the channel.[57,63]

The general arrangement just described achieves the separate linear equalization of the two complementary factors $Y_1(z)$ and $Y_2(z)$ of the channel z-transform. All the available evidence suggests that, for practical purposes, the arrangement minimizes the peak distortion in the equalized signal. It is in general the most suitable design technique for the linear equalization of a known time-invariant baseband channel, where it is required to minimize the peak distortion in the equalized signal.

The technique that has just been described may be further generalized as follows. Let

$$Y(z) = Y_1(z)Y_2(z) \dots Y_n(z) \tag{4.3.63}$$

where $0 < n \leqslant g$, and $Y_i(z)$ for each i is a real (or indeed complex) polynomial *all* of whose roots are either inside or outside the unit circle in the z plane. Determine, for each i, the z-transform $C_i(z)$ of the linear equalizer for $Y_i(z)$, as previously described. The z transform of the linear equalizer for $Y(z)$ is now

$$C(z) = C_1(z)C_2(z) \dots C_n(z) \tag{4.3.64}$$

and the z transform of the channel and linear equalizer is

$$Y(z)C(z) \simeq z^{-h} \tag{4.3.65}$$

where h is an appropriate integer.

The equalizer here does not apply zero forcing to the sampled impulse-response of the equalized channel, except when it becomes identical to the arrangement previously described, nor (with this exception) does it achieve as low a value of the peak distortion in the equalized signal for a given number of taps in the transversal filter, as does the other arrangement. It is also more difficult to evaluate and there is no simple algorithm for holding the tap gains correctly adjusted for a time-varying channel, that operates according to this technique. The arrangement does not therefore appear to be of much practical value. It does however help to clarify an important property of linear equalizers.

It follows from Equation 4.3.63 that

$$Y^{-1}(z) = Y_1^{-1}(z)Y_2^{-1}(z) \ldots Y_n^{-1}(z) \qquad (4.3.66)$$

for *any* selection of the n factors of $Y(z)$. Furthermore, if a practical equalizer is designed to satisfy Equation 4.3.65 *fairly accurately* for *given* and appropriate values of h and m, then it does not matter which of the possible combinations of the factors of $Y(z)$ are used in the design of this equalizer; the resultant equalizer will, for practical purposes, be the same. In other words, the equalizer is effectively determined by $Y(z)$, h and m, and not by the particular design technique. This has an important consequence.

Suppose that in Equation 4.3.63,

$$Y_i(z) = 1 + yz^{-1} \qquad (4.3.67)$$

where $|y| = 1$. It has been shown (near the beginning of this section) that when $y = 1$, $Y_i(z)$ cannot be equalized by a finite linear feedforward transversal filter. This result may readily be extended to the more general case considered here, where y may be a complex number. Thus, when $Y_i(z)$ has a root (zero) on the unit circle in the z plane, it cannot be equalized by a linear feedforward transversal filter. Clearly if $Y_i(z)$ in Equation 4.3.63 cannot be equalized, then neither can $Y(z)$. Furthermore, since the equalizer is effectively determined by $Y(z)$, h and m and *not* by the particular design technique, $Y(z)$ cannot now be equalized through the choice of any *other* set of factors in Equation 4.3.63.

It is clear therefore that if $Y(z)$ contains one or more roots *on* the unit circle in the z plane, it cannot be equalized by a linear feedforward transversal filter.

It is shown in Section 2.4.6 that if $Y(z)$ contains one or more roots (zeros) on the unit circle in the z plane, the channel introduces infinite attenuation at certain discrete frequencies, which means that the corresponding frequency components of the transmitted signal are lost completely. When this happens, no amount of *linear* signal processing at the receiver can restore the lost frequencies. The linear equalization of the channel in fact requires a linear filter that introduces infinite gain at these frequencies, which is not physically realizable. The channel cannot therefore be equalized linearly, no matter what type of linear filter is used. It is not just linear feedforward transversal equalizers that cannot be used.

4.3.4 Example

A baseband channel has a sampled impulse-response whose z transform is

$$Y(z) = 1 - \tfrac{1}{2}z^{-1} + 4z^{-2} - 2z^{-3} \qquad (4.3.68)$$

Design a 9-tap linear feedforward transversal equalizer for this channel.

$$\begin{aligned} Y(z) &= (1 - \tfrac{1}{2}z^{-1})(1 + 4z^{-2}) \\ &= (1 - \tfrac{1}{2}z^{-1})(1 + 2\sqrt{-1}\,z^{-1})(1 - 2\sqrt{-1}\,z^{-1}) \end{aligned} \qquad (4.3.69)$$

Let

$$Y_1(z) = 1 - \tfrac{1}{2}z^{-1} \qquad (4.3.70)$$

and

$$Y_2(z) = 1 + 4z^{-2} \qquad (4.3.71)$$

Clearly, the root of $Y_1(z)$ lies inside the unit circle in the z plane, whereas the two roots of $Y_2(z)$ both lie outside the unit circle. The equalizers for these two factors must therefore be determined separately.

The z-transform $C_1(z)$ of the 5-tap linear equalizer for $Y_1(z)$ is determined through the long division of 1 by $Y_1(z)$, to give

$$C_1(z) = 1 + \tfrac{1}{2}z^{-1} + \tfrac{1}{4}z^{-2} + \tfrac{1}{8}z^{-3} + \tfrac{1}{16}z^{-4} \qquad (4.3.72)$$

which is such that

$$Y_1(z)C_1(z) \simeq 1 \qquad (4.3.73)$$

The z-transform $C_2(z)$ of the linear equalizer for $Y_2(z)$ is determined as follows. The z transform obtained from $Y_2(z)$ by reversing the order of its coefficients is

$$M(z) = 4 + z^{-2} \qquad (4.3.74)$$

and the two roots of $M(z)$ lie inside the unit circle in the z plane.

Thus the z transform of the linear equalizer for $M(z)$ is determined through the long division of 1 by $M(z)$, to give

$$M^{-1}(z) = \tfrac{1}{4} - \tfrac{1}{16}z^{-2} + \tfrac{1}{64}z^{-4} - \cdots \qquad (4.3.75)$$

The z transform obtained from $M^{-1}(z)$ by taking its first three terms and reversing the order of the coefficients is

$$C_2(z) = \tfrac{1}{64} - \tfrac{1}{16}z^{-2} + \tfrac{1}{4}z^{-4} \qquad (4.3.76)$$

and this is the z transform of the 3-tap linear equalizer for $Y_2(z)$, such that

$$Y_2(z)C_2(z) \simeq z^{-6} \qquad (4.3.77)$$

Notice that both $C_1(z)$ and $C_2(z)$ have been selected to have z^{-4} as the highest power of z^{-1}, which means that each of the corresponding transversal filters has effectively 5 taps, although, in the case of the filter with z-transform $C_2(z)$, two of these taps have zero gain. The z-transform $C(z)$ of the 9-tap transversal equalizer for the given baseband channel is now

$$\begin{aligned}
C(z) &= C_1(z)C_2(z) \\
&= \tfrac{1}{64}(1 + \tfrac{1}{2}z^{-1} - 3\tfrac{3}{4}z^{-2} - 1\tfrac{7}{8}z^{-3} + 15\tfrac{1}{16}z^{-4} + 7\tfrac{1}{2}z^{-5} \\
&\qquad + 3\tfrac{3}{4}z^{-6} + 2z^{-7} + z^{-8}) \qquad (4.3.78)
\end{aligned}$$

The z transform of the channel and linear equalizer is

$$\begin{aligned}
Y(z)C(z) &= (1 - \tfrac{1}{2}z^{-1} + 4z^{-2} - 2z^{-3})\tfrac{1}{64}(1 + \tfrac{1}{2}z^{-1} - 3\tfrac{3}{4}z^{-2} \\
&\qquad - 1\tfrac{7}{8}z^{-3} + 15\tfrac{1}{16}z^{-4} + 7\tfrac{1}{2}z^{-5} + 3\tfrac{3}{4}z^{-6} + 2z^{-7} + z^{-8}) \\
&= \tfrac{1}{64}(1 - \tfrac{1}{32}z^{-5} + 64z^{-6} - 2z^{-11}) \\
&\simeq \tfrac{1}{64} + z^{-6} - \tfrac{1}{32}z^{-11} \qquad (4.3.79)
\end{aligned}$$

Thus the ith received signal-element has approximately the z-transform

$$s_i z^{-i}(\tfrac{1}{64} + z^{-6} - \tfrac{1}{32}z^{-11}) = s_i(\tfrac{1}{64}z^{-i} + z^{-i-6} - \tfrac{1}{32}z^{-i-11}) \qquad (4.3.80)$$

and is detected from the sign of the sample value

$$\begin{aligned}
x_{i+6} &\simeq -\tfrac{1}{32}s_{i-5} + s_i + \tfrac{1}{64}s_{i+6} + u_{i+6} \\
&\simeq s_i + u_{i+6} \qquad (4.3.81)
\end{aligned}$$

at the output of the equalizer, at time $t = (i+6)T$. u_{i+6} is a Gaussian noise component.

It would of course have been possible to take

$$C_1(z) = 1 + \tfrac{1}{2}z^{-1} + \tfrac{1}{4}z^{-2} + \tfrac{1}{8}z^{-3} + \tfrac{1}{16}z^{-4} + \tfrac{1}{32}z^{-5} + \tfrac{1}{64}z^{-6} \quad (4.3.82)$$

$$C_2(z) = -\tfrac{1}{16} + \tfrac{1}{4}z^{-2} \quad (4.3.83)$$

or else

$$C_1(z) = 1 + \tfrac{1}{2}z^{-1} + \tfrac{1}{4}z^{-2} \quad (4.3.84)$$

$$C_2(z) = -\tfrac{1}{256} + \tfrac{1}{64}z^{-2} - \tfrac{1}{16}z^{-4} + \tfrac{1}{4}z^{-6} \quad (4.3.85)$$

since each of these combinations of $C_1(z)$ and $C_2(z)$ gives 9 taps for the resultant equalizer with z-transform $C(z)$. However, the evaluation of the z transform of the equalized channel for each of these combinations shows a greater value of the peak distortion in the equalized signal than for the chosen combination. Whenever the tap gains of an equalizer have been evaluated, it is important to check the sampled impulse-response of the equalized channel in order to ensure that the required accuracy of equalization has been achieved.

4.3.5 Feedback transversal filter

The transversal equalizers considered so far are linear feedforward transversal filters of the type normally used in practice. Equalization of a channel may alternatively be achieved by means of a feedback transversal filter, provided that certain conditions are satisfied by the channel response. Suppose that the z transform of the channel is

$$
\begin{aligned}
Y(z) &= y_0 + y_1 z^{-1} + y_2 z^{-2} + \ldots + y_g z^{-g} \\
&= y_0 \left(1 + \frac{y_1}{y_0} z^{-1} + \frac{y_2}{y_0} z^{-2} + \ldots + \frac{y_g}{y_0} z^{-g} \right) \\
&= y_0 (1 + \alpha_1 z^{-1})(1 + \alpha_2 z^{-1}) \ldots (1 + \alpha_g z^{-1}) \quad (4.3.86)
\end{aligned}
$$

where the $\{\alpha_i\}$ may be real or complex and are the negatives of the g roots of $Y(z)$.

The linear feedback transversal equalizer for the channel is as shown in Figure 4.11. The signals shown here are those present at the time instant $t = iT$. It can be seen that if an individual signal-element $s_i \delta(t - iT)$ is fed to the input of the baseband channel in

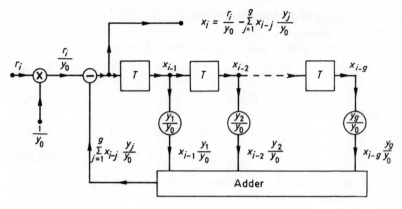

$$x_i = \frac{r_i}{y_0} - \sum_{j=1}^{g} x_{i-j} \frac{y_j}{y_0}$$

Fig. 4.11 *Linear feedback transversal equalizer*

Figure 4.1, the resultant sequence of sample values at the output of the channel, in the absence of noise, are

$$s_i y_0 \quad s_i y_1 \ldots s_i y_g$$

at the time instants iT, $(i+1)T$, . . ., $(i+g)T$, respectively, the preceding and following sample values being all zero. The operation of the feedback transversal filter on the received sequence of sample values is to give at its output a single nonzero sample value $x_i = s_i$, at the time instant iT, this being preceded and followed by zero sample-values, assuming that all signals in the equalizer were initially set to zero. Thus the filter clearly equalizes the channel. Furthermore, the equalization is exact and requires only g taps, which is a far smaller number than that normally required by the corresponding feedforward transversal equalizer.

Consider now the feedback transversal equalizer, shown in Figure 4.12, for a channel with z-transform $1 + \alpha_i z^{-1}$.

Let the z transforms of the sequences of sample values at the input and output of the equalizer in Figure 4.12 be $A(z)$ and $B(z)$, respectively. Then

$$B(z) = A(z) - \alpha_i z^{-1} B(z) \qquad (4.3.87)$$

so that

$$B(z)(1 + \alpha_i z^{-1}) = A(z) \qquad (4.3.88)$$

To solve this equation in terms of $B(z)$ it must be borne in mind that, when $|\alpha_i| > 1$,

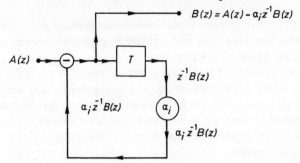

Fig. 4.12 Linear feedback transversal equalizer for a channel with z-transform $1 + \alpha_i z^{-1}$

$$(1 + \alpha_i z^{-1})^{-1} = \alpha_i^{-1} z (1 + \alpha_i^{-1} z)^{-1}$$
$$= \alpha_i^{-1} z (1 - \alpha_i^{-1} z + \alpha_i^{-2} z^2 - \ldots)$$
$$= \alpha_i^{-1} z - \alpha_i^{-2} z^2 + \alpha_i^{-3} z^3 - \ldots \quad (4.3.89)$$

Thus, if Equation 4.3.88 is rewritten as

$$B(z) = A(z)(1 + \alpha_i z^{-1})^{-1} \quad (4.3.90)$$

this is not physically realizable, since the output signal now precedes the input signal, and, in any case, it does *not* describe the actual operation of the equalizer. It is evident from Figure 4.12 that the sampled impulse-response of the equalizer is $1 - \alpha_i \; \alpha_i^2 - \alpha_i^3 \ldots$, with z-transform $1 - \alpha_i z^{-1} + \alpha_i^2 z^{-2} - \alpha_i^3 z^{-3} + \ldots$, so that

$$B(z) = A(z)(1 - \alpha_i z^{-1} + \alpha_i^2 z^{-2} - \alpha_i^3 z^{-3} + \ldots) \quad (4.3.91)$$

In other words, the binomial expansion is applied to $(1 + \alpha_i z^{-1})^{-1}$ in Equation 4.3.90, as though $|\alpha_i| < 1$. Of course, when $|\alpha_i| < 1$, Equation 4.3.90 describes the correct operation of the equalizer.

The z transform of the equalizer, $1 - \alpha_i z^{-1} + \alpha_i^2 z^{-2} - \alpha_i^3 z^{-3} + \ldots$, may alternatively be obtained through the division of 1 by $1 + \alpha_i z^{-1}$, using a normal process of long division, and it can be seen from Figure 4.12 that the process of long division describes exactly the operation of the equalizer. Thus the z transform of the equalizer can be written $1/(1 + \alpha_i z^{-1})$, where this is taken to imply the process of long division.[64] These results apply for both real and complex values of α_i.

The fact that, when $|\alpha_i| > 1$, the operation of the feedback equalizer is given by Equation 4.3.91 and not by Equation 4.3.90, shows that the feedback equalizer is *not* now equivalent to the ideal linear equalizer for $1+\alpha_i z^{-1}$, and in fact it performs a quite different operation on its input signal.

In the absence of noise the equalizer operates perfectly whether $|\alpha_i| < 1$ or $|\alpha_i| > 1$, but whenever $|\alpha_i| > 1$, any noise component added to the data signal at the input to the equalizer, however small its magnitude, results in a sequence of noise samples of steadily increasing magnitudes at the output of the equalizer, so that the system is unstable.

Consider now the feedback transversal equalizer in Figure 4.11. Following the same line of analysis as that just applied to the simpler equalizer and from the analysis in Section 4.6.5, it can be seen that the operation of the equalizer in Figure 4.11 is described exactly by a process of long division, such that its z transform is

$$1 \bigg/ \left(1 + \frac{y_1}{y_0} z^{-1} + \frac{y_2}{y_0} z^{-2} + \ldots + \frac{y_g}{y_0} z^{-g} \right)$$

Now the polynomial obtained through the long division of 1 by

$$1 + \frac{y_1}{y_0} z^{-1} + \frac{y_2}{y_0} z^{-2} + \ldots + \frac{y_g}{y_0} z^{-g}$$

is equal to the *product* of the polynomials obtained through the separate long division of 1 by the g factors

$$\{ 1 + \alpha_i z^{-1} \} \text{ of } 1 + \frac{y_1}{y_0} z^{-1} + \frac{y_2}{y_0} z^{-2} + \ldots + \frac{y_g}{y_0} z^{-g}$$

in Equation 4.3.86. This means that the equalizer in Figure 4.11 may alternatively be implemented as a sequence of g equalizers, with z-transforms $\{ 1/(1 + \alpha_i z^{-1}) \}$, connected in cascade, the resultant equalizer having exactly the same performance as that in Figure 4.11. The $\{ \alpha_i \}$ may, of course, be real or complex. It follows that the equalizer in Figure 4.11 is stable if and only if the equalizer in Figure 4.12 is stable for every value of i in the range 1 to g, that is, if and only if *all* the roots of $Y(z)$ lie *inside* the unit circle in the z plane. When this condition is satisefid, the z-transform

$$1 \bigg/ \left(1 + \frac{y_1}{y_0} z^{-1} + \frac{y_2}{y_0} z^{-2} + \ldots + \frac{y_g}{y_0} z^{-g} \right)$$

of the equalizer in Figure 4.11 becomes equal to

$$\left(1 + \frac{y_1}{y_0}z^{-1} + \frac{y_2}{y_0}z^{-2} + \ldots + \frac{y_g}{y_0}z^{-g}\right)^{-1}$$

and the equalizer is both stable and achieves the accurate equalization of the channel.

To summarize, if one or more roots of the z-transform $Y(z)$ of the baseband channel in Figure 4.1 lie outside or on the unit circle in the z plane, a linear feedback transversal equalizer is unstable and so cannot be used, whether it is implemented as in Figure 4.11 or as a sequence of separate filters, the ith of which is shown in Figure 4.12. When all the roots of $Y(z)$ lie inside the unit circle, the linear feedback transversal equalizer is always stable. It now gives *exact* equalization and is generally much smaller than the equivalent feedforward filter.

When the baseband channel is time invariant and known, the linear equalizer giving the most accurate equalization for a given number of taps is that where the factor $Y_1(z)$ of the z transform of the channel (Equation 4.3.56), whose roots lie inside the unit circle in the z plane, is equalized by means of a feedback transversal filter, while the factor $Y_2(z)$, whose roots lie outside the unit circle, is equalized by a linear feedforward transversal filter.[64] Linear feedback transversal filters are not, however, generally suitable for use as adaptive equalizers for time-varying channels, because of the risk of instability.

4.3.6   *Tolerance to additive Gaussian noise*

Consider the linear feedforward transversal filter with m taps, as shown in Figure 4.4. Assume that the filter equalizes the baseband channel in Figure 4.1 and that the sample value at the output of the equalizer, at time $t = (i+h)T$, is

$$x_{i+h} \simeq s_i + u_{i+h} \tag{4.3.92}$$

where, as before,

$$s_i = \pm k \tag{4.3.93}$$

and u_{i+h} is a noise component. s_i is detected from the sign of x_{i+h}, so that when $x_{i+h} > 0$, s_i is detected as k, and when $x_{i+h} < 0$, s_i is detected as $-k$.

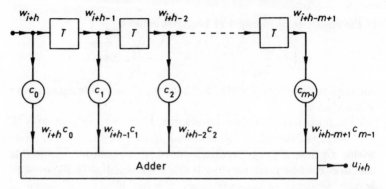

Fig. 4.13 *Noise signals in an m-tap linear transversal equalizer at the time instant $t = (i+h)T$*

The noise components at different points in the transversal equalizer, at the time instant $t = (i+h)T$, are shown in Figure 4.13, from which it can be seen that the output noise signal from the equalizer, at time $t = (i+h)T$, is

$$u_{i+h} = \sum_{j=0}^{m-1} w_{i+h-j} c_j \qquad (4.3.94)$$

The noise components $\{w_i\}$ are jointly Gaussian random variables that are statistically independent, with zero mean and variance σ^2. Thus u_{i+h} is a Gaussian random variable with zero mean and variance

$$\eta^2 = \sum_{j=0}^{m-1} \sigma^2 c_j^2 = \sigma^2 C C^T = \sigma^2 |C|^2 \qquad (4.3.95)$$

where $|C|$ is the *length* of the m-component row vector

$$C = c_0 \ c_1 \ldots c_{m-1} \qquad (4.3.96)$$

An error occurs in the detection of s_i from x_{i+h}, whenever x_{i+h} has the opposite sign to s_i, and therefore whenever u_{i+h} has a magnitude greater than k and the opposite sign to s_i.

The probability density function of the noise sample u_{i+h} is

$$p(u) = \frac{1}{\sqrt{(2\pi\eta^2)}} \exp\left(-\frac{u^2}{2\eta^2}\right) \qquad (4.3.97)$$

writing u in place of u_{i+h}, and this is clearly an even function of u. Thus the probability of error in the detection of s_i from x_{i+h} is the same whether $s_i = k$ or $-k$, and is

$$P_e = \int_k^\infty \frac{1}{\sqrt{(2\pi\eta^2)}} \exp\left(-\frac{u^2}{2\eta^2}\right) du$$

$$= \int_{k/\eta\sqrt{2\pi}}^\infty \frac{1}{\sqrt{2\pi}} \exp(-\tfrac{1}{2}u^2) du$$

$$= Q\left(\frac{k}{\eta}\right) = Q\left(\frac{k}{\sigma|C|}\right) \tag{4.3.98}$$

where $Q(.)$ is the Q-function defined by Equation 3.2.65. The accurate linear equalization of the channel is, of course, assumed here.

Consider now the case where the z transform of the baseband channel in Figure 4.1 is

$$Y(z) = 1 + 2\tfrac{1}{2}z^{-1} + z^{-2} \tag{4.3.99}$$

and the linear transversal equalizer for the channel has the z-transform

$$C(z) = -\tfrac{1}{64}(1 - 2\tfrac{1}{2}z^{-1} + 5\tfrac{1}{4}z^{-2} - 10\tfrac{5}{8}z^{-3} + 21\tfrac{5}{16}z^{-4} - 42\tfrac{21}{32}z^{-5}$$
$$+ 21\tfrac{5}{16}z^{-6} - 10\tfrac{5}{8}z^{-7} + 5\tfrac{1}{4}z^{-8} - 2\tfrac{1}{2}z^{-9} + z^{-10}) \tag{4.3.100}$$

so that the z transform of the equalized channel is

$$Y(z)C(z) \simeq z^{-6} \tag{4.3.101}$$

as in Equation 4.3.47.

s_i is detected from the sign of the sample value

$$x_{i+6} \simeq s_i + u_{i+6} \tag{4.3.102}$$

at the output of the equalizer, at time $t = (i+6)T$. u_{i+6} is a Gaussian random variable with zero mean and variance

$$\eta^2 = \frac{\sigma^2}{64^2}[1^2 + (2\tfrac{1}{2})^2 + (5\tfrac{1}{4})^2 + (10\tfrac{5}{8})^2 + (21\tfrac{5}{16})^2$$
$$+ (42\tfrac{21}{32})^2 + (21\tfrac{5}{16})^2 + (10\tfrac{5}{8})^2 + (5\tfrac{1}{4})^2 + (2\tfrac{1}{2})^2 + 1^2]$$
$$= 0.738\sigma^2 \tag{4.3.103}$$

so that

$$\eta = 0.859\sigma \tag{4.3.104}$$

From Equation 4.3.98, the probability of error in the detection of s_i from x_{i+6} is approximately

$$P_e = Q\left(\frac{k}{\eta}\right) = Q\left(\frac{k}{0.859\sigma}\right) = Q\left(\frac{1.164k}{\sigma}\right) \quad (4.3.105)$$

If now *no* equalization is used, the z transform of the ith received signal-element is

$$s_i z^{-i} Y(z) = s_i(z^{-i} + 2\tfrac{1}{2} z^{-i-1} + z^{-i-2}) \quad (4.3.106)$$

from Equation 4.3.99, so that the components (sample values) of the ith element are s_i, $2\tfrac{1}{2}s_i$ and s_i, at times iT, $(i+1)T$ and $(i+2)T$, respectively.

Clearly, s_i is best detected from the received sample value containing the component $2\tfrac{1}{2}s_i$, that is, from the received sample value

$$r_{i+1} = s_{i-1} + 2\tfrac{1}{2}s_i + s_{i+1} + w_{i+1} \quad (4.3.107)$$

at time $t = (i+1)T$. When $r_{i+1} > 0$, s_i is detected as k, and when $r_{i+1} < 0$, s_i is detected as $-k$.

At high signal/noise ratios with additive Gaussian noise, as assumed here, even a small increase in the distance to the decision threshold, in the detection of a signal, produces a very large reduction in the corresponding error probability. Thus practically *all* the errors occur when

$$s_{i-1} = s_{i+1} = -s_i \quad (4.3.108)$$

since now the signal component in r_{i+1} has its smallest magnitude and is such that

$$r_{i+1} = \tfrac{1}{2}s_i + w_{i+1} \quad (4.3.109)$$

An error occurs here in the detection of s_i when w_{i+1} has a magnitude greater than $\tfrac{1}{2}k$ and the opposite sign to s_i. Of course, w_{i+1} is a Gaussian random variable with zero mean and variance σ^2.

Since there is a probability of $\tfrac{1}{4}$ that Equation 4.3.108 is satisfied, the average probability of error in the detection of s_i from r_{i+1} is approximately equal to a quarter of the conditional probability of error, given that Equation 4.3.108 is satisfied. Thus the average probability of error in the detection of s_i from r_{i+1} is approximately

$$\tfrac{1}{4}\int_{\frac{1}{2}k}^{\infty} \frac{1}{\sqrt{(2\pi\sigma^2)}} \exp\left(-\frac{w^2}{2\sigma^2}\right) dw = \tfrac{1}{4}\int_{\frac{1}{2}k/\sigma}^{\infty} \frac{1}{\sqrt{2\pi}} \exp(-\tfrac{1}{2}w^2) dw$$

$$= \tfrac{1}{4}Q\left(\frac{0.5k}{\sigma}\right) \quad (4.3.110)$$

At error probabilities of around 1 in 10^5 or 1 in 10^6, a reduction in the error probability by a factor of four times is equivalent to an increase in tolerance to additive Gaussian noise of only a little over $\frac{1}{2}$ dB. Although this must obviously be taken into account in an accurate comparison of error probabilities, it may be neglected for our purposes here. Thus the probability of error in the detection of s_i from r_{i+1} can be taken to be

$$P_e = Q\left(\frac{0.5k}{\sigma}\right) \qquad (4.3.111)$$

Comparing Equations 4.3.105 and 4.3.111, it can be seen that for *given* values of P_e and k, the value of σ in the case where the linear equalizer is used is 2.33 times that where no equalizer is used. This means that the linear equalizer gains an advantage in tolerance to additive white Gaussian noise of 7.3 dB over the arrangement with no equalizer. Making an allowance for the omission of the factor $1/4$ in the error probability for no equalizer, it can be seen that the equalizer increases the tolerance to additive white Gaussian noise by about 7 dB. This is a considerable improvement and is quite typical of the advantage that may be gained with a linear equalizer.

4.3.7 Equalizer giving the minimum mean-square error

In the last few years it has become recognized that a linear transversal equalizer that minimizes the mean-square error in its output signal[63-107] generally gives a more useful or effective degree of equalization, for a given number of taps, than an equalizer that minimizes the peak distortion.[53-65] It is also more easily held in the correct adjustment for a slowly time-varying channel, in an adaptive system, particularly when the channel introduces severe signal distortion.[63-107] Such an equalizer *minimizes the mean-square difference* between the *actual* and *ideal* sample values at its output, for the given number of taps, when a continuous stream of data elements is being received in the presence of noise.

Suppose that the sampled impulse-response of the baseband channel in Figure 4.1 is given by the $(g+1)$-component row vector

$$V = y_0 \ y_1 \cdots y_g \qquad (4.3.112)$$

The sampled impulse-response of the m-tap linear feedforward transversal equalizer is given by the m-component row vector

$$C = c_0 \ c_1 \ldots c_{m-1} \tag{4.3.113}$$

and the sampled impulse-response of the channel and equalizer is given by the $(m+g)$-component row vector

$$E = e_0 \ e_1 \ldots e_{m+g-1} \tag{4.3.114}$$

The *ideal* value of the sampled impulse-response of the channel and equalizer, for a total transmission delay of hT seconds, is given by the $(m+g)$-component row vector

$$E_h = \overbrace{0 \ldots 0}^{h} \ 1 \ 0 \ldots 0 \tag{4.3.115}$$

where h is an integer in the range 0 to $m+g-1$. With the accurate equalization of the channel, $E \simeq E_h$.

Under these conditions, the output sample value from the linear equalizer, at time $t = (i+h)T$, is

$$x_{i+h} \simeq s_i + u_{i+h} \tag{4.3.116}$$

where u_{i+h} is the Gaussian noise component. Ideally, exact equality should hold in Equation 4.3.116, but in fact

$$x_{i+h} = \sum_{j=0}^{m+g-1} s_{i+h-j} e_j + u_{i+h} \tag{4.3.117}$$

as can be seen from Equation 4.3.114.

A linear equalizer that minimizes the mean-square error in its output signal, minimizes the mean-square value of $x_{i+h} - s_i$, which is

$$\overline{(x_{i+h} - s_i)^2} = k^2(e_h - 1)^2 + \sum_{\substack{j=0 \\ j \neq h}}^{m+g-1} k^2 e_j^2 + \overline{u_{i+h}^2}$$

$$= k^2 |E - E_h|^2 + \sigma^2 |C|^2 \tag{4.3.118}$$

where $|E - E_h|$ and $|C|$ are the *lengths* of the vectors $E - E_h$ and C, respectively. $k^2 |E - E_h|^2$ is the mean-square error in x_{i+h} due to intersymbol interference (Equation 4.3.11), and $\sigma^2 |C|^2$ is the mean-square error due to the Gaussian noise (Equation 4.3.95).

The derivation of Equation 4.3.118 uses the following facts. The $\{s_i\}$ are statistically independent of each other and equally likely to have either value k or $-k$, so that $\overline{s_i s_j} = 0$ for $i \neq j$, and $\overline{s_i^2} = k^2$.

Also, the $\{s_i\}$ are statistically independent of u_{i+h} which has zero mean, so that $\overline{s_j u_{i+h}} = 0$ for any integer j.

The equalizer here minimizes the mean-square error due to *both* intersymbol interference and noise, and not just the mean-square error $k^2|E-E_h|^2$ due to intersymbol interference alone, as in Equation 4.3.11. The relationship between the two minimization criteria will be considered presently.

Let Y be the $m \times (m+g)$ convolution matrix whose ith row is given by the $(m+g)$-component row vector

$$Y_{i-1} = \overbrace{0 \ldots 0}^{i-1} y_0 \; y_1 \ldots y_g \; 0 \ldots 0 \qquad (4.3.119)$$

Then the sampled impulse-response of the channel and equalizer is given by the $(m+g)$-component vector

$$E = \sum_{i=0}^{m-1} c_i Y_i = CY \qquad (4.3.120)$$

From Equation 4.3.118, the linear equalizer is required to minimize

$$\begin{aligned}
k^2|CY-E_h|^2+\sigma^2|C|^2 &= k^2(CY-E_h)(CY-E_h)^T+\sigma^2 CC^T \\
&= k^2 CYY^T C^T - 2k^2 E_h Y^T C^T + k^2 E_h E_h^T + \sigma^2 CC^T \\
&= C(k^2 YY^T+\sigma^2 I)C^T - 2k^2 E_h Y^T C^T + k^2 \qquad (4.3.121)
\end{aligned}$$

where I is an $m \times m$ identity matrix, and E_h is given by Equation 4.3.115.

The $m \times m$ matrix $k^2 YY^T+\sigma^2 I$ is real, symmetric and positive definite, since the matrices YY^T and I are both real, symmetric and positive definite. It follows that there is an $m \times m$ real nonsingular matrix G such that

$$GG^T = k^2 YY^T+\sigma^2 I \qquad (4.3.122)$$

Now, from Equation 4.3.121,

$$\begin{aligned}
k^2|CY-E_h|^2+\sigma^2|C|^2 &= CGG^T C^T - 2k^2 E_h Y^T C^T + k^2 \\
&= [CG-k^2 E_h Y^T(G^T)^{-1}][CG-k^2 E_h Y^T(G^T)^{-1}]^T \\
&\qquad + k^2[1-k^2 E_h Y^T(G^T)^{-1}G^{-1} Y E_h^T] \\
&= |CG-k^2 E_h Y^T(G^T)^{-1}|^2 + k^2[1-k^2 E_h Y^T(GG^T)^{-1} Y E_h^T] \quad (4.3.123)
\end{aligned}$$

It is required to select C to minimize $k^2|CY-E_h|^2+\sigma^2|C|^2$. In Equation 4.3.123

$$k^2\big[1-k^2E_hY^T(GG^T)^{-1}YE_h^T\big]$$

is independent of C and

$$|CG-k^2E_hY^T(G^T)^{-1}|^2$$

is nonnegative, so that

$$k^2|CY-E_h|^2+\sigma^2|C|^2$$

is minimum when

$$|CG-k^2E_hY^T(G^T)^{-1}|^2 = 0 \qquad (4.3.124)$$

that is, when

$$CG-k^2E_hY^T(G^T)^{-1} = 0 \qquad (4.3.125)$$

or

$$
\begin{aligned}
C &= k^2E_hY^T(G^T)^{-1}G^{-1} \\
&= k^2E_hY^T(GG^T)^{-1} \\
&= k^2E_hY^T(k^2YY^T+\sigma^2I)^{-1} \\
&= E_hY^T\left(YY^T+\frac{\sigma^2}{k^2}I\right)^{-1} \qquad (4.3.126)
\end{aligned}
$$

It is interesting to compare this equation with Equation 3.3.103. Equation 4.3.126 gives the values of the tap gains of the m-tap linear transversal equalizer that minimizes the mean-square error due to both intersymbol interference and noise, in the output signal from the equalizer. From Equation 4.3.123, the minimum value of this mean-square error is

$$
\begin{aligned}
&k^2\big[1-k^2E_hY^T(GG^T)^{-1}YE_h^T\big] \\
&= k^2\left[1-E_hY^T\left(YY^T+\frac{\sigma^2}{k^2}I\right)^{-1}YE_h^T\right] \qquad (4.3.127)
\end{aligned}
$$

At high signal/noise ratios, $k^2\gg\sigma^2$ (assuming that $|V|\simeq 1$ and hence $|Y_i|\simeq 1$), so that, from Equation 4.3.126, the equalizer that minimizes the mean-square error is now given by

$$C \simeq E_hY^T(YY^T)^{-1} \qquad (4.3.128)$$

This equalizer will be considered in more detail later.

At very-low signal/noise ratios, $k^2\ll\sigma^2$, so that the equalizer is now given by

$$C \simeq \frac{k^2}{\sigma^2} E_h Y^T$$

$$= \frac{k^2}{\sigma^2}(y_h \ y_{h-1} \ldots y_0 \ 0 \ldots 0) \qquad (4.3.129)$$

where $y_i = 0$ for $i > g$. The filter is here *matched* to the ith received signal-element, or at least to as much of it as has been received over the m consecutive sampling instants ending at the time instant $t = (i+h)T$, when s_i is detected from the output sample value x_{i+h}. The equalizer therefore maximizes the output signal/noise ratio and hence minimizes the effect of the noise on the signal. It does not however necessarily reduce and may even increase the mean-square error $k^2|E - E_h|^2$ due to intersymbol interference. This is, of course, to be expected since the noise now predominates over the intersymbol interference, so that it is more important to reduce the noise than the intersymbol interference.

Most adaptive equalizers are automatically adjusted to minimize the mean-square error in the equalized signal,[63-107] which means that when the data is received in the presence of additive white Gaussian noise, the equalizer will be adjusted as just described. Unfortunately, over switched telephone circuits, the predominant type of noise is impulsive noise which occurs in short bursts, separated by relatively long intervals that are comparatively noise-free. Under these conditions, the equalizer is normally adjusted so that, for practical purposes, it minimizes the mean-square error $k^2|E - E_h|^2$ due to intersymbol interference alone. Again, when a fixed equalizer is designed for some given time-invariant channel that introduces additive white Gaussian noise, a prior knowledge of the signal/noise ratio k^2/σ^2 is required in order to minimize the mean-square error in the equalized signal, as can be seen from Equation 4.3.126. In practice this prior knowledge is often not available. The best that can be done under these conditions is again to minimize the mean-square error due to intersymbol interference alone.

Consider therefore the m-tap linear transversal equalizer that is designed to minimize $k^2|E - E_h|^2$, which is the mean-square error due to intersymbol interference in its output signal.

The vectors E and E_h may be represented as points in an $(m+g)$-dimensional Euclidean vector space, as illustrated in Figure 4.14, and the equalizer must now be designed to minimize $|E - E_h|$ which is the *distance* between the two vectors in the vector space.

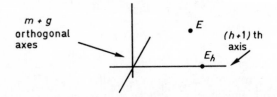

Fig. 4.14   Vectors E and E_h in the $(m+g)$-dimensional vector space

Now, $E = CY$, as can be seen from Equation 4.3.120, and the vector E is a linear combination of the m vectors $\{Y_i\}$, given by Equation 4.3.119. These m vectors are linearly independent, which means that no one of them can be formed by a linear combination of any of the others. Thus they span an m-dimensional subspace of the $(m+g)$-dimensional vector space, and the vector E lies in this subspace. Any vector that cannot be expressed as a linear combination of the m $\{Y_i\}$ does not lie in the subspace. It follows from Equations 4.3.119 and 4.3.120 that whenever the channel introduces *any* intersymbol interference in the sampled signal at the receiver input, E_h cannot lie in the subspace, at least not exactly.

Clearly, $|E - E_h|$ is minimum when E is the point in the m-dimensional subspace at the *minimum distance* from E_h. By the projection theorem, the required vector E is the *orthogonal projection* of E_h on to the m-dimensional subspace. This is illustrated in Figure 4.15.

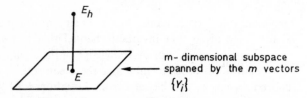

Fig. 4.15   The vector E that minimizes the mean-square error due to intersymbol interference

The vector $E - E_h$ is now *orthogonal* to the m-dimensional subspace and therefore also to every vector that lies in the subspace. But the origin and each vector Y_i must lie in the subspace. Thus the vector $E - E_h$ is orthogonal to each vector Y_i, so that

$$(E - E_h)Y_i^T = 0 \qquad (4.3.130)$$

or
$$(E - E_h)Y^T = 0 \qquad (4.3.131)$$

or
$$(CY - E_h)Y^T = 0 \qquad (4.3.132)$$

or
$$C = E_h Y^T (YY^T)^{-1} \qquad (4.3.133)$$

since the $m \times m$ matrix YY^T is nonsingular. Clearly, C is given by the $(h+1)$th row of the $(m+g) \times m$ matrix $Y^T(YY^T)^{-1}$. To obtain the optimum equalizer, C must be determined for each value of h in the range 0 to $m+g-1$, and the vector C for the required equalizer selected as that which gives the minimum value of $|E - E_h|$.

The minimum mean-square error due to intersymbol interference alone is the minimum value of $k^2 |E - E_h|^2$ and is therefore

$$k^2 |CY - E_h|^2 = k^2 |E_h Y^T (YY^T)^{-1} Y - E_h|^2$$

$$= k^2 E_h [Y^T (YY^T)^{-1} Y - I][Y^T (YY^T)^{-1} Y - I]^T E_h^T$$

$$= k^2 E_h [I - Y^T (YY^T)^{-1} Y] E_h^T \qquad (4.3.134)$$

where I is an $(m+g) \times (m+g)$ identity matrix and $I - Y^T(YY^T)^{-1}Y$ is an $(m+g) \times (m+g)$ real symmetric matrix. It can be seen from Equations 4.3.134 and 4.3.115 that the minimum mean-square error is equal to k^2 times the $(h+1)$th component along the main diagonal of $I - Y^T(YY^T)^{-1}Y$.

Comparing Equations 4.3.128 and 4.3.133, it is clear that a linear equalizer, that minimizes the mean-square error due to intersymbol interference alone, also minimizes the mean-square error due to both intersymbol interference and noise, at high signal/noise ratios. This is, of course, to be expected since, at high signal/noise ratios, the intersymbol interference predominates over the noise, so that it is now more important to reduce the intersymbol interference than the noise.

4.3.8 Example

Consider the case where the z transform of the channel is

$$Y(z) = 1 + 2.5z^{-1} + z^{-2} \qquad (4.3.135)$$

and it is required to equalize the channel by means of an 11-tap linear feedforward transversal filter that minimizes the mean-square error due to intersymbol interference in its output signal.

Let Y be the 11×13 matrix whose ith row is

$$Y_{i-1} = \overbrace{0 \ldots 0}^{i-1} \; 1 \; 2.5 \; 1 \; 0 \ldots 0 \qquad (4.3.136)$$

so that the sampled impulse-response of the channel and equalizer is given by the 13-component row vector $E = CY$, where C is the 11-component row vector whose component values are the tap gains of the required equalizer. From the symmetry of $Y(z)$ it seems reasonable to select the vector C so as to minimize the distance between E and E_6, where E_6 is the 13-component row vector

$$E_6 = \overbrace{0 \ldots 0}^{6} \; 1 \; 0 \ldots 0 \qquad (4.3.137)$$

From Equation 4.3.133, the required vector C is now given by

$$C = \begin{matrix} -0.012 & 0.035 & -0.079 & 0.164 & -0.331 & 0.665 \end{matrix}$$
$$\begin{matrix} -0.331 & 0.164 & -0.079 & 0.035 & -0.012 \end{matrix} \qquad (4.3.138)$$

and from Equation 4.3.120,

$$E = \begin{matrix} -0.012 & 0.006 & -0.003 & 0.001 & -0.001 & 0.000 & 1.000 \end{matrix}$$
$$\begin{matrix} 0.000 & -0.001 & 0.001 & -0.003 & 0.006 & -0.012 \end{matrix} \qquad (4.3.139)$$

The corresponding 11-tap equalizer, designed through the separate equalization of the two factors $1 + \frac{1}{2}z^{-1}$ and $1 + 2z^{-1}$ as described in Section 4.3.3, is given by

$$C = \begin{matrix} -0.016 & 0.039 & -0.082 & 0.166 & -0.334 & 0.667 \end{matrix}$$
$$\begin{matrix} -0.334 & 0.166 & -0.082 & 0.039 & -0.016 \end{matrix} \qquad (4.3.140)$$

as can be seen from Equation 4.3.46, and now

$$E = \begin{matrix} -0.016 & 0 & 0 & 0 & 0 & 0 & 1 \end{matrix}$$
$$\begin{matrix} 0 & 0 & 0 & 0 & 0 & -0.016 \end{matrix} \qquad (4.3.141)$$

The latter equalizer achieves zero forcing in the sampled impulse-response of the equalized channel. Furthermore, it gives a lower peak distortion due to intersymbol interference in its output signal than does the other equalizer but a higher mean-square distortion and a higher mean-square error. There is not, however, much difference between the tap gains of the two equalizers.

For either equalizer, s_i is detected from the sample value x_{i+6} at the output of the equalizer, at time $t = (i+6)T$, where

$$x_{i+6} \simeq s_i + u_{i+6} \qquad (4.3.142)$$

and u_{i+6} is a Gaussian random variable with zero mean and variance $\sigma^2 |C|^2$ (Equation 4.3.95).

In the case of the equalizer that minimizes the mean-square error, the probability of error in the detection of s_i is approximately

$$Q\left(\frac{k}{\sigma|C|}\right) = Q\left(\frac{k}{0.855\sigma}\right) = Q\left(\frac{1.170k}{\sigma}\right) \qquad (4.3.143)$$

whereas, for the equalizer that achieves zero forcing, it is approximately

$$Q\left(\frac{k}{\sigma|C|}\right) = Q\left(\frac{k}{0.859\sigma}\right) = Q\left(\frac{1.164k}{\sigma}\right) \qquad (4.3.144)$$

The slightly lower peak distortion obtained with the latter equalizer reduces the very small advantage gained by the former due to its lower noise variance, so that, for practical purposes, the two arrangements achieve the same tolerance to additive white Gaussian noise, at high signal/noise ratios.

The important fact that has been illustrated by this example is that, with the accurate linear equalization of the channel, the tap gains of the transversal equalizer are effectively fixed by the channel and are not significantly affected by the particular design technique applied to the equalizer. This means that when a channel is equalized linearly, the tolerance to additive white Gaussian noise is essentially determined by the channel.

4.3.9   Equalizer giving the minimum mean-square distortion

At high signal/noise ratios, a linear equalizer should ideally minimize the *intersymbol interference* in the equalized signal. Since the $\{s_i\}$ are statistically independent and equally likely to have either binary value, the mean-square value of the intersymbol interference in a sample value of the equalized signal, expressed as a fraction of the square of the wanted signal, is

$$\frac{1}{e_h^2} \sum_{\substack{i=0 \\ i \neq h}}^{m+g-1} e_i^2$$

This is the *mean-square distortion* in the equalized signal, as given by Equation 4.3.10. Ideally then the equalizer should be designed to minimize this quantity.

Consider for the moment the most general case where the mean-square error and distortion, as given by Equations 4.3.11 and 4.3.10, respectively, are *not* necessarily associated with the impulse-response of an equalized channel. If the component e_h of the vector E in Equation 4.3.8 is greater than 1, the discrepancy between e_h and the corresponding component value 1 in E_h *increases* the mean-square error but *reduces* the mean-square distortion. Again, if $e_h < 1$, the contribution of the error in e_h to the mean-square error is in general much smaller than its contribution to the mean-square distortion. The mean-square error is not therefore *necessarily* equivalent to the mean-square distortion.

Clearly it is important to investigate more precisely the relationship between a linear equalizer designed for the minimum mean-square error and one designed for the minimum mean-square distortion.

Let the m tap gains of the linear equalizer that minimizes the mean-square error due to intersymbol interference be given by the m-component row vector C in Equation 4.3.133. From Equation 4.3.120, the sampled impulse-response of the channel and equalizer is given by the $(m+g)$-component row vector $E = CY$, where Y is given by Equation 4.3.119. The tap gains of the linear equalizer are selected to minimize $|E - E_h|$, where E_h is given by Equation 4.3.115. Thus E is the orthogonal projection of E_h on to the m-dimensional subspace spanned by the m vectors $\{Y_i\}$, and $E_h - E$ is *orthogonal* to E, as shown in Figure 4.16. The vectors are here shown as the points E_h and E in the plane containing the two vectors and the origin O of the $(m+g)$-dimensional Euclidean vector space.

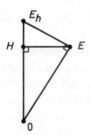

Fig. 4.16   *Relationship between the vectors E and E_h*

The orthogonal projection of E on to the line OE_h is the point H representing the $(m+g)$-component row vector

$$H = \overbrace{0\ldots0}^{h} e_h \, 0\ldots0 \qquad (4.3.145)$$

where e_h is the $(h+1)$th component of E. From Equation 4.3.115 and Figure 4.16, it is clear that

$$e_h < 1 \qquad (4.3.146)$$

so that the situation previously considered, where $e_h > 1$, cannot arise.

Consider now the case where the sampled impulse-response E of the channel and equalizer is *constrained* so that $e_h = 1$ for the same integer h as that in Equation 4.3.115, and the mean-square error due to intersymbol interference in the equalized signal is minimized *subject to this constraint*. The tap gains of the equalizer are here selected to minimize $|E - E_h|$, *given that* $e_h = 1$. Let the resultant sampled impulse-response of the channel and equalizer be given by the $(m+g)$-component row vector

$$F = f_0 \, f_1 \ldots f_{m+g-1} \qquad (4.3.147)$$

where, of course, $f_h = 1$. It now follows that

$$|F - E_h|^2 = \frac{1}{f_h^2} \sum_{\substack{i=0 \\ i \neq h}}^{m+g-1} f_i^2 \qquad (4.3.148)$$

so that the arrangement not only minimizes the mean-square error, *subject to the constraint* that $f_h = 1$, but it also minimizes the mean-square distortion.

If all the tap gains of the linear equalizer are multiplied by a constant d, this changes the components $\{f_i\}$ of the sampled impulse-response of the channel and equalizer to $\{df_i\}$, but it does not change the mean-square distortion in the equalized signal. Thus the equalizer that minimizes the mean-square distortion is not *uniquely* determined by the channel. It is therefore convenient to take as the equalizer that minimizes the mean-square distortion, the equalizer that minimizes the mean-square error subject to the constraint that $f_h = 1$.

Since $f_h = 1$, the orthogonal projection of the vector F on to the line OE_h, which joins the origin and E_h (as in Figure 4.16), is the vector (or point) E_h itself.

It can now be seen that there must be an $(m+g)$-component row vector

$$G_h = \overbrace{0 \ldots 0}^{h} \, g_h \, 0 \ldots 0 \qquad (4.3.149)$$

which is collinear with O and E_h, and such that the orthogonal projection of G_h on to the subspace spanned by the $m \, \{Y_i\}$ is a vector G whose orthogonal projection on to the line OG_h is the vector E_h. The vectors are shown in Figure 4.17. E is here the sampled impulse-response of the channel and equalizer, where the equalizer minimizes the mean-square error due to intersymbol interference. Clearly, $g_h > 1$ and, of course, g_h is in no way related to the integer g.

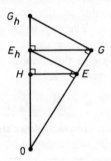

Fig. 4.17   *Relationship between the vectors E and G*

Since O, E_h and G_h are collinear, and since O, E and G are the orthogonal projections of O, E_h and G_h, respectively, on to the m-dimensional subspace, O being the projection of itself, it follows that O, E and G are collinear.

From the projection theorem, G is the vector in the m-dimensional subspace, at the minimum distance from G_h. If now there is another vector in the m-dimensional subspace whose orthogonal projection on to the line OG_h is the vector E_h and which is closer than G to E_h, then this vector must be closer than G to G_h, which is impossible. Thus G is the vector in the m-dimensional subspace, which is at the minimum distance from E_h, *subject to the constraint* that its $(h+1)$th component is fixed at unity. In other words, G is the vector F previously defined.

It follows immediately that O, E and F are collinear, so that

$$\frac{f_i}{e_i} = \frac{1}{e_h} \qquad (4.3.150)$$

for each i, bearing in mind that the $(h+1)$th components of F and E are 1 and e_h, respectively. From the similar triangles OG_hG, OE_hE, OGE_h and OEH, in Figure 4.17, it can be seen that

$$\frac{1}{e_h} = g_h \qquad (4.3.151)$$

It follows that the tap gains of the particular linear equalizer considered here, that minimizes the mean-square distortion, are g_h times those for the linear equalizer that minimizes the mean-square error.

It is now clear that the mean-square distortion corresponding to the vector F is the same as that corresponding to the vector E. Thus a linear equalizer that minimizes the mean-square error (due to intersymbol interference) in the equalized signal also minimizes the mean-square distortion. The converse does not however hold, so that an equalizer designed to minimize the mean-square distortion does not necessarily minimize the mean-square error and may in fact give a very large value for this error. Nevertheless, its tap gains are always in a constant ratio to the corresponding tap gains of the equalizer that minimizes the mean-square error, where this has the same number of taps and introduces the same delay into the equalized signal. Furthermore, since the multiplication of the tap gains of the linear equalizer by some constant d, multiplies both the output signal and the output noise components by d, it does not affect the probability of error in the detection of s_i from x_{i+h} at the output of the equalizer. Thus, for a given channel, an equalizer that minimizes the mean-square distortion gives exactly the same probability of error as the corresponding equalizer that minimizes the mean-square error. The latter equalizer is, of course, a special case of the former.

4.3.10 Linear equalization of phase distortion

Consider the data-transmission system in Figure 4.1 and suppose that the baseband channel introduces pure phase distortion (which may include some delay but no attenuation). Phase distortion is defined and described in some detail in Section 2.4. Pure phase distortion is an orthogonal transformation of the transmitted signal such that the individual received signal-elements remain orthogonal to each other and unchanged in level (energy). It can be considered as a rotation of the orthogonal axes of the Euclidean vector space, in which the vectors corresponding to the individual signal-elements are represented as points. There is therefore no change in the relative positions of these vectors in the vector space.

Suppose that the z transform of the baseband channel is

$$Y(z) = y_0 + y_1 z^{-1} + \ldots + y_g z^{-g} \qquad (4.3.152)$$

where

$$y_0^2 + y_1^2 + \ldots + y_g^2 = 1 \qquad (4.3.153)$$

The z transform of the ith individual received signal-element is now $s_i z^{-i} Y(z)$. It can be seen that any two individual received signal-elements with z-transforms $s_h z^{-h} Y(z)$ and $s_i z^{-i} Y(z)$ are the corresponding two sequences of $g + 1$ sample values $\{y_l\}$, the first multiplied by s_h and the second multiplied by s_i, the second sequence being also shifted $i - h$ places to the right with respect to the first sequence. The condition for any two received signal-elements to be orthogonal is therefore that

$$\sum_{l=0}^{g} y_l y_{l+j} \simeq 0 \qquad (4.3.154)$$

for any nonzero integer $j = i - h$, assuming that $y_l = 0$ for $i < 0$ and $i > g$. Obviously Equation 4.3.154 only holds exactly, for all values of j in the range 1 to g, when $g \to \infty$, but the errors normally become negligible when $g > 8$. With the assumed baseband channel, Equation 4.3.154 holds for *every* nonzero integer j.

If the linear transversal filter at the receiver has a z transform formed from $Y(z)$ by reversing the order of its coefficients, its z transform is

$$y_g + y_{g-1} z^{-1} + \ldots + y_0 z^{-g}$$

and the z transform of the ith individual signal-element at the output of this filter is

$$s_i z^{-i} Y(z)(y_g + y_{g-1} z^{-1} + \ldots + y_0 z^{-g}) \simeq s_i z^{-i-g} \qquad (4.3.155)$$

It follows that the z transform of the *linear equalizer* for the baseband channel is

$$C(z) = y_g + y_{g-1} z^{-1} + \ldots + y_0 z^{-g} \simeq z^{-g} Y^{-1}(z) \qquad (4.3.156)$$

With this equalizer at the receiver, the z transform of the channel and equalizer is

$$Y(z)C(z) \simeq z^{-g} \qquad (4.3.157)$$

so that the sample value at the output of the equalizer, at time $t = (i+g)T$, is

$$x_{i+g} \simeq s_i + u_{i+g} \qquad (4.3.158)$$

where u_{i+g} is a Gaussian random variable with zero mean and variance

$$\sigma^2|C|^2 = \sigma^2 \qquad (4.3.159)$$

as can be seen from Equations 4.3.153 and 4.3.156. Clearly, the $\{u_i\}$ have the same variance as the $\{w_i\}$. s_i is detected from the sign of x_{i+g}, and the probability of error in the detection of s_i is approximately

$$P_e = Q\left(\frac{k}{\sigma|C|}\right) = Q\left(\frac{k}{\sigma}\right) \qquad (4.3.160)$$

The linear equalizer is clearly also a *matched filter*, in the sense that it is matched both to the channel and to each received signal-element, which means that in the presence of additive white Gaussian noise, as assumed here, the output signal/noise ratio from the equalizer is maximized.[141] Since furthermore there is negligible intersymbol interference at the output of the equalizer, it follows that the linear equalizer and detector together achieve the best available tolerance to additive white Gaussian noise. There is no better detection process for the given received signal.

Consider now a received data signal comprising a finite sequence of signal elements that extend over a total of n sample values at the input to the equalizer. The individual signal-elements here can now be represented as vectors in an n-dimensional Euclidean vector space. Furthermore, as is shown in Section 3.2.7, the value of the orthogonal projection of the corresponding noise vector W on to any direction in this vector space, is a Gaussian random variable with zero mean and variance σ^2. Since each individual received signal-element at the input to the equalizer is a vector of length k and having one of two possible values, one of which is the negative of the other, and since the vectors corresponding to different signal-elements are orthogonal, it is evident without further analysis that the probability of error in the optimum detection of each signal-element is given by

P_e in Equation 4.3.160 and is furthermore unaffected by the presence or absence of the other vectors or by the particular directions of any individual vectors in the vector space. If there is *no* distortion, and again no attenuation, each vector lies along a different one of the orthogonal axes of the Euclidean vector space and has just one nonzero component (sample value) which is $\pm k$. No linear equalizer is now required, and the probability of error in the optimum detection of each received signal-element is again given by P_e in Equation 4.3.160. Clearly, so long as the appropriate detection process is used, phase distortion does not change the tolerance to additive white Gaussian noise.

Consider now the two noise components u_i and u_{i+j}, at the time instants iT and $(i+j)T$ at the output of the linear equalizer with z-transform $C(z)$ in Equation 4.3.156. Clearly,

$$u_i = w_i y_g + w_{i-1} y_{g-1} + \ldots + w_{i-g} y_0 \qquad (4.3.161)$$

and

$$u_{i+j} = w_{i+j} y_g + w_{i+j-1} y_{g-1} + \ldots + w_{i+j-g} y_0 \qquad (4.3.162)$$

Since $\overline{w_i} = 0$, $\overline{u_i} = \overline{u_{i+j}} = 0$, where $\overline{x}$ is the expected or mean value of x. Since the $\{w_i\}$ are statistically independent Gaussian random variables, $\overline{w_i w_{i+j}} = 0$ for any $j \neq 0$. Thus, when $j \neq 0$,

$$\overline{u_i u_{i+j}} = \overline{(w_i y_g + \ldots + w_{i-g} y_0)(w_{i+j} y_g + \ldots + w_{i+j-g} y_0)}$$

$$= \sigma^2 \sum_{l=0}^{g} y_l y_{l+j} \simeq 0 \qquad (4.3.163)$$

as can be seen from Equation 4.3.154. Hence the noise components $\{u_i\}$ at the output of the equalizer are uncorrelated and therefore statistically independent Gaussian random variables with zero mean.

The important results may now be summarized as follows. The signal distortion represented by $Y(z)$ is pure phase distortion, with no signal attenuation or gain, and the equalizer with z-transform $C(z)$ is a pure phase equalizer whose output noise components are statistically independent Gaussian random variables with zero mean and variance σ^2, just like the noise components $\{w_i\}$. The pure phase equalizer therefore introduces no change in the level or other statistical properties of the Gaussian noise samples, nor does it change the level of the data signal. Clearly, with a linear equalizer, the presence of any phase distortion does not affect the tolerance of the system to additive white Gaussian noise, since it is removed by the equalizer without affecting either the noise or the level of the signal.

Most practical channels introduce, in addition to phase distortion, some degree of amplitude distortion. Amplitude distortion is defined and described in some detail in Section 2.4. In the presence of amplitude distortion, the linear equalizer for the channel is *not* matched to the channel or received signal, which necessarily implies that the signal/noise ratio at the output of the equalizer is *less* than that at the output of the corresponding matched filter (neglecting here the level of the intersymbol interference). This means that, with a linear equalizer followed by the appropriate detector, the probability of error in the detection of a received signal-element is now *greater* than that where only a single element is received and is detected by a matched-filter detector. In other words, when a channel is equalized linearly at the receiver, *any* amplitude distortion introduced by the channel (with no signal attenuation or gain) always *reduces* the tolerance of the system to additive white Gaussian noise. Furthermore, in the presence of amplitude distortion, a linear equalizer no longer gives the best available tolerance to Gaussian noise, as can be seen from Chapters 2 and 3. Indeed, the tolerance to noise may now fall far below that obtainable with the optimum detection process for the particular received signal, so that a linear equalizer is no longer the most suitable arrangement. With the optimum detection process, the presence of amplitude distortion may or may not noticeably reduce the tolerance of the system to additive white Gaussian noise, depending upon the severity of the distortion.

4.4 NONLINEAR EQUALIZERS

4.4.1 Decision-directed cancellation of intersymbol interference

During the last few years a considerable amount of work has been carried out on nonlinear equalizers and it has been shown that these can sometimes achieve a much better tolerance to additive noise than linear equalizers.[108-135] To understand and hence to be able to analyse the method of operation of nonlinear equalizers, it is necessary first to study the basic principles of decision-directed cancellation of intersymbol interference.

Consider the data-transmission system shown in Figure 4.1, where

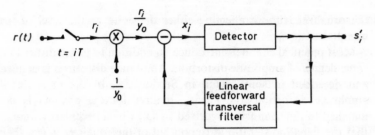

Fig. 4.18 Receiver using nonlinear equalization by decision-directed cancellation of intersymbol interference

the equalizer at the receiver is now the nonlinear equalizer shown in Figure 4.18. The signals shown here are those at the time instant $t = iT$, s_i' being the detected value of s_i. The nonlinear equalizer is implemented as a linear feedforward transversal filter fed from the output of the detector, as shown in Figure 4.19 (where $v_j = y_j/y_0$). The output signal from this filter is subtracted from the corresponding sample value at the output of the multiplier, to give the input signal to the detector. The equalizer therefore operates by *quantized-feedback correction*, removing the intersymbol interference from the detector input signal, as will now be explained.

If the detector is removed from Figure 4.19, the linear feedback transversal equalizer of Figure 4.11 is obtained. The arrangement

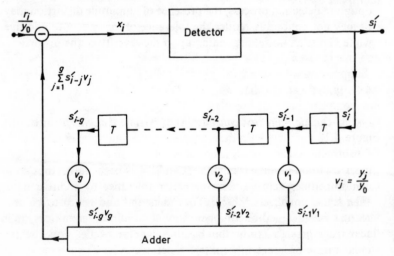

Fig. 4.19 Nonlinear equalizer and detector

in Figure 4.19 is therefore a *nonlinear* feedback transversal equalizer. The equalizer is nonlinear because the detector, which is a nonlinear device, is included in the feedback path of the equalizer.

The sampled impulse-response of the baseband channel is given by the $(g+1)$-component row vector

$$V = y_0 \; y_1 \ldots y_g \qquad (4.4.1)$$

where, as before, $y_0 \neq 0$ and $y_i = 0$ for $i < 0$ and $i > g$. Thus the received sample value at the input to the multiplier in Figure 4.18, at time $t = iT$, is

$$r_i = s_i y_0 + \sum_{j=1}^{g} s_{i-j} y_j + w_i \qquad (4.4.2)$$

and the corresponding sample value at the output of the multiplier is

$$\frac{r_i}{y_0} = s_i + \sum_{j=1}^{g} s_{i-j}\frac{y_j}{y_0} + \frac{w_i}{y_0}$$

$$= s_i + \sum_{j=1}^{g} s_{i-j} v_j + \frac{w_i}{y_0} \qquad (4.4.3)$$

where $v_j = y_j/y_0$. s_i is here the wanted signal, $\sum_{j=1}^{g} s_{i-j} v_j$ is the intersymbol interference and w_i/y_0 is the noise component. The sampled impulse-response of the baseband channel, sampler and multiplier is

$$\frac{1}{y_0}V = 1 \; v_1 \; v_2 \ldots v_g \qquad (4.4.4)$$

and this is assumed to be known at the receiver.

Suppose now that the $\{s_{i-j}\}$, for $j = 1, 2, \ldots, g$, have been correctly detected, so that $s'_{i-j} = s_{i-j}$ for each j. The receiver now has complete prior knowledge of the intersymbol interference in r_i/y_0, as can be seen from Equation 4.4.3. The transversal filter in Figure 4.18 eliminates the intersymbol interference present in r_i/y_0 by subtraction, the signal subtracted from r_i/y_0 being formed as shown in Figure 4.19. This is a process of decision-directed cancellation of intersymbol interference, and operates as follows.

The signals marked in Figure 4.19 are those present at the time instant $t = iT$. The detected element values $\{s'_i\}$ are here fed to a linear transversal filter containing g stores. The filter operates in a similar manner to that in Figure 4.4. Associated with each store in Figure 4.19 is a multiplier that multiplies the output signal s'_{i-j}

from the store by v_j, so that the output signal from the transversal filter, at time $t = iT$, is

$$\sum_{j=1}^{g} s'_{i-j} v_j$$

and the input signal to the detector is

$$x_i = \frac{r_i}{y_0} - \sum_{j=1}^{g} s'_{i-j} v_j = s_i + \frac{w_i}{y_0} \qquad (4.4.5)$$

so long as $s'_{i-j} = s_{i-j}$ for each j. Thus the intersymbol interference in r_i/y_0 is eliminated, and s_i is detected from x_i.

It can be seen from Equations 4.4.3 and 4.4.5 that both r_i/y_0 and x_i have the *same* wanted-signal component s_i and noise component w_i/y_0 so that the nonlinear equalizer removes the intersymbol interference *without changing* the signal/noise ratio.

If s_i is correctly detected, the receiver knows the components $\{s_i v_j\}$ of the signal element in the following g samples $\{r_j/y_0\}$ at the input to the nonlinear filter. Each of these components is now cancelled (removed by subtraction) at the appropriate time instant, so that the intersymbol interference caused by the ith received signal-element in the following g elements is eliminated. Thus the sample value at the output of the nonlinear filter at time $t = (i+1)T$ is

$$x_{i+1} = s_{i+1} + \frac{w_{i+1}}{y_0} \qquad (4.4.6)$$

and s_{i+1} is detected from x_{i+1} with no intersymbol interference. The components $\{s_{i+1} v_j\}$ in the following $\{r_j/y_0\}$ are subsequently cancelled in the nonlinear filter, and so on. It is clear that so long as the received signal-elements are correctly detected, their intersymbol interference in the following elements is eliminated and the channel is accurately equalized.

To start this process of nonlinear equalization, a known sequence of more than g $\{s_i\}$ is transmitted and the intersymbol interference introduced by the received signal-elements is cancelled automatically in the nonlinear filter, without the detection of the corresponding $\{s_i\}$. The channel is now correctly equalized, and the following received elements are detected and cancelled as just described.

It is clear that when a received signal-element is correctly detected and cancelled, the resultant sample values become as though the signal had never been transmitted. Thus, when a received signal-element is being detected, following the correct detection and cancellation of the preceding elements, the received signal is as though the element being detected is the *first* of the received sequence of elements.

In the detector (Figure 4.18), s_i is detected as k or $-k$, depending upon whether x_i is positive or negative, respectively. As can be seen from Equation 4.4.5, an error occurs in the detection of s_i when x_i has the opposite sign to s_i, and therefore when the Gaussian noise component w_i/y_0 has a magnitude greater than k and the opposite sign to s_i. But w_i/y_0 is a Gaussian random variable with zero mean and variance σ^2/y_0^2, so that the probability of error in the detection of a received signal-element, following the correct detection of the preceding g elements, is

$$
\begin{aligned}
P_e &= \int_k^\infty \frac{1}{\sqrt{(2\pi y_0^{-2}\sigma^2)}} \exp\left(-\frac{w^2}{2y_0^{-2}\sigma^2}\right) dw \\
&= \int_{k|y_0|/\sigma}^\infty \frac{1}{\sqrt{2\pi}} \exp(-\tfrac{1}{2}w^2) dw \\
&= Q\left(\frac{k|y_0|}{\sigma}\right)
\end{aligned}
\tag{4.4.7}
$$

When a received signal-element is incorrectly detected, its inter-symbol interference in the following elements, instead of being eliminated, is doubled, and this greatly increases the probability of error in their detection. Errors therefore tend to occur in bursts, and the system suffers from error-extension effects. However, at high signal/noise ratios and for values of the sampled impulse-response of the channel where y_0 is one of the larger components of V, the error extension effects are not normally serious and do not usually increase the average probability of error in the detection of an element by more than ten times. This corresponds to a reduction of up to about 1 dB in tolerance to additive Gaussian noise, so that no serious inaccuracy is introduced here by taking the average error probability as P_e.

An important property of the nonlinear equalizer just described is that it operates correctly even when $Y(z)$, the z transform of the channel, has all its roots outside the unit circle in the z plane. The error extension effects are however now more severe than when some or all of the roots lie inside the unit circle.

When all the roots of $Y(z)$ lie *inside* the unit circle, the linear feedforward transversal equalizer for the channel introduces no delay and has a z transform approximately equal to $Y^{-1}(z)$. The first term of $Y^{-1}(z)$ is clearly $1/y_0$. This means that if the equalizer is as shown in Figure 4.4, its first tap gain is

$$c_0 = \frac{1}{y_0} \qquad (4.4.8)$$

and the sample value r_i at the input to the equalizer is multiplied by $1/y_0$. Since one or more of the remaining tap gains of the linear equalizer must be nonzero, it is clear that

$$|C| > \frac{1}{|y_0|} \qquad (4.4.9)$$

and the noise variance at the equalizer output exceeds $\sigma^2/|y_0|^2$. From Equation 4.3.98, the error probability in the detection of s_i here exceeds $Q\left(\dfrac{k|y_0|}{\sigma}\right)$ and so is greater than P_e in Equation 4.4.7. Thus, whenever all the roots of the z transform of the channel lie *inside* the unit circle in the z plane and the signal/noise ratio is high so that error-extension effects may be neglected, the nonlinear equalizer just described gives a *better* tolerance to additive white Gaussian noise than a linear equalizer.

In practice, $Y(z)$ frequently has roots outside the unit circle, and the nonlinear equalizer now often gives a lower tolerance to additive white Gaussian noise than a linear equalizer. Under these conditions, a better performance is sometimes obtained by detecting s_i, not from the first of the $g+1$ received sample values $\{r_j\}$ that are dependent on s_i, as in Figure 4.19, but from the appropriate one of the following g sample values.

Consider now the receiver shown in Figure 4.20. The multiplier is omitted here and s_i is detected from the sample value x_{i+h} at the output of the nonlinear equalizer, at time $t = (i+h)T$, where h is an integer in the range 0 to g. The received sample value at time $t = (i+h)T$ is

$$r_{i+h} = \sum_{j=0}^{h-1} s_{i+h-j}y_j + s_i y_h + \sum_{j=h+1}^{g} s_{i+h-j}y_j + w_{i+h} \qquad (4.4.10)$$

where $s_i y_h$ is the wanted signal component, and the two terms

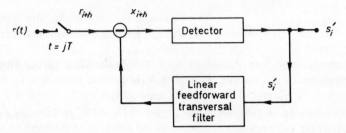

Fig. 4.20 Receiver using only partial nonlinear equalization of the channel

involving $s_{i+h-j}y_j$ represent the intersymbol interference. With the correct detection of the $\{s_j\}$ for $j < i$, s_{i+h-j} is known for $j = h+1, h+2, \ldots, g$, but not for $j = 0, 1, \ldots, h-1$, so that the second (right hand) intersymbol-interference term in Equation 4.4.10 is known, but not the other. The nonlinear filter in Figure 4.20 eliminates the second intersymbol-interference term, but it cannot, of course, eliminate the remaining intersymbol interference, since the corresponding $\{s_j\}$ have not yet been detected. Thus, assuming the correct detection of the preceding $g - h$ signal-elements, the sample value at the output of the nonlinear filter at time $t = (i+h)T$ is

$$x_{i+h} = \sum_{j=0}^{h-1} s_{i+h-j}y_j + s_i y_h + w_{i+h} \qquad (4.4.11)$$

and the channel is only *partially* equalized.

s_i is now detected as k or $-k$, depending upon whether x_{i+h}/y_h is positive or negative, respectively. When $|y_h| \gg |y_j|$, for $j = 0$, $1, \ldots, h-1$, this arrangement usually gives a better tolerance to noise than that where s_i is detected from x_i, with the complete elimination of all intersymbol interference.

In the particular case where $h = 0$ in Figure 4.20, s_i is detected at time $t = iT$ from the sample value

$$x_i = s_i y_0 + w_i \qquad (4.4.12)$$

at the output of the nonlinear equalizer, assuming the correct detection of the g preceding $\{s_j\}$. An error now occurs in the detection of s_i when the noise component w_i has the opposite sign to $s_i y_0$ and a magnitude greater than $k|y_0|$. As before, w_i is a Gaussian random variable with zero mean and variance σ^2. Thus the probability of error in the detection of s_i, given the correct detection of the g preceding $\{s_j\}$, is

$$P_e = Q\left(\frac{k|y_0|}{\sigma}\right) \qquad (4.4.13)$$

As can be seen from Equation 4.4.7, this is the same as the error probability of the receiver in Figure 4.18, so that when $h = 0$, the receivers in Figures 4.18 and 4.20 are equivalent. The reason for describing the arrangement in Figure 4.18 is to emphasize the relationship between this and some of the nonlinear equalizers to be considered later.

An important property of the nonlinear equalizers studied here is that they can be used to equalize a channel whose z transform has one or more roots *on* the unit circle in the z plane. Such a channel cannot be equalized linearly.

Nonlinear equalizers of the type just considered will sometimes be referred to as *pure* nonlinear equalizers, to distinguish them from the more complex equalizers to be considered presently.

4.4.2 Example

Consider the case where the z transform of the baseband channel in Figure 4.1 is

$$Y(z) = 1 + 2.5z^{-1} + z^{-2} \qquad (4.4.14)$$

so that the z transform of the ith received signal-element, at the output of the sampler, is

$$s_i z^{-i} Y(z) = s_i z^{-i} + 2.5 s_i z^{-i-1} + s_i z^{-i-2} \qquad (4.4.15)$$

Assume that the receiver is as in Figure 4.20. The received sample value at time $t = iT$, at the input to the equalizer, is now

$$r_i = s_i + 2.5s_{i-1} + s_{i-2} + w_i \qquad (4.4.16)$$

where w_i is a Gaussian random variable with zero mean and variance σ^2, as before.

If $h = 0$ in Figure 4.20, s_i is detected from the sample value x_i at the output of the nonlinear equalizer, at time $t = iT$. With the correct detection of s_{i-1} and s_{i-2}, all intersymbol interference is eliminated from x_i and

$$x_i = r_i - 2.5s_{i-1} - s_{i-2} = s_i + w_i \qquad (4.4.17)$$

When $x_i > 0$, s_i is detected as k, and when $x_i < 0$, s_i is detected as $-k$. An error occurs in the detection of s_i when w_i has a magnitude greater than k and the opposite sign to s_i. Thus the probability of error in the detection of s_i from x_i is

$$P_e = Q\left(\frac{k}{\sigma}\right) \qquad (4.4.18)$$

The received sample value at time $t = (i+1)T$ is

$$r_{i+1} = s_{i-1} + 2.5s_i + s_{i+1} + w_{i+1} \qquad (4.4.19)$$

If $h = 1$ in Figure 4.20, s_i is detected from the sample value x_{i+1} at the output of the nonlinear equalizer, at time $t = (i+1)T$. With the correct detection of s_{i-1} its intersymbol interference in x_{i+1} is eliminated, and

$$x_{i+1} = r_{i+1} - s_{i-1} = 2.5s_i + s_{i+1} + w_{i+1} \qquad (4.4.20)$$

When $x_{i+1} > 0$, s_i is detected as k, and when $x_{i+1} < 0$, s_i is detected as $-k$. At high signal/noise ratios and with the correct detection of s_{i-1}, practically all errors occur when

$$s_{i+1} = -s_i \qquad (4.4.21)$$

and now

$$x_{i+1} = 1.5s_i + w_{i+1} \qquad (4.4.22)$$

Under these conditions, an error occurs in the detection of s_i from x_{i+1} when w_{i+1} has a magnitude greater than $1.5k$ and the opposite sign to s_i. Since there is a probability of $\frac{1}{2}$ that $s_{i+1} = -s_i$, the average probability of error in the detection of s_i from x_{i+1} is approximately equal to half the conditional probability of error, given that $s_{i+1} = -s_i$. Thus the probability of error in the detection of s_i is approximately $\frac{1}{2}Q(1.5k/\sigma)$. Since, at high signal/noise ratios with additive Gaussian noise, a change in the error probability by a factor of 2 corresponds to a change of less than $\frac{1}{2}$ dB in the signal/noise ratio, which is not very significant, the probability of error can be taken to be

$$P_e = Q\left(\frac{1.5k}{\sigma}\right) \qquad (4.4.23)$$

It can be seen from Equations 4.4.18 and 4.4.23 that, for a given error probability and a given value of k, the second of the two arrangements just considered has a noise level of 3.5 dB above that of the other arrangement. It is therefore clearly the better system.

The error probabilities given by Equations 4.4.18 and 4.4.23 assume the correct detection of the immediately preceding elements. As before, when a signal element is incorrectly detected, there is a considerable increase in the probability of error in the detection of the immediately following elements, so that errors tend to occur in bursts. However, at high signal/noise ratios with additive Gaussian noise, the error extension effects cause a reduction in tolerance to noise of only a fraction of 1 dB and so can be neglected here.

Comparing Equations 4.3.143, 4.3.144, 4.4.18 and 4.4.23, it can be seen that for the particular channel considered here, the non-linear equalizer in which s_i is detected from x_i gives a tolerance to additive white Gaussian noise approximately 1.3 dB below that of a linear equalizer, whereas the nonlinear equalizer in which s_i is detected from x_{i+1} gives a tolerance to additive white Gaussian noise approximately 2.2 dB above that of a linear equalizer.

Clearly, the arrangement of nonlinear equalization using decision-directed cancellation of intersymbol interference can sometimes gain an advantage in tolerance to Gaussian noise over a linear equalizer, an advantage being always obtained at high signal/noise ratios, when the z transform of the channel has all its roots inside the unit circle and the nonlinear equalizer fully equalizes the channel. Furthermore, the nonlinear equalizer is much simpler to implement than the linear equalizer, since it generally requires far fewer taps in the transversal filter. However, when the channel introduces nearly pure phase distortion, the linear equalizer gives a much better tolerance to white Gaussian noise, and over typical channels there is probably not very much to choose between the two techniques, as far as their performances are concerned.

4.4.3   Zero forcing

The improved tolerance to noise obtained by detecting s_i from x_{i+1} rather than from x_i, in the example just considered, suggests that a further improvement in tolerance to noise may be obtained with this arrangement by adding a linear filter at the input to the non-linear equalizer, to remove the intersymbol-interference component in x_{i+1} that is not eliminated by the nonlinear equalizer. A linear

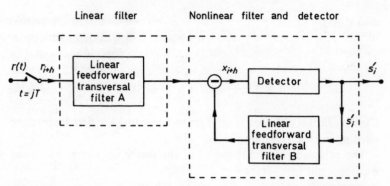

Fig. 4.21   Nonlinear equalizer containing both a linear and a nonlinear filter

filter is therefore added at the input to the equalizer in Figure 4.20 to give the arrangement shown in Figure 4.21. The signals shown here are those at the time instant $t = (i+h)T$.

The received sample values $\{r_j\}$ are fed through the linear filter which modifies the sampled impulse-response of the channel by setting to zero the components preceding that of the largest magnitude in this response, and changing the remaining components by a constant factor. Thus the linear filter partially equalizes the channel and the nonlinear filter then completes the equalization process. The nonlinear filter operates on the sampled impulse-response of the channel and linear filter in exactly the same way as that in which the nonlinear equalizer in Figures 4.18 and 4.19 operates on the sampled impulse-response of the channel and multiplier. Furthermore, the basic structure of the nonlinear filter in Figure 4.21 is the same as that in Figure 4.19, the first tap of the linear feedforward transversal filter B occurring after a delay of T seconds. The linear and nonlinear filters in Figure 4.21 are together assumed to achieve the accurate equalization of the channel.

Suppose that y_l is the component of V (Equation 4.4.1) of the greatest magnitude, and let the z transform of the required linear filter with n taps be

$$D(z) = C(z)E(z) \qquad (4.4.24)$$

where $C(z)$ is the z transform of the m-tap linear equalizer for the baseband channel, such that

$$Y(z)C(z) \simeq z^{-h} \qquad (4.4.25)$$

and where

$$E(z) = \frac{1}{y_l}(y_l + y_{l+1}z^{-1} + \ldots + y_g z^{-g+l})$$

$$= 1 + \frac{y_{l+1}}{y_l}z^{-1} + \ldots + \frac{y_g}{y_l}z^{-g+l} \qquad (4.4.26)$$

Clearly, $E(z)$ is obtained from $Y(z)$ by removing the first l terms and multiplying the remaining terms by z^l/y_l.

The sampled impulse-response of the linear filter is given by the n-component row vector

$$D = d_0 \ d_1 \ldots d_{n-1} \qquad (4.4.27)$$

with z-transform

$$D(z) = d_0 + d_1 z^{-1} + \ldots + d_{n-1}z^{-n+1} \qquad (4.4.28)$$

satisfying Equation 4.4.24. Clearly

$$n = m+g-l \qquad (4.4.29)$$

From Equations 4.4.24 and 4.4.25, the z transform of the channel and linear filter is

$$Y(z)D(z) = Y(z)C(z)E(z) \simeq z^{-h}E(z) \qquad (4.4.30)$$

which is the required response together with a delay of hT seconds. Clearly, the sampled impulse-response of the channel and linear filter is given by the $(g-l+1)$-component row vector

$$E = 1 \ \frac{y_{l+1}}{y_l} \ \frac{y_{l+2}}{y_l} \ldots \frac{y_g}{y_l} \qquad (4.4.31)$$

the delay of hT seconds being neglected here.

The z transform of the ith received signal-element at the output of the linear filter in Figure 4.21 is $s_i z^{-i-h}E(z)$. Whenever $E(z)$ has more than one nonzero term, as is usually the case, there is inter-symbol interference between the signal elements at the output of the linear filter. This intersymbol interference is removed in the nonlinear filter, which operates exactly as the nonlinear equalizer in Figure 4.18, except that it equalizes a response with z-transform $z^{-h}E(z)$ instead of the response with z-transform $y_0^{-1}Y(z)$. The linear and nonlinear filters in Figure 4.21 together achieve the accurate equalization of the baseband channel in Figure 4.1.

Following the correct detection of s_{i-j}, for $j = 1, 2, \ldots, g-l$, the sample value at time $t = (i+h)T$ at the input to the detector in Figure 4.21 is

$$x_{i+h} \simeq s_i + u_{i+h} \qquad (4.4.32)$$

where u_{i+h} is the noise component at time $t = (i+h)T$ at the output of the linear filter. s_i is detected as k or $-k$ depending upon whether x_{i+h} is positive or negative, respectively, and the components of the ith received signal-element, in the following sample values at the output of the linear filter, are then eliminated by cancellation in the nonlinear filter.

u_{i+h} in Equation 4.4.32 is a Gaussian random variable with zero mean and variance

$$\eta^2 = \sigma^2 \sum_{i=0}^{n-1} d_i^2 = \sigma^2 |D|^2 \qquad (4.4.33)$$

where $|D|$ is the length of the vector D. Thus the probability of error in the detection of a received signal-element, following the correct detection of the preceding $g-l$ elements, is

$$P_e = \int_k^\infty \frac{1}{\sqrt{(2\pi\eta^2)}} \exp\left(-\frac{u^2}{2\eta^2}\right) du$$

$$= Q\left(\frac{k}{\eta}\right) = Q\left(\frac{k}{\sigma|D|}\right) \qquad (4.4.34)$$

The error extension effects are not normally serious here, so that no significant inaccuracy is introduced by taking the average probability of error as P_e.

This combination of linear and nonlinear filters often gives a useful improvement in tolerance to additive Gaussian noise over both the linear and nonlinear equalizers previously described. In the particular case where the z transform of the channel is $1 + 2.5z^{-1} + z^{-2}$, it gains an advantage of 2.9 dB over the linear equalizer.

4.4.4 *Linear and nonlinear equalization of complementary factors of the channel response*

When the linear filter in Figure 4.21 sets to zero the components preceding the largest in the sampled impulse-response of the channel, it is not necessary that the remaining components should be changed only by a constant factor. An alternative technique is as follows.

The z transform of the sampled impulse-response of the channel is

$$Y(z) = Y_1(z)Y_2(z) \tag{4.4.35}$$

where all the roots (zeros) of $Y_1(z)$ lie inside the unit circle in the z plane, and all the roots of $Y_2(z)$ lie outside the unit circle. Let

$$Y_1(z) = 1 + p_1 z^{-1} + p_2 z^{-2} + \ldots + p_{g-f} z^{-g+f} \tag{4.4.36}$$

and

$$Y_2(z) = q_0 + q_1 z^{-1} + q_2 z^{-2} + \ldots + q_f z^{-f} \tag{4.4.37}$$

where f is an integer in the range 0 to g. All coefficients here are real.

The n-tap linear filter in Figure 4.21 now equalizes the factor $Y_2(z)$ of the channel z-transform, and the nonlinear filter equalizes the factor $Y_1(z)$. The z-transform $D(z)$ of the linear filter therefore satisfies

$$Y_2(z)D(z) \simeq z^{-h} \tag{4.4.38}$$

where $h = f + n - 1$, so that the z transform of the channel and linear filter is

$$Y(z)D(z) \simeq z^{-h}Y_1(z) \tag{4.4.39}$$

This is equalized by the nonlinear filter. From Equation 4.4.36, the sampled impulse-response of the channel and linear filter is given by the $(g - f + 1)$-component row vector

$$E = 1 \quad p_1 \quad p_2 \cdots p_{g-f} \tag{4.4.40}$$

the delay of hT seconds being neglected here. Thus the first nonzero component in the sampled impulse-response of the channel and linear filter is unity and is one of the larger components, although it need not now be the largest as in Equation 4.4.31.

As before, a received signal-element is detected from the sample value containing the first nonzero component of that element at the output of the linear filter, and the components of the element in the following sample values at the output of the linear filter are then removed in the nonlinear filter.

The sample value at time $t = (i + h)T$ at the input to the detector in Figure 4.21 is given by x_{i+h} in Equation 4.4.32, where u_{i+h} is again a Gaussian random variable with zero mean and variance η^2. η^2 is given by Equation 4.4.33, where D is the appropriate sampled impulse-response. As before, the average probability of error in the detection of a received signal-element can be taken to be P_e in Equation 4.4.34, with no significant inaccuracy.

The linear filter here normally has fewer taps than that in the arrangement of zero forcing or than the corresponding linear equalizer for the channel, and the tolerance of the system to additive white Gaussian noise is more often than not a little better than that of zero forcing (Section 4.4.3). This technique is therefore preferable to that of zero forcing.

If $Y_1(z)$ and $Y_2(z)$ are equalized by the linear and nonlinear filters, respectively, the arrangement always has a lower tolerance to additive white Gaussian noise than that just described. This can be seen intuitively from the fact that if $Y(z) = Y_1(z)$, the nonlinear equalizer in Figure 4.18 gives a better tolerance to Gaussian noise than a linear equalizer, whereas if $Y(z) = Y_2(z)$, a linear equalizer generally gives a better tolerance to noise.

4.4.5 Example

Consider the case where the z transform of the baseband channel in Figure 4.1 is

$$Y(z) = 1 + 2\tfrac{1}{2}z^{-1} + z^{-2} = (1 + \tfrac{1}{2}z^{-1})(1 + 2z^{-1}) \quad (4.4.41)$$

In the preferred arrangement of pure nonlinear equalization by decision-directed cancellation of intersymbol interference (Figure 4.20), s_i is detected from the sample value x_{i+1} (Equation 4.4.20) at the output of the nonlinear filter, at time $t = (i+1)T$. The intersymbol interference of s_{i-1} has here been eliminated from x_{i+1} but not the intersymbol interference of s_{i+1}. The uncancelled intersymbol-interference may, if required, be removed by a linear filter, as in the arrangement of zero forcing described in Section 4.4.3. The resultant z-transform of the channel and linear filter (Figure 4.21) is now $z^{-h}(2.5 + z^{-1})$, and this is equalized by the nonlinear filter. Alternatively, the factor $1 + 2z^{-1}$ of $Y(z)$ may be equalized linearly and the factor $(1 + \tfrac{1}{2}z^{-1})$ may be equalized nonlinearly. This is the linear and nonlinear equalization of complementary factors of the channel response, and will now be considered in some detail.

If the linear filter has 7 taps, its z transform is

$$D(z) = \tfrac{1}{128} - \tfrac{1}{64}z^{-1} + \tfrac{1}{32}z^{-2} - \tfrac{1}{16}z^{-3} + \tfrac{1}{8}z^{-4} - \tfrac{1}{4}z^{-5} + \tfrac{1}{2}z^{-6} \quad (4.4.42)$$

Then the z transform of the channel and linear filter is

$$(1+2\tfrac{1}{2}z^{-1}+z^{-2})(\tfrac{1}{128}-\tfrac{1}{64}z^{-1}+\tfrac{1}{32}z^{-2}-\tfrac{1}{16}z^{-3}$$
$$+\tfrac{1}{8}z^{-4}-\tfrac{1}{4}z^{-5}+\tfrac{1}{2}z^{-6})$$
$$= \tfrac{1}{128}+\tfrac{1}{256}z^{-1}+z^{-7}+\tfrac{1}{2}z^{-8}$$
$$\simeq z^{-7}+\tfrac{1}{2}z^{-8} \qquad (4.4.43)$$

and the z transform of the ith received signal-element at the output of the linear filter is approximately

$$s_i z^{-i}(z^{-7}+\tfrac{1}{2}z^{-8}) = s_i z^{-i-7}+\tfrac{1}{2}s_i z^{-i-8} \qquad (4.4.44)$$

Thus the sample value at the output of the linear filter, at time $t = (i+7)T$, is

$$v_{i+7} \simeq s_i+\tfrac{1}{2}s_{i-1}+u_{i+7} \qquad (4.4.45)$$

where u_{i+7} is a Gaussian random variable with zero mean and variance

$$\eta^2 = \sigma^2[(\tfrac{1}{128})^2+(\tfrac{1}{64})^2+(\tfrac{1}{32})^2+(\tfrac{1}{16})^2+(\tfrac{1}{8})^2+(\tfrac{1}{4})^2+(\tfrac{1}{2})^2]$$
$$= 0.333\sigma^2 \qquad (4.4.46)$$

so that

$$\eta = 0.577\sigma \qquad (4.4.47)$$

With the correct detection of s_{i-1} and the cancellation of its intersymbol interference in v_{i+7} by the nonlinear filter, the sample value at the input to the detector, at time $t = (i+7)T$, is

$$x_{i+7} = v_{i+7}-\tfrac{1}{2}s_{i-1} \simeq s_i+u_{i+7} \qquad (4.4.48)$$

It can be seen that the nonlinear filter has removed the intersymbol interference term in v_{i+7} to give x_{i+7}, without changing either the wanted-signal component s_i or the noise component u_{i+7}. Thus the nonlinear filter does not change the signal/noise ratio.

The probability of error in the detection of s_i is now

$$P_e = Q\left(\frac{k}{\eta}\right) = Q\left(\frac{k}{0.577\sigma}\right) = Q\left(\frac{1.733k}{\sigma}\right) \qquad (4.4.49)$$

From Equations 4.3.143 and 4.3.144, the advantage in tolerance to additive white Gaussian noise, gained by the combination of linear and nonlinear equalizers over a linear equalizer, for the given channel at high signal/noise ratios, is approximately 3.4 dB. This compares with an advantage of 2.9 dB gained by the corresponding nonlinear equalizer using zero forcing, and an advantage of 2.2 dB gained by the better of the two arrangements of pure nonlinear equalization, where there is no linear filter.

Suppose now that, in the arrangement of the linear and nonlinear equalization of complementary factors of the channel response, the factor $1 + \frac{1}{2}z^{-1}$ of the channel z-transform is equalized linearly and the factor $1 + 2z^{-1}$ is equalized nonlinearly.

If the linear filter has 7 taps, its z transform is

$$D(z) = 1 - \tfrac{1}{2}z^{-1} + \tfrac{1}{4}z^{-2} - \tfrac{1}{8}z^{-3} + \tfrac{1}{16}z^{-4} - \tfrac{1}{32}z^{-5} + \tfrac{1}{64}z^{-6} \quad (4.4.50)$$

The z transform of the channel and linear filter is approximately $1 + 2z^{-1}$, so that the sample value at the output of the linear filter, at time $t = iT$, is

$$v_i \simeq s_i + 2s_{i-1} + u_i \quad (4.4.51)$$

where u_i is a Gaussian random variable with zero mean and variance

$$\eta^2 = \sigma^2(1^2 + (\tfrac{1}{2})^2 + (\tfrac{1}{4})^2 + (\tfrac{1}{8})^2 + (\tfrac{1}{16})^2 + (\tfrac{1}{32})^2 + (\tfrac{1}{64})^2)$$
$$= 1.333\sigma^2 \quad (4.4.52)$$

so that

$$\eta = 1.155\sigma \quad (4.4.53)$$

With the correct detection of s_{i-1} and the cancellation of its intersymbol interference in v_i by the nonlinear filter, the sample value at the input to the detector, at time $t = iT$, is

$$x_i = v_i - 2s_{i-1} \simeq s_i + u_i \quad (4.4.54)$$

The probability of error in the detection of s_i is now

$$P_e = Q\left(\frac{k}{\eta}\right) = Q\left(\frac{k}{1.155\sigma}\right) = Q\left(\frac{0.866k}{\sigma}\right) \quad (4.4.55)$$

The tolerance to noise is thus 6 dB below that of the previous nonlinear equalizer, where the factor $1 + 2z^{-1}$ is equalized linearly and the factor $1 + \frac{1}{2}z^{-1}$ is equalized nonlinearly. It is also 2.6 dB below that of a linear equalizer. Clearly, the previous nonlinear equalizer is much to be preferred.

4.4.6 Optimum equalizer

It is clear from the previous analysis that when the channel is equalized by a linear filter, the filter is effectively determined by the channel. This, of course, follows from the fact that if the channel has the z-transform $Y(z)$, the corresponding linear equalizer has a z transform approximately equal to $z^{-h}Y^{-1}(z)$, where h is an appropriate integer, and for the given value of h, $z^{-h}Y^{-1}(z)$ is uniquely determined by $Y(z)$. With the accurate equalization of the channel, the tolerance to noise of the resultant system is determined

Sampled impulse-response
of channel and linear
equalizer:

Sampled impulse-response
of channel and linear
filter in nonlinear
equalizer:

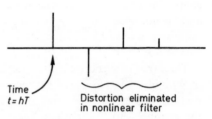

Time
$t = hT$

Distortion eliminated
in nonlinear filter

Fig. 4.22   Relationship between linear and nonlinear equalizers

by the noise variance at the equalizer output. This, in turn, is determined by the sum of the squares of the tap gains of the linear equalizer and so by the channel itself. Thus the tolerance to noise is not significantly affected by the particular design technique applied to the linear equalizer.

However, when the channel is equalized by a nonlinear equalizer, containing both a linear and a nonlinear filter, the channel is equalized partly by the linear filter and partly by the nonlinear filter, the two filters together achieving the accurate equalization of the channel. Under these conditions the equalization of the channel may be achieved by an infinite number of different combinations of linear and nonlinear filters. Thus there must be one or more particular combinations of the two filters that give the best tolerance to additive white Gaussian noise. The situation here is illustrated in Figure 4.22.

With a nonlinear equalizer, the tolerance of the system to additive white Gaussian noise at high signal/noise ratios is essentially determined by the noise variance at the output of the linear filter and therefore by the sum of the squares of the tap gains of this filter. The noise variance is not affected by the nonlinear filter. Furthermore, the signal component at the detector input is the first nonzero component of the corresponding signal-element at the output of the linear filter. Thus the wanted signal component and the noise component, at the output of the linear filter, are passed *unchanged* to the detector input and are therefore unaffected by the nonlinear filter. This merely removes the unwanted signal components that form the residual intersymbol interference at the output of the linear filter. To maximize the signal/noise ratio at the detector input, it is necessary to maximize the ratio of the magnitude of the first nonzero component in the sampled impulse-response of the channel and linear filter to the sum of the squares of the linear-filter tap gains. The linear filter can clearly be selected to maximize this ratio, and the nonlinear filter is now completely determined by the sampled impulse-response of the channel and linear filter, as can be seen in Figure 4.22.

The error-extension effects, caused by the incorrect cancellation of wrongly detected signal-elements, are dependent entirely on the nonlinear filter and therefore on the sampled impulse-response of the channel and linear filter. They may cause the error rate to be typically up to about ten times that with the corresponding idealized system, where there is always the correct cancellation of intersymbol interference, regardless of the errors in detection. However, at high signal/noise ratios, in the presence of additive Gaussian noise, this increase in error rate represents a reduction in tolerance to noise of up to only about 1 dB, which is not very serious and for our purposes may be neglected. Thus, if the signal/noise ratio at the detector input is maximized, then, for practical purposes, the error rate in the detection of the received signal-elements is minimized, to give the optimum nonlinear equalizer.

With 4- or 8-level signals, the increase in the error rate due to error-extension effects may sometimes be considerably more than 2 or 4 times that for the 2-level signals assumed here, and in the case of 8-level signals, the error-extension effects begin to become significant. These signals are not, however, considered here.

It is important to notice that the optimization process just outlined is subject (constrained) to the accurate equalization of the channel. To minimize (without constraint) the error rate in the detected signal, at very low signal/noise ratios, it becomes more important to minimize the level of the noise than the level of the intersymbol interference, in the signal at the detector input. Thus a significant level of intersymbol interference must now be accepted in the equalized signal, so that the channel is not accurately equalized. On the other hand to minimize (without constraint) the error rate in the detected signal, at high signal/noise ratios, it becomes more important to minimize the level of the intersymbol interference than the level of the noise, so that the channel is now accurately equalized. Thus, at high signal/noise ratios, the tolerance to additive white Gaussian noise of the equalizer that minimizes the error rate, subject to the accurate equalization of the channel, is not noticeably inferior to that of the optimum equalizer, that has not been constrained to the accurate equalization of the channel. The analysis in Chapter 3 and Section 4.3.7 clearly supports the above conclusion.

A study will now be made of the nonlinear equalizer that maximizes the signal/noise ratio at the detector input, subject to the accurate equalization of the channel. For practical purposes, this is also the nonlinear equalizer that minimizes the error rate in the detected signal, at high signal/noise ratios.

Consider the nonlinear equalizer shown in Figure 4.21. Let the linear filter have n taps and a sampled impulse-response

$$D = d_0 \ d_1 \dots d_{n-1} \qquad (4.4.56)$$

with z-transform

$$D(z) = d_0 + d_1 z^{-1} + \dots + d_{n-1} z^{-n+1} \qquad (4.4.57)$$

Also let

$$D(z) = C(z)E(z) \qquad (4.4.58)$$

where

$$E(z) = 1 + e_1 z^{-1} + e_2 z^{-2} + \dots + e_{n-m} z^{-n+m} \qquad (4.4.59)$$

and $C(z)$ is the z transform of the m-tap linear equalizer for the baseband channel with z-transform $Y(z)$, such that

$$C(z) \simeq z^{-h} Y^{-1}(z) \qquad (4.4.60)$$

The z transform of the channel and linear filter is

$$Y(z)D(z) \simeq z^{-h} E(z) \qquad (4.4.61)$$

so that the sampled impulse-response of the channel and linear filter is given approximately by the $(n-m+1)$-component row vector

$$E = 1 \ e_1 \ e_2 \dots e_{n-m} \qquad (4.4.62)$$

neglecting the delay of hT seconds. The first component of E is fixed at unity, whereas its remaining $n-m$ components $\{e_i\}$ may be selected appropriately to optimize the tolerance of the system to noise.

The channel and linear filter together are equalized by the nonlinear filter in Figure 4.21. Thus the nonlinear filter equalizes $E(z)$, and the linear feedforward transversal filter B that forms part of the nonlinear filter in Figure 4.21 has $n-m$ taps whose gains are equal respectively to the $\{e_i\}$. The nonlinear filter is therefore given by some simple modifications to Figure 4.19 and the channel is now accurately equalized.

The binary value s_i of the ith received signal-element is detected from the sign of the sample value

$$x_{i+h} \simeq s_i + u_{i+h} \tag{4.4.63}$$

at the input to the detector, at time $t = (i+h)T$. u_{i+h} is a Gaussian random variable with zero mean and variance

$$\eta^2 = \sigma^2 |D|^2 \tag{4.4.64}$$

where $|D|$ is the length of the vector D. Thus the probability of error in the detection of a received signal-element, following the correct detection of the preceding $n-m$ elements, is

$$P_e = \int_k^\infty \frac{1}{\sqrt{(2\pi\eta^2)}} \exp\left(-\frac{u^2}{2\eta^2}\right) \mathrm{d}u$$

$$= Q\left(\frac{k}{\eta}\right) = Q\left(\frac{k}{\sigma|D|}\right) \tag{4.4.65}$$

The error-extension effects are not normally serious here, so that no significant inaccuracy is introduced by taking the average probability of error as P_e.

It is clear from Equation 4.4.65, that to minimize the error rate in the detected element values $\{s_i'\}$, it is necessary to minimize $|D|$. The minimization is, of course, subject to the constraint imposed by Equations 4.4.58 and 4.4.59, which imply the accurate equalization of the channel by the nonlinear equalizer.

Let G be the $(n-m+1)\times n$ convolution matrix whose ith row is given by the n-component row vector

$$G_{i-1} = \overbrace{0 \ldots 0}^{i-1} c_0 \; c_1 \ldots c_{m-1} \; 0 \ldots 0 \tag{4.4.66}$$

where c_i is the coefficient of z^{-i} in $C(z)$ (Equation 4.4.60). Then, from Equation 4.4.58,

$$D = EG = G_0 - LM \qquad (4.4.67)$$

where

$$L = -e_1 - e_2 \ldots - e_{n-m} \qquad (4.4.68)$$

and M is the $(n-m) \times n$ matrix whose ith row is G_i. Clearly, the removal of the first row from G gives the matrix M.

Thus, to minimize the error probability, it is necessary to minimize

$$|D| = |G_0 - LM| \qquad (4.4.69)$$

For given values of $C(z)$ and n, G_0 and M are fixed, leaving L as the only variable in $|G_0 - LM|$. Thus L must be chosen to minimize this quantity.

G_0 and LM are n-component vectors and so can be represented as points in an n-dimensional Euclidean vector space. Furthermore,

$$LM = \sum_{i=1}^{n-m} -e_i G_i \qquad (4.4.70)$$

and the $n-m$ vectors $\{G_i\}$ are linearly independent. Thus LM is a point in the $n-m$ dimensional subspace spanned by the $n-m$ $\{G_i\}$, and $|G_0 - LM|$ is the *distance* between G_0 and LM in the n-dimensional vector space. There are no constraints on L, which may be selected such that LM is any required point in the $(n-m)$-dimensional subspace. It follows that $|G_0 - LM|$ is minimum when LM is the point in the subspace at the *minimum distance* from G_0. By the projection theorem, LM is now the orthogonal projection of G_0 on to the subspace. Thus each vector G_i, for $i = 1, 2, \ldots, n-m$, is orthogonal to the vector $G_0 - LM$, so that

$$(G_0 - LM)M^T = 0 \qquad (4.4.71)$$

or

$$L = G_0 M^T (MM^T)^{-1} \qquad (4.4.72)$$

The $(n-m) \times (n-m)$ matrix MM^T is of course nonsingular, since the matrix M has rank $n-m$. From Equation 4.4.67,

$$D = G_0(I - M^T(MM^T)^{-1}M) \qquad (4.4.73)$$

where I is an $n \times n$ identity matrix. D now gives the tap gains of the required linear filter in Figure 4.21.

For a sufficiently large value of n, the number of taps in fact required for the linear filter is frequently smaller than m, and the tap gains are essentially independent of n. The reason for this will be made clear presently.

The sampled impulse-response of the channel and linear filter is given approximately by the vector E in Equation 4.4.62, where the $\{e_i\}$ satisfy Equations 4.4.68 and 4.4.72. The nonlinear filter in Figure 4.21 equalizes the distortion represented by E, and its transversal filter therefore has $n-m$ taps with gains equal respectively to the negatives of the $n-m$ components of the vector L in Equation 4.4.72.

At high signal/noise ratios and with the accurate equalization of the channel, the system often gives a significantly better tolerance to additive white Gaussian noise than any of the other equalizers described here. In no case does it have a lower tolerance.

An alternative derivation of the optimum nonlinear equalizer just described, that leads to the same result but expressed in a quite different form, is as follows.

Consider the nonlinear equalizer shown in Figure 4.21 and suppose that the z transform of the baseband channel and linear filter is

$$Y(z)D(z) \simeq z^{-h}V_1(z)V_2(z) \tag{4.4.74}$$

where $V_2(z)$ is *any* real factor of $Y(z)D(z)$ such that *all* the roots of $V_2(z)$ lie *outside* the unit circle in the z plane. No assumptions are made about the roots of $V_1(z)$, except, of course, that any complex roots occur in complex-conjugate pairs.

$$V_2(z) = v_0 + v_1 z^{-1} + \ldots + v_j z^{-j} \tag{4.4.75}$$

where j is a positive integer equal to the number of roots of $V_2(z)$. The first term of $z^{-h}V_1(z)V_2(z)$ is taken to be $e_0 z^{-h}$. As before, $Y(z)$ and $D(z)$ are the z transforms of the channel and linear filter, respectively, and h is a nonnegative integer.

Suppose now that there is added, at the input of the linear filter in Figure 4.21, another linear filter that is a phase equalizer with z-transform

$$A(z) \simeq z^{-l}V_2^{-1}(z)V_3(z) \tag{4.4.76}$$

where $$V_3(z) = v_j + v_{j-1}z^{-1} + \ldots + v_0 z^{-j} \tag{4.4.77}$$

l is the positive integer needed to make $z^{-l}V_2^{-1}(z)$ physically realizable, when expressed to a sufficient number of terms for the required degree of accuracy, and $V_3(z)$ is obtained from $V_2(z)$ by reversing the order of its coefficients. As is shown in Section 2.4.3, $V_2^{-1}(z)V_3(z)$ is the z transform of the phase equalizer for the pure phase distortion $V_2(z)V_3^{-1}(z)$. This is because $V_2^{-1}(z)V_3(z)$ is the *inverse* of $V_2(z)V_3^{-1}(z)$ and is also obtained from the latter by *reversing* the order of its coefficients (pivoted about the term containing z^0), so that $V_2^{-1}(z)V_3(z)$ is *matched* to $V_2(z)V_3^{-1}(z)$. Thus $A(z)$ represents an orthogonal transformation with no attenuation or gain.

The z transform of the channel and two linear filters is

$$Y(z)A(z)D(z) \simeq z^{-h-l}V_1(z)V_3(z) \qquad (4.4.78)$$

where the first term of $z^{-h-l}V_1(z)V_3(z)$ is taken to be $f_0 z^{-h-l}$. Since the roots of $V_3(z)$ are the reciprocals of the roots of $V_2(z)$, the action of the additional linear filter is to replace the roots of $V_2(z)$ by their reciprocals, and the latter, of course, all lie *inside* the unit circle in the z plane. Again, since all the roots of $V_2(z)$ lie outside the unit circle, it can be seen from Equation 4.4.75 that $|v_j| > |v_0|$. But, from Equation 4.4.78, the sampled impulse-response of the channel and two linear filters is given by the coefficients in $V_1(z)V_3(z)$, and, from Equation 4.4.74, the sampled impulse-response of the channel and original linear filter is given by the coefficients in $V_1(z)V_2(z)$. Thus the channel and two linear filters have a sampled impulse-response with a first nonzero component f_0, whose magnitude is *greater* than that of e_0, the first nonzero component of the sampled impulse-response of the channel and original linear filter.

It is assumed that the nonlinear equalizer in Figure 4.21, with the additional linear filter at its input, achieves the accurate equalization of the baseband channel. Thus the nonlinear filter in Figure 4.21 sets to zero all but the first of the components of the sampled impulse-response of the channel and two linear filters, to give a resultant sampled impulse-response containing only the one nonzero component f_0. Furthermore, the nonlinear filter does not change the noise variance.

The binary element value s_i of the ith received signal-element is now detected from the sign of the sample value

$$x_{i+h+l} \simeq s_i f_0 + u_{i+h+l} \qquad (4.4.79)$$

at the detector input, at time $t = (i+h+l)T$. The noise component u_{i+h+l} is a Gaussian random variable with zero mean and variance dependent on the two linear filters. The correct detection of the preceding signal-elements is, of course, assumed here.

It is clear from Section 4.3.10 that the noise components at the output of the linear filter with z-transform $A(z)$, which is connected between the sampler and the linear filter with z-transform $D(z)$, are statistically independent Gaussian random variables with zero mean and variance σ^2, just like the noise components at the input of this filter. It follows that the variance of u_{i+h+l} is

$$\eta^2 = \sigma^2 |D|^2 \qquad (4.4.80)$$

where $|D|^2$ is the sum of the squares of the tap gains of the linear filter with z-transform $D(z)$. Thus the probability of error in the detection of s_i can be taken to be

$$P_e = Q\left(\frac{k|f_0|}{\eta}\right) \qquad (4.4.81)$$

For the case where the nonlinear equalizer in Figure 4.21 contains just the single linear filter with z-transform $D(z)$, the nonlinear filter being appropriately modified to achieve the accurate equalization of the channel and linear filter, s_i is detected from the sign of the sample value

$$x_{i+h} \simeq s_i e_0 + u_{i+h} \qquad (4.4.82)$$

at the detector input, at time $t = (i+h)T$. The noise component u_{i+h} is a Gaussian random variable with zero mean and variance $\eta^2 = \sigma^2 |D|^2$, as before. Thus the probability of error in the detection of s_i can now be taken to be

$$P_e = Q\left(\frac{k|e_0|}{\eta}\right) \qquad (4.4.83)$$

and this is always *greater* than the error probability P_e in Equation 4.4.81 since $|e_0| < |f_0|$.

Clearly, whenever the z-transform $Y(z)D(z)$ of the channel and linear filter, for the nonlinear equalizer in Figure 4.21, contains one or more roots outside the unit circle in the z plane, an improved tolerance to additive white Gaussian noise is always achieved by adding, at the input to the linear filter, a pure phase equalizer that replaces the roots outside the unit circle by their reciprocals, the nonlinear filter being appropriately modified to maintain the accurate equalization of the channel.

Suppose now that the linear filter in Figure 4.21 comprises two linear feedforward transversal filters in cascade, the first filter being a pure phase equalizer with z-transform $A(z)$ and the second filter having a z-transform $B(z)$. Thus the resultant z transform of the linear filter is

$$D(z) = A(z)B(z) \qquad (4.4.84)$$

The nonlinear equalizer in Figure 4.21 is otherwise exactly as shown.

The z transform of the baseband channel in Figure 4.1 is assumed to be

$$Y(z) = Y_1(z)Y_2(z) = y_0 + y_1 z^{-1} + \ldots + y_g z^{-g} \qquad (4.4.85)$$

where all the roots of $Y_1(z)$ lie inside the unit circle in the z plane and all the roots of $Y_2(z)$ lie outside the unit circle. $Y_1(z)$ and $Y_2(z)$ are given by Equations 4.4.36 and 4.4.37, respectively. Let

$$Y_3(z) = q_f + q_{f-1} z^{-1} + \ldots + q_0 z^{-f} \qquad (4.4.86)$$

where f is an integer in the range 0 to g, so that $Y_3(z)$ is obtained from $Y_2(z)$ by reversing the order of its coefficients. The roots of $Y_3(z)$ are the reciprocals of those of $Y_2(z)$ and therefore lie *inside* the unit circle.

It is assumed that

$$A(z) \simeq z^{-h} Y_2^{-1}(z) Y_3(z) \qquad (4.4.87)$$

so that the z transform of the channel and phase equalizer is

$$Y(z)A(z) \simeq z^{-h} Y_1(z) Y_3(z) \qquad (4.4.88)$$

all of whose roots lie *inside* the unit circle.

The roots of $Y(z)A(z)$ comprise the roots of $Y(z)$ that lie inside the unit circle together with the reciprocals of the roots of $Y(z)$ that lie outside the unit circle. Thus, in $Y(z)A(z)$, the phase equalizer replaces the roots of $Y(z)$ that lie outside the unit circle by their reciprocals and leaves the remaining roots unchanged. When all the roots of $Y(z)$ lie inside the unit circle, $A(z) = 1$. The first term in $Y(z)A(z)$ is $q_f z^{-h}$.

Suppose that the second linear filter has the z-transform

$$B(z) = q_f^{-1} + b_1 z^{-1} + b_2 z^{-2} + \ldots + b_j z^{-j} \qquad (4.4.89)$$

where j is a nonnegative integer, $B(z)$ being such that the resultant z transform of the channel and two linear filters is

$$Y(z)A(z)B(z) \simeq z^{-h}(1 + a_1 z^{-1} + a_2 z^{-2} + \ldots + a_l z^{-l}) \qquad (4.4.90)$$

with all its roots *inside* the unit circle, but otherwise with no restrictions on the values of the $\{a_i\}$ which are, of course, all real. l may be any nonnegative integer. Clearly, over the possible values of the $\{a_i\}$ and of the integer l, j may be any nonnegative integer and the $\{b_i\}$ may take on all real values, subject to the constraint that all the roots of $B(z)$ must lie *inside* the unit circle.

The second linear filter is a minimum delay network whose main tap is its first tap and has a fixed gain of q_f^{-1}. It may introduce additional roots into the z-transform $Y(z)A(z)B(z)$ that are not present in $Y(z)A(z)$, or it may be such that there are roots in $Y(z)A(z)$ that are not present in $Y(z)A(z)B(z)$, or again it may satisfy both of these conditions. Thus $B(z)$ can be selected to give any number of roots in $Y(z)A(z)B(z)$, the roots being located anywhere within the unit circle. In an extreme case, for instance, $B(z)$ could be such that $Y(z)A(z)B(z)$ has no roots (other than those at infinity, corresponding to the term z^{-h} in Equation 4.4.90) so that the linear filter in Figure 4.21 becomes the linear equalizer for the baseband channel. The nonlinear filter in Figure 4.21 is assumed to be appropriately adjusted to maintain the accurate equalization of the channel for any value of $B(z)$.

Roots of $Y(z)A(z)B(z)$ *outside* the unit circle are not considered, since it has already been shown that the presence of any such roots necessarily implies that the nonlinear equalizer in Figure 4.21 does not have the best available tolerance to additive white Gaussian noise. If the z-transform $A(z)$ of the phase equalizer is now adjusted to replace these roots in $Y(z)A(z)B(z)$ by their reciprocals, this always improves the tolerance of the system to additive white Gaussian noise and gives the arrangement considered here where $Y(z)A(z)B(z)$ has *no* roots outside the unit circle.

Since the first term in $Y(z)A(z)$ is $q_f z^{-h}$, it can be seen from Equation 4.4.90 that the gain of the first tap of the second linear filter must be q_f^{-1} and, except in the particular case where $B(z) = q_f^{-1}$ and the linear filter becomes a simple multiplier, the linear filter must have one or more *additional* taps with nonzero gains. Since the noise components at the output of the phase equalizer are statistically independent Gaussian random variables with zero mean and variance σ^2, the noise component u_i, at the output of the second linear filter at time $t = iT$, is a Gaussian random variable with zero mean and variance

$$\eta^2 = \sigma^2(q_f^{-2} + b_1^2 + b_2^2 + \ldots + b_j^2) \qquad (4.4.91)$$

The sample value at the input to the detector in Figure 4.21, at time $t = (i+h)T$, is

$$x_{i+h} \simeq s_i + u_{i+h} \qquad (4.4.92)$$

so that, at high signal/noise ratios, the probability of error in the detection of s_i can be taken to be

$$P_e = Q\left(\frac{k}{\eta}\right) \qquad (4.4.93)$$

To minimize the probability of error, η must be minimized. From Equation 4.4.91, the minimum value of η is $\sigma/|q_f|$ and this is obtained when $b_i = 0$ for $i = 1, 2, \ldots, j$. Thus, for the minimum error probability,

$$B(z) = q_f^{-1} \qquad (4.4.94)$$

and the second linear filter becomes a multiplier that multiplies each sample value at its input by q_f^{-1}. The corresponding error probability, at high signal/noise ratios, can be taken to be

$$P_e = Q\left(\frac{k|q_f|}{\sigma}\right) \qquad (4.4.95)$$

The phase equalizer together with the multiplier can of course be replaced by a single linear filter with z-transform $D(z)$ (Equation 4.4.84). If now all the tap gains of this linear filter are multiplied by some constant, this produces the same change in both the signal and the noise components at its output and so does not affect either the signal/noise ratio or the error probability. Thus the gain of q_f^{-1} that has been assumed to be introduced by the multiplier, does not affect the performance of the system.

It can be seen from Equations 4.4.84, 4.4.87 and 4.4.94 that in the optimum nonlinear equalizer (Figure 4.21), that minimizes the error rate in the detected element values at high signal/noise ratios, the linear filter is a phase equalizer with the z-transform

$$D(z) \simeq q_f^{-1} z^{-h} Y_2^{-1}(z) Y_3(z) \qquad (4.4.96)$$

The z-transform of the channel and linear filter is now

$$Y(z)D(z) \simeq q_f^{-1} z^{-h} Y_1(z) Y_3(z) \qquad (4.4.97)$$

and the first nonzero component of the sampled impulse-response of the channel and linear filter has the value unity, as in Equation 4.4.62.

The linear feedforward transversal filter B, that forms part of the nonlinear filter in Figure 4.21, has g taps whose gains are given by the coefficients (other than the first) in $q_f^{-1} z^{-h} Y_1(z) Y_3(z)$. With the correct detection of the received signal-elements, the nonlinear filter removes the intersymbol interference represented by $q_f^{-1} z^{-h} Y_1(z) Y_3(z)$, so that the sample value at the input to the detector, at time $t = (i+h)T$, is x_{i+h} in Equation 4.4.92, and s_i is detected from the sign of x_{i+h}. Since $z^{-h} Y_2^{-1}(z) Y_3(z)$ is formed from $Y_2(z) Y_3^{-1}(z)$ by reversing the order of its coefficients, and since the product of the two polynomials is equal to z^{-h}, it is clear that the sum of the squares of the coefficients of $z^{-h} Y_2^{-1}(z) Y_3(z)$ is equal to unity. It follows from Equation 4.4.96 that the sum of the squares of the tap gains of the linear filter in Figure 4.21 is q_f^{-2}, so that the Gaussian noise component u_{i+h} has a variance of σ^2/q_f^2, as shown before. Thus the probability of error in the detection of s_i can be taken to be P_e in Equation 4.4.95.

It is interesting to observe that in the optimum nonlinear equalizer, both the linear and the nonlinear filters perform the particular operations for which each of them always gives a greater output signal/noise ratio than can be achieved by the other, leading to the optimum sharing of the equalization of the channel between the linear and nonlinear filters. Thus the linear filter equalizes pure phase distortion and the nonlinear filter equalizes a z transform all of whose roots lie inside the unit circle in the z plane.

With the linear and nonlinear equalization of complementary factors of the channel response, the linear equalizer merely equalizes the factor $Y_2(z)$ of the channel z-transform, as shown by Equation 4.4.38. In the optimum nonlinear equalizer, the linear filter not only equalizes $Y_2(z)$ but also introduces the factor $q_f^{-1} Y_3(z)$, as shown by Equation 4.4.97.

It may readily be shown that if $Y(z)$ contains one or more roots *on* the unit circle in the z plane, the optimum nonlinear equalizer is obtained by taking $Y_1(z)$ as the factor of $Y(z)$ that includes all the roots *on or inside* the unit circle.[123]

When the baseband channel introduces pure phase distortion, the optimum nonlinear equalizer degenerates into a pure linear equalizer which, of course, now achieves the best available tolerance to additive white Gaussian noise. Again, when all the roots of $Y(z)$ lie on or inside the unit circle, the optimum nonlinear equalizer becomes a pure nonlinear equalizer as in Figure 4.18, that removes all the signal distortion by decision-directed cancellation of inter-symbol interference.

4.4.7   Example

Consider the case where the z transform of the baseband channel in Figure 4.1 is

$$Y(z) = 1 + 2.5z^{-1} + z^{-2} \tag{4.4.98}$$

and it is required to design the nonlinear equalizer that minimizes the error rate in the detected element values at high signal/noise ratios. The receiver has a linear transversal filter with up to 16 taps, followed by a nonlinear filter, and is as shown in Figure 4.21.

From Equation 4.3.138, $Y(z)$ is equalized *linearly* by an 11-tap feedforward transversal filter whose tap gains are given by the first 11 components of the 16-component vector

$$G_0 = \begin{array}{cccc} -0.012 & 0.035 & -0.079 & 0.164 \\ & -0.331 & 0.665 & -0.331 & 0.164 \\ -0.079 & 0.035 & -0.012 & 0.000 \\ & 0.000 & 0.000 & 0.000 & 0.000 \end{array} \tag{4.4.99}$$

From Equation 4.4.73, the 16 tap gains of the required linear transversal filter in the nonlinear equalizer are given by the 16 components of the vector

$$D = G_0(I - M^T(MM^T)^{-1}M) \tag{4.4.100}$$

where M is the 5×16 matrix whose ith row is

$$G_i = \overbrace{0\ldots0}^{i} \overbrace{-0.012 \; 0.035\ldots -0.012}^{11} \; 0\ldots0 \tag{4.4.101}$$

the eleven nonzero components of G_i being the same as those of G_0 in Equation 4.4.99. Thus

$$D = \begin{array}{cccc} -0.012 & 0.023 & -0.047 & 0.094 \\ & -0.188 & 0.376 & 0.249 & 0.001 \\ -0.002 & 0.003 & -0.003 & 0.000 \\ & -0.005 & 0.001 & 0.000 & 0.000 \end{array} \tag{4.4.102}$$

It is clear that only the first seven taps of the linear filter need in fact be used, the remaining taps having only a small effect on the sampled impulse-response of the channel and linear filter. Thus the linear filter in the nonlinear equalizer is *smaller* than the corresponding linear equalizer for the channel, which requires some eleven taps, as can be seen from Equation 4.4.99. The design of the linear filter does not appear to be significantly affected by the number of the components of the vector D, so long as this number is not less than $m+2$, where m is the number of nonzero components of G_0 and hence the number of taps of the linear equalizer that achieves the reasonably accurate equalization of the channel.

From Equation 4.4.72,

$$L = G_0 M^T (MM^T)^{-1} = -1.00 \quad -0.25 \quad 0.00 \quad 0.00 \quad 0.00 \quad (4.4.103)$$

so that, from Equations 4.4.62 and 4.4.68, the sampled impulse-response of the channel and linear filter is given approximately by the 6-component vector

$$E = 1.00 \quad 1.00 \quad 0.25 \quad 0.00 \quad 0.00 \quad 0.00 \quad (4.4.104)$$

neglecting the delay in transmission. The corresponding z-transform is

$$E(z) = 1 + z^{-1} + 0.25z^{-2} \quad (4.4.105)$$

The linear *equalizer* for the channel, given by Equation 4.4.99, has the z-transform

$$C(z) \simeq z^{-6} Y^{-1}(z) \quad (4.4.106)$$

The z transform of the linear filter, in the nonlinear equalizer, is

$$D(z) = C(z)E(z) \simeq z^{-6} Y^{-1}(z)E(z) \quad (4.4.107)$$

so that the z transform of the channel and linear filter is

$$Y(z)D(z) \simeq z^{-6}E(z) \quad (4.4.108)$$

and the z transform of the ith received signal-element at the output of the linear filter is approximately

$$s_i z^{-i-6} E(z) = s_i z^{-i-6} (1 + z^{-1} + 0.25z^{-2}) \quad (4.4.109)$$

The nonlinear filter has two taps, with gains 1.00 and 0.25. It equalizes $E(z)$ by removing all but its first term to give at the detector input, at time $t = (i+6)T$, the sample value

$$x_{i+6} \simeq s_i + u_{i+6} \quad (4.4.110)$$

where u_{i+6} is a Gaussian random variable with zero mean and variance

$$\eta^2 = \sigma^2|D|^2 = 0.25\sigma^2 \qquad (4.4.111)$$

Thus the probability of error in the detection of s_i can be taken to be

$$P_e = Q\left(\frac{k}{\eta}\right) = Q\left(\frac{2k}{\sigma}\right) \qquad (4.4.112)$$

the error-extension effects not being important.

The optimum nonlinear equalizer for the given channel may alternatively be derived as follows.

The z transform of the channel is

$$Y(z) = (1 + \tfrac{1}{2}z^{-1})(1 + 2z^{-1}) \qquad (4.4.113)$$

so that, in Equation 4.4.85, $Y_1(z) = 1 + \tfrac{1}{2}z^{-1}$ and $Y_2(z) = 1 + 2z^{-1}$. From Equation 4.4.96, the z-transform $D(z)$ of the linear filter in the nonlinear equalizer must be

$$D(z) \simeq \tfrac{1}{2}z^{-h}(1 + 2z^{-1})^{-1}(2 + z^{-1}) \qquad (4.4.114)$$

where h is the appropriate positive integer. The z transform of the channel and linear filter is now

$$Y(z)D(z) \simeq \tfrac{1}{2}z^{-h}(1 + \tfrac{1}{2}z^{-1})(2 + z^{-1}) \qquad (4.4.115)$$

in which the root of $Y(z)$ outside the unit circle in the z plane has been replaced by its reciprocal.

Suppose that the linear filter has seven taps. Then

$$\begin{aligned}
D(z) &= \tfrac{1}{2}(-\tfrac{1}{64} + \tfrac{1}{32}z^{-1} - \tfrac{1}{16}z^{-2} + \tfrac{1}{8}z^{-3} - \tfrac{1}{4}z^{-4} \\
&\qquad\qquad + \tfrac{1}{2}z^{-5})(2 + z^{-1}) \\
&= -\tfrac{1}{64} + \tfrac{3}{128}z^{-1} - \tfrac{3}{64}z^{-2} + \tfrac{3}{32}z^{-3} - \tfrac{3}{16}z^{-4} \\
&\qquad\qquad + \tfrac{3}{8}z^{-5} + \tfrac{1}{4}z^{-6} \quad (4.4.116)
\end{aligned}$$

so that

$$\begin{aligned}
D = -0.016 \quad 0.023 \quad -0.047 \quad 0.094 \\
-0.188 \quad 0.375 \quad 0.250 \quad (4.4.117)
\end{aligned}$$

which approximates quite well to D in Equation 4.4.102.

The z transform of the channel and linear filter is now

$$Y(z)D(z) \simeq z^{-6}(1 + z^{-1} + 0.25z^{-2}) \qquad (4.4.118)$$

which is the same as that in Equation 4.4.108. The nonlinear filter is therefore the same as before, and s_i is detected from x_{i+6}, given by Equation 4.4.110.

From Equation 4.4.95, the probability of error in the detection of s_i can be taken to be

$$P_e = Q\left(\frac{k|2|}{\sigma}\right) = Q\left(\frac{2k}{\sigma}\right) \tag{4.4.119}$$

as before.

The second method of evaluating the nonlinear equalizer is clearly the simpler of the two, and the method also gives the more accurate optimization of the equalizer, when only a limited number of taps can be used for the linear filter.

It can be seen from Equations 4.3.143 and 4.4.119 that the optimum nonlinear equalizer gains an advantage of 4.7 dB in tolerance to additive white Gaussian noise over a linear equalizer, for the given channel at high signal/noise ratios. This compares with an advantage of 3.4 dB gained by the nonlinear equalizer in which linear and nonlinear equalization is applied to complementary factors of the channel z-transform.

4.5 ADAPTIVE EQUALIZERS

When the sampled impulse-response of the channel varies with time, it is necessary to keep readjusting the tap gains of a transversal equalizer in order to hold it correctly set for the channel. The most cost-effective method of doing this is to derive the information needed for the correct setting of the tap gains from the received data signal itself.[53-135] So long as the element values $\{s_i\}$ are reasonably uncorrelated, the method works quite well. The basic principles of the more important adaptive equalizers will now be described very briefly and without proof.

Consider first the linear transversal equalizer that minimizes the mean-square error in its output signal,[70] as shown in Figure 4.23. The signals shown here are those present at the time instant $t = iT$. With the reasonably accurate equalization of the channel, the output signal from the linear equalizer at this time instant is

$$x_i \simeq s_{i-h} + u_i \tag{4.5.1}$$

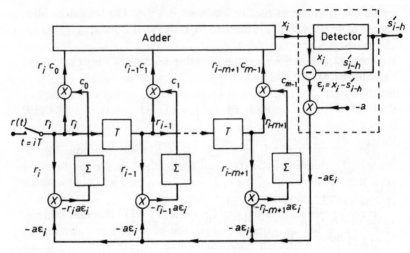

Fig. 4.23   Adaptive linear equalizer

where h is an appropriate positive integer and u_i is the noise component. The error in x_i is $x_i - s_{i-h}$, and at low error rates, when usually $s'_{i-h} = s_{i-h}$, the error in x_i can be taken to be

$$\varepsilon_i = x_i - s'_{i-h} \qquad (4.5.2)$$

where, of course, s'_{i-h} is the detected value of s_{i-h}. When $s'_{i-h} = s_{i-h}$, ε_i is composed partly of noise and partly of intersymbol interference. The mean-square error in x_i can thus be taken to be

$$\overline{\varepsilon_i^2} = \overline{(x_i - s'_{i-h})^2} \qquad (4.5.3)$$

It has been shown[70] that the mean-square error is minimized by an arrangement in which, at each time interval of T seconds when a new signal-element is received, the tap gain c_j of the $(j+1)$th tap of the m-tap transversal equalizer is incremented by

$$\delta c_j = -a\varepsilon_i r_{i-j} \qquad (4.5.4)$$

where a is a suitable small positive constant, and $j = 0, 1, \ldots, m-1$. Over a number of l received signal-elements, this arrangement effectively measures the cross-correlation between the corresponding l error signals $\{\varepsilon_i\}$ and the l received samples $\{r_{i-j}\}$, for any given value of j. The resultant change in the tap gain c_j is now

$$\Delta c_j = -la\,\overline{\varepsilon_i r_{i-j}} \qquad (4.5.5)$$

where the average value $\overline{\varepsilon_i r_{i-j}}$ of $\varepsilon_i r_{i-j}$ is determined over the l values of i.[69,70,121] The values of the $\{c_j\}$ now converge towards the values at which the mean-square error in x_i is minimum, and as the channel varies with time the tap gains are appropriately adjusted to hold the equalizer correctly set for the channel. The conditions for the correct convergence and adaptive operation have been widely studied.[63-107] The smaller the value of a in Equation 4.5.5, the greater the value of l for a given magnitude of Δc_j. Hence the slower the response of the tap gains and the greater the effective averaging or integration period of the system.

A necessary condition for the correct operation of the arrangement is that the $\{s_i\}$ are uncorrelated, but even now a small value of a is required to obtain a sufficiently accurate measure of the true average value of $\varepsilon_i r_{i-j}$. A small value of a is also necessary to reduce the effects of noise on the settings of the $\{c_i\}$.

Since the arrangement minimizes $\overline{\varepsilon_i^2}$ in Equation 4.5.3, it minimizes the mean-square error due to *both* intersymbol interference and noise. Thus even with an extremely small value of a (and so the effective averaging over a very large number of received samples), the settings of the $\{c_i\}$ are affected by the noise variance σ^2. For instance, an increase in σ^2 causes a reduction in the magnitudes of *all* $\{c_i\}$. However, at high signal/noise ratios, where the predominant source of error in the $\{x_i\}$ is the intersymbol interference, the tap gains are effectively adjusted to minimize the mean-square error caused by intersymbol interference alone.

The adaptive equalizer may be modified in various different ways. For instance, either or both ε_i and the $\{r_{i-j}\}$ may be quantized, as may the $\{c_j\}$. Again, the changes in the $\{c_j\}$ may only be applied after every lT seconds or alternatively only when the magnitudes of the changes exceed a given threshold level.[63-107] However, the method of operation is still basically similar to that just described.

Consider next the nonlinear transversal equalizer that minimizes the mean-square error in its output signal. The nonlinear filter of this equalizer is shown in Figure 4.24. The linear filter, which has n taps with tap gains $\{d_i\}$, for $i = 0, 1, \ldots, n-1$, is similar to the portion of Figure 4.23 outside the dotted rectangle. When correctly adjusted at high signal/noise ratios, the nonlinear equalizer maximizes the signal/noise ratio at the detector input, and, with additive white Gaussian noise, this minimizes the error rate in the detected element values.

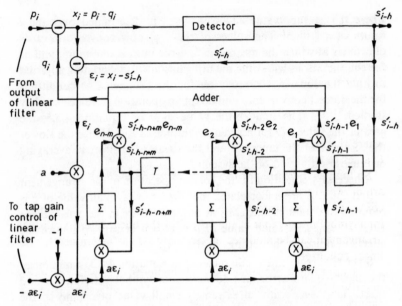

Fig. 4.24   Nonlinear filter of adaptive nonlinear equalizer

Suppose that the z transform of the linear filter is $D(z)$ and the z transform of the channel and linear filter is

$$Y(z)D(z) \simeq z^{-h}E(z) \qquad (4.5.6)$$

where

$$E(z) = e_0 + e_1 z^{-1} + \ldots + e_{n-m} z^{-n+m} \qquad (4.5.7)$$

For the minimum mean-square error in the equalized signal x_i, it is again necessary to minimize $\overline{\varepsilon_i^2}$ in Equation 4.5.3. Since the tap gains of the nonlinear filter do not affect the noise level in x_i, it is clear that, for the minimum mean-square error in x_i, the nonlinear filter must equalize $E(z)$ *exactly*, so that its ith tap gain (from the right, in Figure 4.24) is e_i, there being $n-m$ taps altogether.

It has been shown[111,114,121] that the required adjustment of the equalizer is maintained by an arrangement in which, at the arrival of each new signal-element, the tap gain e_j of the jth tap is incremented by

$$\delta e_j = a\varepsilon_i s'_{i-h-j} \qquad (4.5.8)$$

where a is a suitably small positive constant, and $j = 1, 2, \ldots,$ $n - m$. Over a number of l received signal-elements, this arrangement effectively measures the cross-correlation between the corresponding l error signals $\{\varepsilon_i\}$ and the l detected element values $\{s'_{i-h-j}\}$, for any given value of j. The resultant change in the tap gain e_j is now

$$\Delta e_j = la \, \overline{\varepsilon_i s'_{i-h-j}} \qquad (4.5.9)$$

where the average value $\overline{\varepsilon_i s'_{i-h-j}}$ of $\varepsilon_i s'_{i-h-j}$ is determined over the l values of i.

So long as there are not too many errors in the $\{s'_i\}$, so that these are mostly the same as the $\{s_i\}$, the tap gains of the nonlinear filter converge to the correct values $\{e_i\}$ that eliminate the residual intersymbol interference at the output of the linear filter. Again, the arrangement can be modified in various ways, as mentioned for the linear equalizer.

The adjustment of the tap gains of the linear filter is as previously described for the linear equalizer, except that now the last $n - m$ samples of the sampled impulse-response of the channel and linear filter are set to zero by the nonlinear filter. Thus the tap gains of the linear filter are adjusted to minimize the mean-square error in x_i, given the partial equalization achieved by the nonlinear filter. This is achieved by incrementing the tap gain d_j of the $(j+1)$th tap by

$$\delta d_j = -a\varepsilon_i r_{i-j} \qquad (4.5.10)$$

at each received element, so that over l received elements the tap gain changes by

$$\Delta d_j = -la \, \overline{\varepsilon_i r_{i-j}} \qquad (4.5.11)$$

where the average value $\overline{\varepsilon_i r_{i-j}}$ of $\varepsilon_i r_{i-j}$ is determined over the l values of i.

At high signal/noise ratios, where the effects of intersymbol interference tend to be more serious than the effects of noise, the tap gains of the linear filter are adjusted to minimize the noise variance in x_i, subject to the accurate equalization of the channel, and the nonlinear equalizer becomes the optimum nonlinear equalizer described in Section 4.4.6. Thus in Equation 4.5.7, $e_0 = 1$ and $n - m$ need not exceed g. It is assumed here that a suitable value of h is used in the equalizer.

4.6 EQUALIZATION PROCESS SHARED BETWEEN TRANSMITTER AND RECEIVER

4.6.1 *Data transmission over a time-invariant channel*

In all the systems considered so far, the equalization of the channel is carried out at the receiver. This is generally the most cost-effective arrangement in those applications where the channel characteristics vary with time, since the information needed to hold the equalizer correctly set for the channel is derived from the received signal. If all or part of the equalizer is located at the transmitter, the information needed to hold the equalizer correctly adjusted must now be fed back from the receiver to the transmitter, thus introducing additional equipment complexity. On the other hand, with a time-invariant channel whose sampled impulse-response is known, both the transmitter and receiver have at all times the necessary prior knowledge of the channel characteristics, so that no additional equipment complexity need be involved in sharing the equalization process between the transmitter and receiver, and such an arrangement may in some cases be more cost-effective than that where the whole of the channel equalization is carried out at the receiver.[32-34] The typical magnitude of the improvement obtainable in this way was first evaluated by Shum.

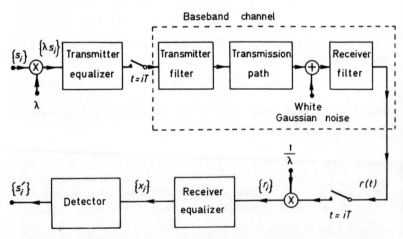

Fig. 4.25 Data-transmission system with the equalization of the channel shared between the transmitter and receiver

A known time-invariant baseband channel is assumed throughout the following discussion, and the data-transmission system is now as shown in Figure 4.25. This data-transmission system is the same as that in Figure 4.1, except that there are now two equalizers, one at the transmitter and the other at the receiver. Various combinations of transmitter and receiver equalizers will be considered, including of course the important case where all the equalization of the channel is carried out at the transmitter. The transmitter and receiver equalizers together achieve the accurate equalization of the channel, and a high signal/noise ratio is assumed, as before.

4.6.2 Linear equalization

Each binary element value s_i at the input to the transmitter in Figure 4.25 is multiplied by the positive scalar quantity λ, before feeding it to the transmitter equalizer. The sequence of values at the output of the equalizer are sampled once for each element-value s_i, at the time instants $\{iT\}$, to give the corresponding sequence of impulses which are fed to the baseband channel.

The transmitter and receiver equalizers are linear feedforward transversal filters whose sampled impulse-responses are given by the l-component row vectors

$$A = a_0 \; a_1 \ldots a_{l-1} \tag{4.6.1}$$

and

$$B = b_0 \; b_1 \ldots b_{l-1} \tag{4.6.2}$$

respectively, with z-transforms

$$A(z) = a_0 + a_1 z^{-1} + \ldots + a_{l-1} z^{-l+1} \tag{4.6.3}$$

and

$$B(z) = b_0 + b_1 z^{-1} + \ldots + b_{l-1} z^{-l+1} \tag{4.6.4}$$

Where A or B has fewer than l components, the appropriate number of zero components are added to follow the last component of the vector and bring the number up to l. All components of A and B are real. In every case,

$$A(z)B(z) \simeq z^{-\theta} C(z) \tag{4.6.5}$$

where θ is an appropriate nonnegative integer and

$$C(z) = c_0 + c_1 z^{-1} + \ldots + c_{m-1} z^{-m+1} \tag{4.6.6}$$

is the z transform of the m-tap linear equalizer for the channel, that minimizes the mean-square error due to intersymbol interference in the equalized signal. As before, the z transform of the baseband channel is

$$Y(z) = y_0 + y_1 z^{-1} + \ldots + y_g z^{-g} \qquad (4.6.7)$$

and the z transform of the channel and single linear equalizer is

$$Y(z)C(z) \simeq z^{-h} \qquad (4.6.8)$$

When $A(z)$ and $B(z)$ are simple factors of $C(z)$, Equation 4.6.5 holds exactly and $\theta = 0$. Now, in Equations 4.6.3 and 4.6.4, $l \geqslant \frac{1}{2}(m+1)$. When $A(z)$ and $B(z)$ are not simple factors of $C(z)$, then $\theta \geqslant 0$, $l \gg \frac{1}{2}(m+1)$ and Equation 4.6.5 holds only approximately, the approximation however improving as l increases.

The transfer function of the transmitter filter in Figure 4.25 is assumed to be such that a unit impulse at its input gives a waveform of unit energy at its output, and such that orthogonal output waveforms are obtained from any two input impulses separated by a multiple of T seconds. These conditions are satisfied by the transmitter filter assumed for the data-transmission system in Figure 4.1 and described in Section 4.2. The ith transmitted signal-element, at the input to the baseband channel in Figure 4.25, is

$$\lambda s_i \sum_{j=0}^{l-1} a_j \delta[t-(i+j)T]$$

and the energy of this individual signal-element, at the input to the transmission path, is

$$\lambda^2 s_i^2 \sum_{j=0}^{l-1} a_j^2 = \lambda^2 k^2 |A|^2 \qquad (4.6.9)$$

since $s_i = \pm k$. $|A|$ is here the length of the vector A. Since the $\{s_i\}$ are statistically independent and have zero mean, they are statistically orthogonal, so that the expected energy of a sequence of n signal-elements, at the input to the transmission path, is equal to the sum of the expected energies of the individual elements. Thus the average transmitted energy per signal element, at the input to the transmission path in Figure 4.25, is $\lambda^2 k^2 |A|^2$. The value of λ is selected to set the average transmitted energy per signal element to k^2, as in Figure 4.1, so that

$$|A| = \frac{1}{\lambda} \qquad (4.6.10)$$

The receiver filter in Figure 4.25 has all the properties assumed for this filter in the data-transmission system of Figure 4.1 and described in Section 4.2.

After sampling at the receiver, the received sample values are multiplied by $1/\lambda$ to give the sequence of sample values $\{r_i\}$ that are fed to the receiver equalizer.

Since, in Figure 4.25, the signal is sampled at the time instants $\{iT\}$, both between the transmitter equalizer and baseband channel and between the baseband channel and receiver equalizer, the z transform of the transmitter equalizer, baseband channel and receiver equalizer, in cascade, is

$$A(z)Y(z)B(z) \simeq z^{-h-\theta} \qquad (4.6.11)$$

as can be seen from Equations 4.6.5 and 4.6.8. Thus the z transform of the ith received signal-element, at the output of the receiver equalizer, is

$$s_i z^{-i}\lambda A(z)Y(z)\frac{1}{\lambda}B(z) \simeq s_i z^{-i-h-\theta} \qquad (4.6.12)$$

From Equation 4.6.8, this is the same as that obtained with the corresponding single linear equalizer at the receiver, in the data-transmission system of Figure 4.1, except for the additional delay of θT seconds.

4.6.3 Optimum combination of transmitter and receiver linear-equalizers

From Equation 4.6.12, the sample value at the output of the receiver equalizer, at time $t = (i+h+\theta)T$ is

$$x_{i+h+\theta} \simeq s_i + u_{i+h+\theta} \qquad (4.6.13)$$

where $u_{i+h+\theta}$ is a Gaussian random variable with zero mean and variance

$$\eta^2 = \frac{\sigma^2}{\lambda^2}\sum_{i=0}^{l-1} b_i^2 = \frac{\sigma^2}{\lambda^2}|B|^2 = \sigma^2|A|^2|B|^2 \qquad (4.6.14)$$

as can be seen from Equations 4.6.2 and 4.6.10, bearing in mind that the noise components $\{w_i\}$ at the input to the receiver equalizer are statistically independent Gaussian random variables with zero mean and variance σ^2/λ^2.

The detector detects s_i from the sign of $x_{i+h+\theta}$, so that, from Equation 4.6.13, the probability of error in the detection of s_i is

$$Q\left(\frac{k}{\eta}\right) = Q\left(\frac{k}{\sigma|A|\,|B|}\right) \tag{4.6.15}$$

Clearly, to minimize the probability of error in a detection process, it is necessary to minimize $|A|\,|B|$.

Let the n-component discrete Fourier transforms (DFT's) of A and B (added zeros extending A and B to n components) be

$$E = e_0\ e_1 \ldots e_{n-1} \tag{4.6.16}$$

and

$$F = f_0\ f_1 \ldots f_{n-1} \tag{4.6.17}$$

respectively, where $n = 2l-1$. The components of E and F are in general complex, whereas the components of A, B and C are always real. Furthermore, from Section 2.3.10,

$$|A|^2 = \frac{1}{n}\sum_{i=0}^{n-1}|e_i|^2 = \frac{1}{n}|E|^2 \tag{4.6.18}$$

and

$$|B|^2 = \frac{1}{n}\sum_{i=0}^{n-1}|f_i|^2 = \frac{1}{n}|F|^2 \tag{4.6.19}$$

so that

$$|A|\,|B| = \frac{1}{n}|E|\,|F| \tag{4.6.20}$$

Thus to minimize the probability of error in a detection process, it is necessary to minimize $|E|\,|F|$.

Let the DFT of the sampled impulse-response of the transmitter and receiver equalizers in cascade be the n-component row vector

$$G = g_0\ g_1 \ldots g_{n-1} \tag{4.6.21}$$

Then, from Section 2.3.3,

$$g_i = e_i f_i \tag{4.6.22}$$

for $i = 0, 1, \ldots, n-1$.

Let E_0 and F_0 be the real nonnegative n-component row vectors whose $(i+1)$th components are $|e_i|$ and $|f_i|$, respectively. Clearly,

$$|E| |F| = \left(\sum_{i=0}^{n-1} |e_i|^2 \right)^{\frac{1}{2}} \left(\sum_{i=0}^{n-1} |f_i|^2 \right)^{\frac{1}{2}} = |E_0| |F_0| \qquad (4.6.23)$$

Also,

$$E_0 F_0^T = \sum_{i=0}^{n-1} |e_i| |f_i| = \sum_{i=0}^{n-1} |g_i| \qquad (4.6.24)$$

since $|g_i| = |e_i| |f_i|$, from Equation 4.6.22.

But C, and therefore also G, are *fixed* by the baseband channel, if we neglect the delay of θT seconds introduced by the transmitter and receiver equalizers relative to the corresponding single equalizer (Equation 4.6.5). It can be seen from Section 2.3.5 that this delay in fact affects the $\{g_i\}$ but not the $\{|g_i|\}$, so that it does not affect Equation 4.6.24 and need not therefore be considered for our purposes here. Thus $E_0 F_0^T$ has a *fixed* nonnegative value. Now, by the Schwarz inequality,[140] or simply by using the fact that the angle ϕ between the vectors E_0 and F_0 satisfies the relationship $\cos\phi = E_0 F_0^T / (|E_0| |F_0|)$ where $E_0 F_0^T \geqslant 0$, it is clear that

$$|E_0| |F_0| \geqslant E_0 F_0^T \qquad (4.6.25)$$

with equality when

$$E_0 = \varepsilon F_0 \qquad (4.6.26)$$

where ε is a positive real scalar. Thus the minimum value of $|E_0| |F_0|$ and hence of $|E| |F|$ is given by $E_0 F_0^T$ in Equation 4.6.24, and this is obtained when

$$|e_i| = \varepsilon |f_i| = (\varepsilon |g_i|)^{\frac{1}{2}} \qquad (4.6.27)$$

for $i = 0, 1, \ldots, n-1$.

It now follows from Equations 4.6.20 and 4.6.24 that the minimum value of $|A| |B|$ is

$$v = \frac{1}{n} \sum_{i=0}^{n-1} |g_i| \qquad (4.6.28)$$

where $|g_i|$ is the modulus of the $(i+1)$th component of the n-component DFT of the sampled impulse-response of the single linear equalizer for the baseband channel, the z transform of this equalizer being $C(z)$ in Equation 4.6.6. Finally, from Equation 4.6.15, the minimum value of the probability of error in the detection of s_i is

$$P_e = Q\left(\frac{k}{\sigma v}\right) \qquad (4.6.29)$$

It is clear from Equation 4.6.27 that there is not in general a *unique* combination of transmitter and receiver equalizers that minimizes the probability of error. A particular combination of some interest is that where $E = F$ and $A = B$, so that the transmitter and receiver equalizers are the same. Now

$$A(z) = B(z) \simeq z^{-\tau}C^{\frac{1}{2}}(z) \qquad (4.6.30)$$

where 2τ is an appropriate nonnegative integer. From Equation 4.6.5, $\theta = 2\tau$. A possible method of evaluating $C^{\frac{1}{2}}(z)$ is described in Section 2.4.7.

It has been shown by Tint that if the conventional algebraic techniques for evaluating the square root of a polynomial are applied to $C(z)$,[136] these usually fail when $C(z)$ has any roots outside the unit circle in the z plane. The correct technique for evaluating $A(z)$ and $B(z)$ is therefore as follows. Let

$$C(z) = C_1(z)C_2(z) \qquad (4.6.31)$$

where all the roots of $C_1(z)$ lie inside the unit circle and all the roots of $C_2(z)$ lie outside. $C_1^{\frac{1}{2}}(z)$ can now readily be evaluated from $C_1(z)$ by conventional techniques,[136] a sufficient number of terms of the infinite expansion of $C_1^{\frac{1}{2}}(z)$ being taken to obtain the required degree of accuracy. To evaluate $C_2^{\frac{1}{2}}(z)$, let $C_3(z)$ be the polynomial obtained from $C_2(z)$ by reversing the order of its coefficients. The roots of $C_3(z)$ are the reciprocals of the roots of $C_2(z)$ and therefore all lie inside the unit circle. $C_3^{\frac{1}{2}}(z)$ can now readily be evaluated from $C_3(z)$,[136] the evaluation being taken to a sufficient number of terms to give the required approximation to $C_3^{\frac{1}{2}}(z)$. The polynomial obtained from the approximation to $C_3^{\frac{1}{2}}(z)$, by reversing the order of its coefficients, is now an approximation to $z^{-\tau}C_2^{\frac{1}{2}}(z)$. Finally, $A(z)$ and $B(z)$ are each given by the product of the approximations to $C_1^{\frac{1}{2}}(z)$ and $z^{-\tau}C_2^{\frac{1}{2}}(z)$, so that

$$A(z) = B(z) \simeq C_1^{\frac{1}{2}}(z)z^{-\tau}C_2^{\frac{1}{2}}(z) = z^{-\tau}C^{\frac{1}{2}}(z) \qquad (4.6.32)$$

To ensure that the coefficients of $A(z)$ and $B(z)$ are real, it may be necessary to let $A(z) = -B(z)$.

Suppose next that $A(z) \neq \pm B(z)$, but now

$$A(z) = M_1(z)N_1(z) \qquad (4.6.33)$$

and
$$B(z) = M_2(z)N_2(z) \qquad (4.6.34)$$

where
$$M_1(z) = \mu_0 + \mu_1 z^{-1} + \ldots + \mu_{j-1} z^{-j+1} \qquad (4.6.35)$$

and
$$M_2(z) = \mu_{j-1} + \mu_{j-2} z^{-1} + \ldots + \mu_0^- z^{-j+1} \qquad (4.6.36)$$

$A(z)$ and $B(z)$ also satisfy Equations 4.6.3 and 4.6.4, as before, and j is any nonnegative integer not greater than l. M_1 and M_2 are the j-component row vectors corresponding to $M_1(z)$ and $M_2(z)$, respectively. N_1 and N_2 are the $(l-j+1)$-component row vectors corresponding to $N_1(z)$ and $N_2(z)$, respectively. M_1, M_2, N_1 and N_2 are all *real* vectors, added zeros extending these to n components.

The DFT's of M_1 and M_2 are the n-component row vectors whose $(i+1)$th components are α_i and β_i, respectively, and the DFT's of N_1 and N_2 are the n-component row vectors whose $(i+1)$th components are γ_i and δ_i, respectively. As before, $n = 2l-1$.

Since M_2 is obtained from M_1 by reversing the order of its components, it can be seen from the reversal theorem (Section 2.3.7) that

$$|\alpha_i| = |\beta_i| \qquad (4.6.37)$$

for $i = 0, 1, \ldots, n-1$.

Using the fact that the ith component of the DFT of the convolution of two vectors is simply the *product* of the ith components of the DFT's of the individual vectors, it follows from Equations 4.6.16, 4.6.17, 4.6.33 and 4.6.34 that

$$|E|^2 = \sum_{i=0}^{n-1} |e_i|^2 = \sum_{i=0}^{n-1} |\alpha_i|^2 |\gamma_i|^2 \qquad (4.6.38)$$

and
$$|F|^2 = \sum_{i=0}^{n-1} |f_i|^2 = \sum_{i=0}^{n-1} |\beta_i|^2 |\delta_i|^2 \qquad (4.6.39)$$

So that, if now $A(z) = M_2(z)N_1(z)$ and $B(z) = M_1(z)N_2(z)$, $|E| |F|$ remains unchanged, and hence, from Equations 4.6.15 and 4.6.20, the probability of error in the detection of a received signal-element remains unchanged.

It can be seen that the $j-1$ roots of $M_2(z)$ are the reciprocals of the $j-1$ roots of $M_1(z)$. Thus, if i of the roots of $B(z)$ are the reciprocals of the corresponding i roots of $A(z)$, $A(z)$ and $B(z)$ may be modified by interchanging one or more of these roots of $A(z)$ with the roots of $B(z)$ that are their reciprocals, and this will not affect the tolerance of the system to additive white Gaussian noise. Complex roots must here be interchanged in complex-conjugate pairs. If a factor

$$V_1(z) = v_0 + v_1 z^{-1} + \ldots + v_h z^{-h} \qquad (4.6.40)$$

of the z-transform $Y(z)$ of the baseband channel is equalized by a linear filter with z-transform $M_1(z)$, given by Equation 4.6.35, then a factor

$$V_2(z) = v_h + v_{h-1} z^{-1} + \ldots + v_0 z^{-h} \qquad (4.6.41)$$

is equalized by a linear filter with z-transform $M_2(z)$, given by Equation 4.6.36. h is here an integer in the range 1 to g and all coefficients are real. Furthermore, the roots of $V_2(z)$ and $M_2(z)$ are the reciprocals of the roots of $V_1(z)$ and $M_1(z)$, respectively. It follows that if the factor of $Y(z)$ equalized by $B(z)$ contains i roots which are reciprocals of the corresponding i roots of the factor of $Y(z)$ equalized by $A(z)$, $A(z)$ and $B(z)$ may be modified to interchange one or more of these i roots equalized by $A(z)$ with the corresponding roots equalized by $B(z)$, without affecting the tolerance of the system to additive white Gaussian noise. In other words, the same interchange of factors containing reciprocal roots may be carried out between the portions of $Y(z)$ that are equalized by $A(z)$ and $B(z)$, as may be carried out between $A(z)$ and $B(z)$ themselves, without affecting the tolerance of the system to noise. Again, complex roots must be interchanged in complex-conjugate pairs.

Consider now the case where the z transform of the channel is

$$Y(z) = Y_1(z)Y_2(z) = y_0 + y_1 z^{-1} + \ldots + y_g z^{-g} \qquad (4.6.42)$$

and satisfies the following conditions. All the roots of $Y_1(z)$ lie inside the unit circle in the z plane, and all the roots of $Y_2(z)$ lie outside the unit circle. $Y(z)$ has an odd number of terms and its $g+1$ coefficients are symmetric in the sense that $y_i = y_{g-i}$, for $i = 0, 1, \ldots, g$. Now $Y_1(z)$ and $Y_2(z)$ each have $\frac{1}{2}g + 1$ terms, with real coefficients, and the coefficients of $Y_2(z)$ are the same as those of $Y_1(z)$ but in the reverse order. Clearly, the roots of $Y_2(z)$ are the reciprocals of the roots of $Y_1(z)$. Let the roots of $Y_1(z)$ be ϕ_1, $\phi_2, \ldots, \phi_{\frac{1}{2}g}$, and let the roots of $Y_2(z)$ be $\psi_1, \psi_2, \ldots, \psi_{\frac{1}{2}g}$. With the appropriate ordering of the roots of $Y_1(z)$ and $Y_2(z)$, these must now satisfy

$$\phi_i = \frac{1}{\psi_i} \qquad (4.6.43)$$

for each i. Evidently, if $A(z)$ equalizes $Y_1(z)$ and $B(z)$ equalizes $Y_2(z)$, $B(z)$ is obtained from $A(z)$ by reversing the order of its coefficients, so that the roots of $B(z)$ are the reciprocals of the roots of $A(z)$. From the reversal theorem (Section 2.3.7) it follows that

$$|e_i| = |f_i| \qquad (4.6.44)$$

for each of the corresponding pair of components of the n-component DFT's of the vectors A and B. From Equation 4.6.27, this satisfies the condition for the minimum probability of error in the detection of a received signal-element.

If we now lift the restriction that the roots of $Y_1(z)$ lie inside the unit circle in the z plane and the roots of $Y_2(z)$ lie outside the unit circle, it can be seen that, so long as the coefficients of $Y_1(z)$ and $Y_2(z)$ remain real, any one or more pairs of reciprocal roots of $Y_1(z)$ and $Y_2(z)$ may be interchanged, without affecting Equations 4.6.43 and 4.6.44 and therefore without affecting the tolerance of the system to additive white Gaussian noise. Strictly speaking, when one of the two interchanged factors is a constant times the reverse of the other, Equation 4.6.43 is still satisfied but Equation 4.6.44 is replaced by Equation 4.6.27.

Thus, whenever the sampled impulse-response of the channel is symmetric about the central sample, the tolerance of the system to additive white Gaussian noise is optimized by equalizing, at the transmitter, any real factor of $Y(z)$ whose roots are the reciprocals of the roots of the remaining factor. The latter is equalized at the receiver.

It is clear from Equation 4.6.15 that for *any* combination of transmitter and receiver equalizers, in the data-transmission system of Figure 4.25, the error probability in the detection of a received signal-element is unaffected by an interchange of transmitter and receiver equalizers. In the particular case where only a single equalizer is used, the error probability is the same whether the channel is equalized at the receiver or at the transmitter.

If, in Equation 4.6.21,

$$|g_i| = d \qquad (4.6.45)$$

for $i = 0, 1, \ldots, n-1$, where d is any positive real constant, then, for the optimum tolerance to noise, the channel may either be equalized entirely at the transmitter or receiver, or else by any combination of transmitter and receiver equalizers satisfying

Equation 4.6.27. But it is shown in Section 2.4.2 that Equation 4.6.45 represents pure phase distortion, with possibly some attenuation. Thus the transmitter and receiver equalizers here combine to form a phase equalizer, which also appropriately adjusts the signal level. This means, of course, that the channel must introduce pure phase distortion, with possibly some attenuation. It can be seen from Equation 4.6.27 that in any combination of transmitter and receiver equalizers that minimize the probability of error under these conditions, both the transmitter and receiver equalizers must themselves be phase equalizers, although the way in which the phase equalization of the channel is shared between the transmitter and receiver does not affect the performance of the system. In the general case, where the channel introduces both amplitude and phase distortions, the transmitter or receiver equalizer in an optimum combination of these equalizers may introduce any required amount of phase distortion, so long as the resultant phase distortion of the channel and this equalizer is corrected by the other equalizer.

Finally, it can be seen from Equation 4.6.27 and the definition of amplitude distortion (Section 2.4.4) that in an optimum combination of linear equalizers, the equalization of the amplitude distortion must be shared *equally* between the transmitter and receiver equalizers, although these need not introduce the same attenuation or gain. This means that in the particular case where the channel introduces pure amplitude distortion and neither equalizer introduces any phase distortion, the tap gains of the transmitter equalizer are in a constant ratio to the corresponding tap gains of the receiver equalizer. In an arrangement of the optimum linear equalization of a channel introducing pure amplitude distortion and some delay, where the roots of the factor of $Y(z)$ equalized by the transmitter equalizer are the reciprocals of the roots of the factor equalized by the receiver equalizer, each equalizer introduces some phase distortion, the phase distortion introduced by one of these being such as to correct the phase distortion introduced by the other. One of the two equalizers is now the reverse of the other, so that the two equalizers are not the same or in a constant ratio. Clearly, by permitting the transmitter and receiver equalizers to introduce the appropriate levels of phase distortion, the linear equalizers that minimize the error probability in the detected element values can sometimes be implemented more simply than when the two equalizers are constrained to be the same or in a constant ratio.

It is reasonable to conclude from the preceding analysis that the smaller the fraction of the signal distortion corresponding to pure phase distortion, the greater the advantage in tolerance to additive white Gaussian noise which may be achieved when a single equalizer at the receiver is replaced by an optimum combination of transmitter and receiver equalizers.

4.6.4 Optimum combination of a linear equalizer at the transmitter and a nonlinear equalizer at the receiver

When a nonlinear equalizer, containing both a linear and a nonlinear filter, is used at the receiver, the linear filter with z-transform $D(z)$ partially equalizes the channel, the equalization process being completed by the nonlinear filter. For any given combination of linear and nonlinear filters, the linear filter may be shared between the transmitter and receiver, the nonlinear filter remaining at the receiver.

It can be seen from Equations 4.3.98 and 4.4.65 that where the whole of the equalization process is carried out at the receiver, the probability of error in the detection of a received signal-element under the assumed conditions is the same function of the sampled impulse-response of the linear filter, whether the linear filter itself equalizes the channel or whether an additional nonlinear filter is used, that is, regardless of whether linear or nonlinear equalization is used. It follows from Equations 4.6.5 and 4.6.15 that the optimum sharing of the linear filter between the transmitter and receiver is unaffected by whether or not a nonlinear filter is used at the receiver.

If now the z transforms of the transmitter and receiver linear-filters are $A(z)$ and $B(z)$, respectively, then

$$A(z)B(z) \simeq z^{-\theta}D(z) \qquad (4.6.46)$$

where θ is an appropriate nonnegative integer, and the condition for the minimum probability of error in the detection of a received signal-element is given by Equation 4.6.27, as before. The corresponding value of the error probability is given by Equation 4.6.29, if error-extension effects are neglected, bearing in mind that $|g_i|$ in Equation 4.6.28 is now the modulus of the $(i+1)$th component in the n-component DFT of the sampled impulse-response of the linear filter with z-transform $D(z)$.

4.6.5   Pure nonlinear equalizer at the transmitter

An interesting technique has recently been proposed whereby it is possible to equalize a channel nonlinearly at the transmitter.[112,120] Consider the arrangement where the channel is equalized entirely by a pure nonlinear equalizer at the transmitter. The data-transmission system is now as shown in Figure 4.26, where the baseband channel is exactly the same as that in Figure 4.25.

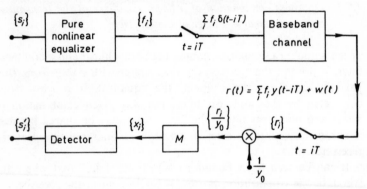

Fig. 4.26   *Equalization of the channel by a pure nonlinear equalizer at the transmitter*

The nonlinear equalizer converts the sequence of binary element values $\{s_i\}$ into the sequence of values $\{f_i\}$ which are sampled and transmitted as the corresponding impulses $\{f_i \delta(t - iT)\}$. As before, $s_i = \pm k$, but the signal level at the input of the baseband channel is *not* now assumed to be such that the average transmitted energy per signal element, $\overline{f_i^2}$, at the input to the transmission path, is necessarily equal to k^2. Again, the noise components $\{w_i\}$ at the output of the sampler in the receiver, are statistically independent Gaussian random variables with zero mean and variance σ^2. Thus the tolerance of the data-transmission system to additive white Gaussian noise is given by the signal/noise ratio $\overline{f_i^2}/\sigma^2$ for a specified average error rate in the detected element values $\{s_i'\}$. Clearly, the z transform of the baseband channel is $Y(z)$ in Equation 4.6.7, where $y_0 \neq 0$, and the sample value at the output of the sampler in the receiver, at time $t = iT$, is

$$r_i = \sum_{j=0}^{g} f_{i-j} y_j + w_i \qquad (4.6.47)$$

The corresponding sample value at the output of the multiplier is r_i/y_0.

The square marked M in Figure 4.26 is a nonlinear network that performs a 'modulo-m' operation on its input signal, as follows. Suppose that the input signal to the network has the value q, where q may be any positive or negative real number. Then the output signal from the network is

$$M[q] = q(\text{modulo } 4k) - 2k \qquad (4.6.48)$$

where

$$q(\text{modulo } 4k) = q - 4jk \qquad (4.6.49)$$

and j is the most positive integer such that $q - 4jk \geqslant 0$. Clearly,

$$0 \leqslant q(\text{modulo } 4k) < 4k \qquad (4.6.50)$$

so that

$$-2k \leqslant M[q] < 2k \qquad (4.6.51)$$

If $-2k \leqslant q < 2k$, $M[q] = q$. The nonlinear transformation M is implemented quite simply by means of a quantizer and subtraction circuit.[120]

The data-transmission system of Figure 4.26 can be represented in terms of the z transforms of the signals, as in Figure 4.27. The

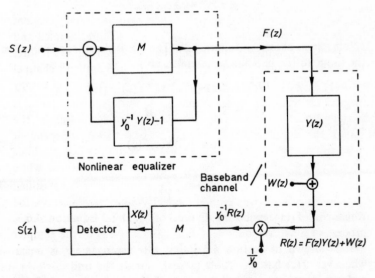

Fig. 4.27 *Mathematical model of the data-transmission system in Figure 4.26*

nonlinear network M at the transmitter is here identical to that at the receiver, and the baseband channel includes the transmitter and receiver samplers. Every signal in Figure 4.27 is a sequence of sample values at the time instants $\{iT\}$.

The nonlinear equalizer in Figure 4.27 is the same as the nonlinear equalizer in Figure 4.19, except that in the former the detector is replaced by the nonlinear network M. The linear feedforward transversal filter in each of these nonlinear equalizers has a sampled impulse-response with z-transform $y_0^{-1} Y(z) - 1$. Thus the transversal filter in Figure 4.27 has g taps, the ith of which has a gain $v_i = y_i/y_0$, as in Figure 4.19. Let the signal at the output of the nonlinear equalizer be the sequence of sample values $\{f_i\}$ with z-transform $F(z)$.

Suppose first that the nonlinear network M is removed from the nonlinear equalizer in the transmitter of Figure 4.27, to give the

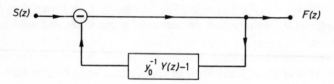

Fig. 4.28   Modified equalizer

arrangement shown in Figure 4.28. The z transform of the signal at the input to the baseband channel is now

$$F(z) = S(z) - F(z)(y_0^{-1}Y(z) - 1)$$

$$= S(z) - y_0^{-1}Y(z)F(z) + F(z) \qquad (4.6.52)$$

so that

$$S(z) = y_0^{-1}Y(z)F(z) \qquad (4.6.53)$$

or

$$F(z) = \frac{y_0 S(z)}{Y(z)} \qquad (4.6.54)$$

where $S(z)/Y(z)$ signifies the division of $S(z)$ by $Y(z)$, using a normal process of long division.

It is shown in Section 4.3.5 that this arrangement is unstable whenever $Y(z)$ has any roots (zeros) outside the unit circle in the z plane. However, when all the roots of $Y(z)$ lie inside the unit circle,

$$\frac{S(z)}{Y(z)} = S(z)Y^{-1}(z) \tag{4.6.55}$$

The arrangement is now stable and achieves the accurate linear equalization of the channel. The nonlinear network M at the receiver is no longer needed in the data-transmission system of Figure 4.27, and, as is shown in Section 4.6.3, the tolerance of the resultant system to additive white Gaussian noise is the same as that of the corresponding system with a linear equalizer at the receiver.

Consider next the data-transmission system in Figure 4.27. The inclusion of the nonlinear network M in the equalizer ensures that the value (area) f_i of a transmitted impulse $f_i\delta(t-iT)$ satisfies $-2k \leqslant f_i < 2k$. By this means it prevents instability in the equalizer regardless of $Y(z)$. If $M[A(z)]$ signifies that the transformation M is applied to the *coefficients* in $A(z)$, the z transform of the signal at the input to the baseband channel is

$$F(z) = M[S(z)-F(z)(y_0^{-1}Y(z)-1)] \tag{4.6.56}$$

so that

$$M[F(z)] = M[S(z)+F(z)-y_0^{-1}Y(z)F(z)] \tag{4.6.57}$$

since, of course, $M[F(z)] = F(z)$. From Equations 4.6.48 and 4.6.57,

$$M[y_0^{-1}Y(z)F(z)] = M[S(z)] \tag{4.6.58}$$

The z transform of the signal at the output of the baseband channel is

$$R(z) = F(z)Y(z)+W(z) \tag{4.6.59}$$

and the z transform of the signal at the input to the nonlinear network M in the receiver is $y_0^{-1}R(z)$.

The z transform of the signal at the output of the nonlinear network M in the receiver is

$$\begin{aligned}
X(z) &= M[y_0^{-1}R(z)] \\
&= M[y_0^{-1}Y(z)F(z)+y_0^{-1}W(z)] \\
&= M[S(z)+y_0^{-1}W(z)]
\end{aligned} \tag{4.6.60}$$

as can be seen from Equations 4.6.59 and 4.6.58. Equation 4.6.60 can be expressed in terms of the sample values at time $t = iT$, to give

$$x_i = M[y_0^{-1}r_i] = M[s_i+y_0^{-1}w_i] \tag{4.6.61}$$

where x_i is the sample value at the detector input at this time instant. Clearly, all intersymbol interference has been eliminated in x_i, so that the channel is accurately equalized.

In the detector, s_i is detected as k or $-k$ depending upon whether x_i is positive or negative, respectively, to give s_i', the detected value of s_i. The z transform of the sequence of the $\{s_i'\}$ is designated $S'(z)$.

It can be seen from Equations 4.6.48 and 4.6.61 that an error occurs in the detection of s_i whenever

$$(4j-3)k < |y_0^{-1} w_i| < (4j-1)k \tag{4.6.62}$$

for all positive integers $\{j\}$. $y_0^{-1} w_i$ is a Gaussian random variable with zero mean and variance $y_0^{-2}\sigma^2$, and $|y_0^{-1} w_i|$ is the modulus or magnitude of $y_0^{-1} w_i$.

At practical signal/noise ratios, the probability of error in the detection of s_i is approximately equal to the probability that $|y_0^{-1} w_i| > k$ and is therefore approximately

$$2\int_k^\infty \frac{1}{\sqrt{(2\pi y_0^{-2}\sigma^2)}} \exp\left(-\frac{w^2}{2y_0^{-2}\sigma^2}\right) dw = 2Q\left(\frac{k|y_0|}{\sigma}\right) \tag{4.6.63}$$

There are no error-extension effects here. At high signal/noise ratios, with error probabilities around 1 in 10^5 or 1 in 10^6, the error probability in Equation 4.6.63 can be taken to be

$$P_e = Q\left(\frac{k|y_0|}{\sigma}\right) \tag{4.6.64}$$

with an inaccuracy of only a small fraction of 1 dB in the corresponding signal/noise ratio.

When the baseband channel in Figure 4.27 introduces no signal distortion, so that $Y(z) = y_0$, bearing in mind that the transmission delay is neglected here, the average transmitted energy per signal element, at the input to the transmission path, is

$$\overline{f_i^2} = k^2 \tag{4.6.65}$$

Suppose now that the baseband channel introduces signal distortion. The value (area) of the impulse at the input to the baseband channel, at time $t = iT$, is now

$$f_i = M[s_i + a_i] \tag{4.6.66}$$

where a_i is a function of the $\{s_j\}$, for $j < i$. This is clear from the structure of the nonlinear equalizer. But the $\{s_i\}$ are statistically independent and equally likely to have either value k or $-k$. Thus, if $|f_i|$ can have the value $k-b$, where $0 \leqslant b \leqslant k$, then it is equally likely to have the value $k+b$. This necessarily implies that, when the channel introduces signal distortion, the average transmitted energy per signal element, at the input to the transmission path, must *exceed* k^2. Hence, for any channel,

$$\overline{f_i^2} \geqslant k^2 \tag{4.6.67}$$

But $-2k \leqslant f_i < 2k$, and when $|f_i|$ can have its maximum value $2k$, it is equally likely to have the value 0. Thus

$$\overline{f_i^2} \leqslant 2k^2 \tag{4.6.68}$$

which means that

$$k^2 \leqslant \overline{f_i^2} \leqslant 2k^2 \tag{4.6.69}$$

Clearly, the average transmitted energy per signal element *must* lie in the range 0–3 dB above k^2. When the channel introduces severe signal distortion, a good estimate of $\overline{f_i^2}$ is obtained by assuming that f_i is equally likely to have *any* value in the range $-2k$ to $2k$. Now $\overline{f_i^2} = 1.333k^2$, which is 1.25 dB above its minimum value of k^2. It seems likely therefore that $\overline{f_i^2}$ will often have a value in the range 0–1.5 dB above k^2.

The above analysis suggests that, in the general case, a reasonable approximation for the average transmitted energy per signal element is $\overline{f_i^2} = k^2$, bearing in mind that it is likely to exceed this value by a little over 1 dB when there is severe distortion. Since the tolerance of the system to noise is measured by the value of the signal/noise ratio $\overline{f_i^2}/\sigma^2$, for a given error rate in the $\{s_i'\}$, it follows that the actual tolerance to noise of the system may be typically up to about 1.5 dB below that evaluated on the assumption that $\overline{f_i^2} = k^2$.

Consider now the data-transmission system shown in Figure 4.1, where the receiver is as shown in Figure 4.18 and contains the pure nonlinear equalizer shown in Figure 4.19 (where $v_i = y_i/y_0$). The linear feedforward transversal filter here has the z-transform $y_0^{-1} Y(z) - 1$, and is therefore the same as the transversal filter in the nonlinear equalizer of Figure 4.27. It is clear that the data-transmission system with the pure nonlinear equalizer at the receiver is obtained from the system in Figure 4.27 by transferring the nonlinear equalizer from the transmitter to the receiver and removing the two nonlinear networks M. In the data-transmission system with the pure non-linear equalizer at the receiver, the average transmitted energy per signal element is *exactly* k^2 and so the signal/noise ratio is exactly k^2/σ^2. Also, from Equation 4.4.7, the probability of error in the detection of s_i can be taken to be

$$P_e = Q\left(\frac{k|y_0|}{\sigma}\right) \qquad (4.6.70)$$

when error-extension effects are neglected. At high signal/noise ratios, the error-extension effects tend to reduce the tolerance to noise by typically up to 1 dB and sometimes a little more, say up to 1.5 dB.

From Equations 4.6.64 and 4.6.70, the two systems have theoretically the same tolerance to additive white Gaussian noise, when it is assumed that the average transmitted energy per element is k^2 in each case, and bearing in mind the other approximations that have been made in these equations.

At high signal/noise ratios, the reduction in tolerance to additive white Gaussian noise caused by error-extension effects, in the data-transmission system with a pure nonlinear equalizer at the receiver, is of the same order as the reduction in tolerance to noise of the system with a pure nonlinear equalizer at the transmitter, resulting from the following facts. Firstly, the actual error probability is twice that in Equation 4.6.64, and secondly, the transmitted signal level here generally exceeds the assumed value of k^2. Thus, at high signal/noise ratios, the two systems have similar tolerances to additive white Gaussian noise, these being normally within about 1 dB of each other. The two systems can be considered as duals of each other, since they use the same equalization process, which is located in one case at the transmitter and in the other case at the receiver.

4.6.6 Optimum nonlinear equalizer at the transmitter

Just as the data-transmission system, with a pure nonlinear filter (equalizer) at the receiver, can be modified by the addition of a linear filter and the appropriate change in the nonlinear filter, to achieve the optimum nonlinear equalization of the channel, so also can the data-transmission system with a pure nonlinear filter (equalizer) at

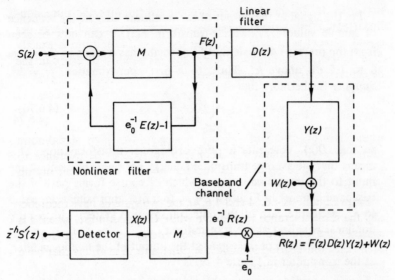

Fig. 4.29 Equalization of the channel by the optimum nonlinear equalizer at the transmitter

the transmitter, shown in Figure 4.27. The resultant system is shown in Figure 4.29 and operates as follows.

The linear filter added at the transmitter is an *n*-tap feedforward transversal filter with z-transform

$$D(z) = d_0 + d_1 z^{-1} + \ldots + d_{n-1} z^{-n+1} \qquad (4.6.71)$$

The output signal from this filter is fed to the baseband channel, which is that in Figure 4.26 together with the two samplers.

The *z* transform of the linear filter and baseband channel in Figure 4.29 is

$$D(z)Y(z) \simeq z^{-h}E(z) \qquad (4.6.72)$$

where

$$E(z) = e_0 + e_1 z^{-1} + \ldots + e_l z^{-l} \qquad (4.6.73)$$

h and *l* are appropriate positive integers. The linear filter and baseband channel, with z-transform $z^{-h}E(z)$, are equalized by the nonlinear filter at the transmitter in Figure 4.29, in a manner similar to that for the equalization of the baseband channel in Figure 4.27. Thus the *z* transform of the transversal filter, that forms part of the nonlinear filter, is $e_0^{-1}E(z) - 1$.

Let the signal at the output of the nonlinear filter be the sequence of sample values $\{f_i\}$ with z-transform $F(z)$. It can now be seen from the previous analysis that $\overline{f_i^2}$ normally has a value in the range 0 to 1.5 dB above k^2. Thus, to a first approximation, $\overline{f_i^2} = k^2$. Suppose furthermore that

$$\sum_{i=0}^{n-1} d_i^2 = 1 \tag{4.6.74}$$

so that $D(z)$ represents a pure orthogonal transformation. This means that the average transmitted energy per signal element, at the input to the transmission path (which of course forms part of the baseband channel in Figure 4.29) is, to a first approximation, equal to k^2, as in the case of the data-transmission system with a pure nonlinear equalizer at the transmitter.

The z transform of the signal at the output of the nonlinear filter, at the transmitter in Figure 4.29, is

$$F(z) = M[S(z) - F(z)(e_0^{-1}E(z) - 1)] \tag{4.6.75}$$

so that

$$M[F(z)] = M[S(z) + F(z) - e_0^{-1}E(z)F(z)] \tag{4.6.76}$$

or

$$M[e_0^{-1}E(z)F(z)] = M[S(z)] \tag{4.6.77}$$

The z transform of the signal at the input to the baseband channel is $F(z)D(z)$ and the z transform of the signal at the receiver input is

$$R(z) = F(z)D(z)Y(z) + W(z)$$

$$\simeq z^{-h}E(z)F(z) + W(z) \tag{4.6.78}$$

from Equation 4.6.72. This signal is multiplied by e_0^{-1} and fed to the nonlinear network M.

The z transform of the signal at the input to the detector is

$$X(z) = M[e_0^{-1}R(z)]$$

$$\simeq M[e_0^{-1}z^{-h}E(z)F(z) + e_0^{-1}W(z)]$$

$$= M[z^{-h}S(z) + e_0^{-1}W(z)] \tag{4.6.79}$$

from Equation 4.6.77. It follows that the sample value at the detector input, at time $t = (i+h)T$, is

$$x_{i+h} = M[s_i + e_0^{-1}w_{i+h}] \tag{4.6.80}$$

s_i is detected from the sign of x_{i+h}. At high signal/noise ratios, the probability of error in the detection of s_i can be taken to equal the probability that $e_0^{-1} w_{i+h} > k$, which is

$$P_e = Q\left(\frac{k|e_0|}{\sigma}\right) \qquad (4.6.81)$$

Clearly, to minimize P_e, $|e_0|$ must be maximized and this can be achieved by the appropriate choice of $D(z)$.

The value of $D(z)$ that maximizes $|e_0|$, given that $\sum_{i=0}^{n-1} d_i^2 = 1$, is the same as (or a constant times) the value of $D(z)$ that minimizes $\sum_{i=0}^{n-1} d_i^2$, given e_0, and this is determined in Section 4.4.6. The fact that e_0 is set to unity in Section 4.4.6 does not affect the minimization process other than by multiplying the terms of both $D(z)$ and $E(z)$ by the appropriate constant. Furthermore, it does not affect the tolerance of the system to noise, since both the signal and the noise are affected equally by the chosen value of e_0.

Suppose that the z transform of the baseband channel is

$$Y(z) = Y_1(z)Y_2(z) = y_0 + y_1 z^{-1} + \ldots + y_g z^{-g} \qquad (4.6.82)$$

where all the roots of $Y_1(z)$ lie inside the unit circle in the z plane, and all the roots of $Y_2(z)$ lie outside the unit circle. Let $Y_3(z)$ be the polynomial obtained from $Y_2(z)$ by reversing the order of its coefficients. $Y_1(z)$, $Y_2(z)$ and $Y_3(z)$ are given by Equations 4.4.36, 4.4.37 and 4.4.86, respectively.

From Section 4.4.6, the linear filter that minimizes $\sum_{i=0}^{n-1} d_i^2$, given that $e_0 = 1$, has the z-transform

$$D_0(z) \simeq q_f^{-1} z^{-h} Y_2^{-1}(z) Y_3(z) \qquad (4.6.83)$$

where q_f is the first term of $Y_3(z)$. Thus the linear filter that maximizes $|e_0|$, given that $\sum_{i=0}^{n-1} d_i^2 = 1$, has the z-transform

$$D(z) \simeq z^{-h} Y_2^{-1}(z) Y_3(z) \qquad (4.6.84)$$

since $z^{-h} Y_2^{-1}(z) Y_3(z)$ represents an orthogonal transformation with no attenuation or gain. The linear filter equalizes a signal with z-transform $Y_2(z) Y_3^{-1}(z)$, which represents pure phase distortion, so that the linear filter is a pure phase equalizer.

The z transform of the linear filter and baseband channel is

$$D(z)Y(z) \simeq z^{-h}Y_1(z)Y_3(z) \qquad (4.6.85)$$

and, from Equation 4.6.72,

$$E(z) = Y_1(z)Y_3(z) \qquad (4.6.86)$$

The first term of $Y_1(z)$ is 1, the first term of $Y_3(z)$ is q_f, and $Y_1(z)Y_3(z)$ has $g+1$ terms. Thus $e_0 = q_f$ and $l = g$. From Equation 4.6.81, the probability of error in the detection of s_i can now be taken to be

$$P_e = Q\left(\frac{k|q_f|}{\sigma}\right) \qquad (4.6.87)$$

In practice, the average transmitted energy per signal element is likely to be up to about 1.5 dB above the value k^2 assumed (for the reasons explained in Section 4.6.5), giving the corresponding reduction in tolerance to noise.

The linear filter with z transform $D(z)$ is a pure phase equalizer such that, in the z transform of the channel and linear filter, all the roots of $Y(z)$ lying outside the unit circle in the z plane are replaced by their reciprocals. The linear filter partially equalizes the channel, the equalization process being completed by the nonlinear filter that comprises the feedback transversal filter and the nonlinear network M.

If the nonlinear network at the transmitter is removed, the baseband channel is now equalized linearly at the transmitter and the nonlinear network M at the receiver may also be removed. The linear equalizer is stable since all the roots of $E(z)$ lie inside the unit circle in the z plane. This system, of course, has the same tolerance to noise as that where the channel is equalized linearly at the receiver.

Consider now the data-transmission system shown in Figure 4.1, where the receiver is as shown in Figure 4.21 and uses the optimum nonlinear equalizer described in Section 4.4.6. The linear feedforward transversal filter, that forms part of the nonlinear filter here, has the z-transform $e_0^{-1}E(z)-1$ and is therefore the same as the transversal filter in the nonlinear equalizer of Figure 4.29. In Section 4.4.6 of course, the first term of $E(z)$ is 1, and not e_0 as it is here, so that $e_0^{-1}E(z)-1$ becomes $E(z)-1$. It is clear that the data-transmission system with the optimum nonlinear equalizer at the receiver is obtained from the system in Figure 4.29 by transferring the nonlinear equalizer from the transmitter to the receiver and removing the two nonlinear networks M. Since the first term of $E(z)$ is now taken to be e_0, the multiplier in the receiver of Figure 4.29 must be retained.

From Equations 4.4.95 and 4.6.87, the two systems have theoretically the same tolerance to additive white Gaussian noise, given the various approximations that have been made. In practice, the tolerance to noise of each system at high signal/noise ratios is reduced a little, for the same reasons as those given in Section 4.6.5 for the corresponding two systems using pure nonlinear equalizers, so that the two systems should in fact have similar tolerances to additive white Gaussian noise. The two systems can be considered as duals of each other, since they use the same equalization process, which is located in one case at the transmitter and in the other at the receiver.

Clearly the two suboptimum nonlinear equalizers that are described in Sections 4.4.3 and 4.4.4 can be implemented at the transmitter instead of at the receiver, in a manner similar to that just described for the optimum nonlinear equalizer. The tolerances to additive white Gaussian noise of the two resultant systems are not significantly different from the tolerances of the systems from which they are derived, and their performances are therefore inferior to that of the optimum nonlinear equalizer, located at the transmitter.

It may readily be shown that if $Y(z)$ contains one or more roots *on* the unit circle in the z plane, the design of the optimum nonlinear equalizer, located at the transmitter, is obtained by taking $Y_1(z)$ in Equation 4.6.82 as the factor of $Y(z)$ that includes all the roots *on or inside* the unit circle.

4.7 COMPARISON OF THE DIFFERENT EQUALIZERS

A comparison has been made of the tolerances to additive white Gaussian noise of the different equalizers studied here, using the results of the theoretical analysis in this chapter.[135] It is assumed that the data-transmission system is as shown in Figure 4.1 or 4.25 and as described in Section 4.2 or 4.6, respectively. The systems have been tested over eight different channels and the results of these tests are given in Table 4.1. The meanings of the system numbers

Table 4.1 Number of dB reduction in tolerance to additive white Gaussian noise, at high signal/noise ratios, when the given channel replaces one that introduces no distortion or attenuation

Channel No.	Sampled impulse-response of channel					System								
						1A	2A	3A	4A	5A	1B	3B	4B	5B
1	0.348	0.870	0.348	0	0	7.9	9.2	5.0	4.4	3.1	5.6	4.1	3.8	3.1
2	0.348	0.870	−0.348	0	0	0.1	9.2	0.8	0.6	0.1	0.1	0.4	0.4	0.1
3	0.625	0.750	0.219	0	0	11.4	4.1	9.0	4.1	4.1	8.0	6.5	4.1	4.1
4	0.219	0.750	0.625	0	0	11.4	13.2	4.7	11.4	4.1	8.0	4.4	8.0	4.1
5	0.167	0.471	0.707	0.471	0.167	20.5	15.5	16.1	13.5	9.9	17.2	13.6	12.0	9.9
6	0.182	−0.571	0.642	−0.428	0.214	18.7	14.9	15.0	19.8	8.9	15.1	12.1	15.0	8.9
7	0.265	−0.486	0.728	−0.368	0.169	13.9	11.5	11.0	10.1	7.1	11.3	9.2	8.8	7.1
8	0.152	0.597	0.643	−0.429	−0.152	2.1	16.4	4.5	3.2	0.9	1.5	3.1	2.1	0.9

Table 4.2   Meanings of the symbols in the system numbers

Symbol	Meaning of the symbol
A	Equalization process located either entirely at the receiver or entirely at the transmitter
B	Equal sharing of the linear equalization process between the transmitter and receiver
1	Linear equalizer (Section 4.3)
2	Pure nonlinear equalizer achieving the accurate equalization of the channel (Sections 4.4.1 and 4.6.5)
3	Nonlinear equalizer using zero forcing (Section 4.4.3)
4	Nonlinear equalizer using linear and nonlinear equalization of complementary factors of the channel response (Section 4.4.4)
5	Optimum nonlinear equalizer (Sections 4.4.6 and 4.6.6)

in Table 4.1 are given in Table 4.2. In every case the channel is accurately equalized and there is a high signal/noise ratio. Each entry in Table 4.1 gives the number of decibels increase in the noise level, which is necessary to maintain a given low error probability, when a channel with the corresponding sampled impulse-response is replaced by one introducing no distortion or attenuation, and the transmitted signal level is held unchanged. All systems have the same tolerance to noise in the presence of no distortion or attenuation. The 5-component row vector, giving the sampled impulse-response of the channel, has unit length in every case, so that each channel introduces distortion but no attenuation or gain. The different values of the sampled impulse-response of the channel have been selected to give a fairly wide range of different signal distortions.

The roots of the z transforms of the Channels 1 and 5 are in reciprocal pairs, all the roots of Channel 3 lie inside the unit circle in the z plane, and all the roots of Channel 4 lie outside the unit circle. Channels 1 and 5 introduce pure amplitude distortion, and Channel 2 introduces nearly pure phase distortion, one of its two roots being the negative of the reciprocal of the other. Channels 3, 4, 6, 7 and 8 introduce different combinations of amplitude and phase distortions.

The tolerance to additive white Gaussian noise, at high signal/noise ratios, of any system using a nonlinear equalizer, either at the transmitter or at the receiver, may be up to about 1.5 dB inferior to the value quoted in Table 4.1, due to the various reasons previously discussed. However, where the nonlinear equalizer is at the receiver, the reduction in tolerance to noise is most often no more than $\frac{1}{2}$ dB.

The results in Table 4.1 show that the poorest tolerance to additive white Gaussian noise is likely to be obtained with System 2A, closely followed by System 1A. Thus the use of pure linear or nonlinear equalization at the receiver, with no equalizer at the transmitter, will usually (but not always) give a lower tolerance to additive white Gaussian noise than any of the arrangements of combined linear and nonlinear equalization studied here. Of course, in the case of pure phase distortion (the nearest approach to which is Channel 2) System 1A is at least as good as or better than any other system, and when all the roots of the channel lie inside the unit circle, as for Channel 3, System 2A is at least as good as or better than any other system.

System 1B gains an advantage of up to between 3 and 4 dB over System 1A. This advantage is generally (but not always) greater than that obtained by System 3B over 3A or by System 4B over 4A. System 5B gains no advantage over System 5A, since the linear equalizer is here a pure phase equalizer, which gives no improvement in tolerance to noise when shared between the transmitter and receiver. Systems 1B, 3B and 4B are consistently as good as or better than 1A, 3A and 4A, respectively.

The best tolerance to additive white Gaussian noise is consistently obtained with Systems 5A and 5B, with often a considerable advantage over the other systems. It appears therefore that the most cost-effective system is 5A.

REFERENCES

1. Savage, J. E. 'Some simple self-synchronizing digital data scramblers', *Bell Syst. Tech. J.*, **46**, 449–487 (1967)
2. Nakamura, K. and Iwadare, Y. 'Data scramblers for multilevel pulse sequences', *Electron. Commun. Japan*, **55-A**, No. 6, 8–16 (1972)
3. Leeper, D. G. 'A universal digital data scrambler', *Bell Syst. Tech. J.*, **52**, 1851–1865 (1973)
4. Kasai, H., Senmoto, S. and Matsushita, M. 'PCM jitter suppression by scrambling', *IEEE Trans. Commun.*, **COM-22**, 1114–1122 (1974)
5. Saltzberg, B. R. 'Error probabilities for a binary signal perturbed by intersymbol interference and Gaussian noise', *IEEE Trans. Commun. Syst.*, **CS-12**, 117–120 (1964)
6. Marinides, H. F. and Reijns, G. L. 'Influence of bandwidth restriction on the signal-to-noise performance of a PCM/NRZ signal', *IEEE Trans. Aerospace Electron. Syst.*, **AES-4**, 35–40 (1968)

7. Saltzberg, B. R. 'Intersymbol interference error bounds with application to ideal bandlimiting signalling', *IEEE Trans. Inform. Theory*, **IT-14**, 563–568 (1968)

8. Saltzberg, B. R. and Simon, M. K. 'Data transmission error probabilities in the presence of low-frequency removal and noise', *Bell Syst. Tech. J.*, **48**, 225–273 (1969)

9. Lugannani, R. 'Intersymbol interference and probability of error in digital systems', *IEEE Trans. Inform. Theory*, **IT-15**, 682–688 (1969)

10. Ho, E. Y. and Yeh, Y. S. 'A new approach for evaluating the error probability in the presence of intersymbol interference and additive Gaussian noise', *Bell Syst. Tech. J.*, **49**, 2249–2266 (1970)

11. Ho, E. Y. and Yeh, Y. S. 'Error probability of a multilevel digital system with intersymbol interference and Gaussian noise', *Bell Syst. Tech. J.*, **50**, 1017–1023 (1971)

12. Shimbo, O. and Celebiler, M. I. 'The probability of error due to intersymbol interference and Gaussian noise in digital communication systems', *IEEE Trans. Commun. Technol.*, **COM-19**, 113–119 (1971)

13. Hill, F. S. 'The computation of error probability for digital transmission', *Bell Syst. Tech. J.*, **50**, 2055–2077 (1971)

14. Yeh, Y. S. and Ho, E. Y. 'Improved intersymbol interference error bounds in digital systems', *Bell Syst. Tech. J.*, **50**, 2585–2598 (1971)

15. Prabhu, V. K. 'Some considerations of error bounds in digital systems', *Bell Syst. Tech. J.*, **50**, 3127–3151 (1971)

16. Forney, G. D. 'Lower bounds on error probability in the presence of large intersymbol interference', *IEEE Trans. Commun.*, **COM-20**, 76–77 (1972)

17. Falconer, D. D. and Gitlin, R. D. 'Bounds on error-pattern probabilities for digital communications systems', *IEEE Trans. Commun. Technol.*, **COM-20**, 132–139 (1972)

18. Glave, F. E. 'An upper bound to the probability of error due to intersymbol interference for correlated digital signals', *IEEE Trans. Inform. Theory*, **IT-18**, 356–363 (1972)

19. Yao, K. 'On minimum average probability of error expression for a binary pulse communication system with intersymbol interference', *IEEE Trans. Inform. Theory*, **IT-18**, 528–531 (1972)

20. Benedetto, S., de Vincentis, G. and Luvison, A. 'Error probability in the presence of intersymbol interference and additive noise for multilevel digital signals', *IEEE Trans. Commun.*, **COM-21**, 181–190 (1973)

21. Hill, F. S. and Blanco, M.A. 'Random geometric series and intersymbol interference', *IEEE Trans. Inform. Theory*, **IT-19**, 326–335 (1973)

22. Matthews, J. W. 'Sharp error bounds for intersymbol interference', *IEEE Trans. Inform. Theory*, **IT-19**, 440–447 (1973)

23. McGee, W. F. 'A modified intersymbol interference error bound', *IEEE Trans. Commun.*, **COM-21**, 862–863 (1973)

24. Korn, I. 'Probability of error in binary communication systems with causal band-limited filters—Part 1: Nonreturn-to-zero signal', *IEEE Trans. Commun.*, **COM-21**, 878–890 (1973)

25. Korn, I. 'Probability of error in binary communication systems with causal band-limited filters—Part 2: Split-phase signal', *IEEE Trans. Commun.*, **COM-21**, 891–898 (1973)

26. Bozic, S. M. and Tahim, K. S. 'Dependence of intersymbol interference upon the signalling rate', *Int. J. Electron.*, **35**, 523–528 (1973)

27. Korn, I. 'Bounds to probability of error in binary communication systems

with intersymbol interference and dependent or independent symbols', *IEEE Trans. Commun.*, **COM-22**, 251–254 (1974)

28. Vanelli, J. C. and Shehadeh, N. M. 'Computation of bit-error probability using the trapezoidal integration rule', *IEEE Trans. Commun.*, **COM-22**, 331–334 (1974)

29. Korn, I. 'Improvement to sharp error bounds for intersymbol interference', *Proc. IEE*, **122**, 265–267 (1975)

30. Aein, J. M. and Hancock, J. C. 'Reducing the effects of intersymbol interference with correlation receivers', *IEEE Trans. Inform. Theory*, **IT-9**, 167–175 (1963)

31. George, D. A. 'Matched filters for interfering symbols', *IEEE Trans. Inform. Theory*, **IT-11**, 153–154 (1965)

32. Tufts, D. W. 'Nyquist's problem—the joint optimization of transmitter and receiver in pulse amplitude modulation', *Proc. IEEE*, **53**, 248–259 (1965)

33. Smith, J. W. 'The joint optimization of transmitted signal and receiving filter for data transmission systems', *Bell Syst. Tech. J.*, **44**, 2363–2392 (1965)

34. Aaron, M. R. and Tufts, D. W. 'Intersymbol interference and error probability', *IEEE Trans. Inform. Theory*, **IT-12**, 26–34 (1966)

35. Deighton, P. D. 'Optimisation of realisable receiver in pulse-amplitude modulation', *Electronics Letters*, **3**, 129–130 (1967)

36. Berger, T. and Tufts, D. W. 'Optimum pulse amplitude modulation, Part 1: Transmitter-receiver design and bounds from information theory', *IEEE Trans. Inform. Theory*, **IT-13**, 196–208 (1967)

37. Berger, T. and Tufts, D. W. 'Optimum pulse amplitude modulation, Part 2: Inclusion of timing jitter', *IEEE Trans. Inform. Theory*, **IT-13**, 209–216 (1967)

38. Deighton, P. D. 'Joint optimisation of realisable receiver and transmitter in data-transmission systems', *Electronics Letters*, **3**, 342–344 (1967)

39. Chang, R. W. and Freeny, S. L. 'Hybrid digital transmission systems—Part 1: Joint optimization of analogue and digital repeaters', *Bell Syst. Tech. J.*, **47**, 1663–1686 (1968)

40. Davisson, L. E. 'Steady state error in adaptive mean-square minimization', *IEEE Trans. Inform. Theory*, **IT-16**, 382–385 (1970)

41. Chang, R. W. 'Joint optimization of automatic equalization and carrier acquisition for digital communication', *Bell Syst. Tech. J.*, **49**, 1069–1104 (1970)

42. Ericson, T. 'Structure of optimum receiving filters in data transmission systems', *IEEE Trans. Inform. Theory*, **IT-17**, 352–353 (1971)

43. Ho, E. Y. 'Optimum equalization and the effect of timing and carrier phase on synchronous data systems', *Bell Syst. Tech. J.*, **50**, 1671–1689 (1971)

44. Hansler, E. 'Some properties of transmission systems with minimum mean-square error', *IEEE Trans. Commun. Technol.*, **COM-19**, 576–579 (1971)

45. Ericson, T. and Johansson, U. 'Digital transmission over coaxial cables', *Ericsson Technics*, **27**, No. 4, 191–272 (1971)

46. Cho, Y. S. 'Optimal equalization of wideband coaxial cable channels using "bump" equalizers', *Bell Syst. Tech. J.*, **51**, 1327–1345 (1972)

47. Moore, J. B. and Hetrakul, P. 'Optimal demodulation of PAM signals', *IEEE Trans. Inform. Theory*, **IT-19**, 188–196 (1973)

48. Gardner, W. A. 'The structure of least mean square linear estimators for synchronous M-ary signals', *IEEE Trans. Inform. Theory*, **IT-19**, 240–243 (1973)

49. Mark, J. W. and Budihardjo, P. S. 'Joint optimization of receive filter and equalizer', *IEEE Trans. Commun.*, **COM-21**, 264–266 (1973)

50. Ericson, T. 'Optimum PAM filters are always band limited', *IEEE Trans. Inform. Theory*, **IT-19**, 570–572 (1973)
51. Mark, J. W. 'Relationship between source coding, channel coding and equalization in data transmission', *Proc. IEEE*, **61**, 1657–1659 (1973)
52. Benedetto, S. and Biglieri, E. 'On linear receivers for digital transmission systems', *IEEE Trans, Commun.*, **COM-22**, 1205–1215 (1974)
53. Rappeport, M. A. 'Automatic equalization of data transmission facility distortion using transversal equalizers', *IEEE Trans. Commun. Technol.*, **COM-12**, 65–73 (1964)
54. Schreiner, K. E., Funk, H. L. and Hopner, E. 'Automatic distortion correction for efficient pulse transmission', *IBM J. Res. Develop.*, **9**, 20–30 (1965)
55. Becker, F. K., Holzman, L. N., Lucky, R. W. and Port, E. 'Automatic equalization for digital communication', *Proc. IEEE*, **53**, 96–97 (1965)
56. Lucky, R. W. 'Automatic equalization for digital communication', *Bell Syst. Tech. J.*, **44**, 547–588 (1965)
57. Lucky, R. W. 'Techniques for adaptive equalization of digital communication systems', *Bell Syst. Tech. J.*, **45**, 255–286 (1966)
58. Rudin, H. R. 'Automatic equalization using transversal filters', *IEEE Spectrum*, **4**, 53–59 (1967)
59. Lytle, D. W. 'Convergence criteria for transversal equalizers', *Bell Syst. Tech. J.*, **47**, 1775–1800 (1968)
60. Hirsh, D. and Wolf, W. J. 'A simple adaptive equalizer for efficient data transmission', *IEEE Trans. Commun. Technol.*, **COM-18**, 5–12 (1970)
61. Newhall, E. E., Qureshi, S. U. H. and Simone, C. F. 'A technique for finding approximate inverse systems and its application to equalization', *IEEE Trans. Commun. Technol.*, **COM-19**, 1116–1127 (1971)
62. Guida, A. 'Optimum tapped delay line for digital signals', *IEEE Trans. Commun.*, **COM-21**, 277–283 (1973)
63. Lucky, R. W., Salz, J. and Weldon, E. J. *Principles of Data Communication*, pp. 93–165, McGraw-Hill, New York (1968)
64. Salomonsson, G. 'An equalizer with feedback filter', *Ericsson Technics*, **28**, No. 2, 57–101 (1972)
65. Gitlin, R. D. and Mazo, J. E. 'Comparison of some cost functions for automatic equalization', *IEEE Trans. Commun.*, **COM-21**, 233–237 (1973)
66. Lucky, R. W. and Rudin, H. R. 'Generalized automatic equalization for communication channels', *Proc. IEEE*, **54**, 439–440 (1966)
67. Lucky, R. W. and Rudin, H. R. 'An automatic equalizer for general-purpose communication channels', *Bell Syst. Tech. J.*, **46**, 2179–2208 (1967)
68. Di Toro, M. J. 'Communication in time-frequency spread media using adaptive equalization', *Proc. IEEE*, **56**, 1653–1679 (1968)
69. Gersho, A. 'Adaptive equalization of highly dispersive channels for data transmission', *Bell Syst. Tech. J.*, **48**, 55–70 (1969)
70. Proakis, J. G. and Miller, J. H. 'An adaptive receiver for digital signalling through channels with intersymbol interference', *IEEE Trans. Inform. Theory*, **IT-15**, 484–497 (1969)
71. Rudin, H. R. 'A continuously adaptive equalizer for general purpose communication channels', *Bell Syst. Tech. J.*, **48**, 1865–1884 (1969)
72. Potter, J. B. 'The adaptive equalization of communication systems, particularly those using waveform fidelity criteria', *Proc. Inst. Radio Electron. Eng. Aust.*, **30**, 326–336 (1969)
73. Davisson, L. E. and Schwartz, S. C. 'Analysis of a decision-directed receiver with unknown priors', *IEEE Trans. Inform. Theory*, **IT-16**, 270–276 (1970)

74. Proakis, J. G. 'Adaptive digital filters for equalization of telephone channels', *IEEE Trans. Audio Electroacoustics*, **AU-18**, 195–200 (1970)
75. Niessen, C. W. and Willim, D. K. 'Adaptive equalizer for pulse transmission', *IEEE Trans. Commun. Technol.*, **COM-18**, 377–395 (1970)
76. Lender, A. 'Decision-directed digital adaptive equalization technique for high-speed data transmission', *IEEE Trans. Commun. Technol.*, **COM-18**, 625–632 (1970)
77. De, S. and Davies, A. C. 'Convergence of adaptive equaliser for data transmission', *Electronics Letters*, **6**, 858–861 (1970)
78. Schonfield, T. J. and Schwartz, M. 'Rapidly converging first order training algorithm for an adaptive equalizer', *IEEE Trans. Inform. Theory*, **IT-17**, 431–439 (1971)
79. Chang, R. W. 'A new equalizer structure for fast start-up digital communication', *Bell Syst. Tech. J.*, **50**, 1969–2014 (1971)
80. Schonfield, T. J. and Schwartz, M. 'Rapidly converging second order tracking algorithm for adaptive equalization', *IEEE Trans. Inform. Theory*, **IT-17**, 572–579 (1971)
81. Lawrence, R. E. and Kaufman, H. 'The Kalman filter for the equalization of a digital communications channel', *IEEE Trans. Commun. Technol.*, **COM-19**, 1137–1141 (1971)
82. Mark, J. W. and Haykin, S. S. 'Adaptive equalization for digital communication', *Proc. IEE*, **118**, 1711–1720 (1971)
83. Brewster, R. L. 'A digital adaptive equaliser for high-speed voice channel modems', *IERE Conf. Proc.*, No. 23, 373–382 (1972)
84. Stuttard, E. B. 'Automatic adaptive equalisation in data modems', *IERE Conf. Proc.*, No. 23, 383–390 (1972)
85. Richman, S. H. and Schwartz, M. 'Dynamic programming training period for an MSE adaptive equalizer', *IEEE Trans. Commun.*, **COM-20**, 857–864 (1972)
86. Ungerboeck, G. 'Theory on the speed of convergence in adaptive equalizers for digital communication', *IBM J. Res. Develop.*, **16**, 546–555 (1972)
87. Sha, R. T. and Tang, D. T. 'A new class of automatic equalizers', *IBM J. Res. Develop.*, **16**, 556–566 (1972)
88. Chang, R. W. and Ho, E. Y. 'On fast start-up data communication systems using pseudo-random training sequences', *Bell Syst. Tech. J.*, **51**, 2013–2027 (1972)
89. Walzman, T. and Schwartz, M. 'Automatic equalization using the discrete frequency domain', *IEEE Trans. Inform. Theory*, **IT-19**, 59–68 (1973)
90. Potter, J. B. 'Application of time-series algebra to the adaptive equalisation of band-limited waveform-transmission systems', *Proc. IEE*, **120**, 191–196 (1973)
91. Gitlin, R. D., Ho, E. Y. and Mazo, J. E. 'Passband equalization of differentially phase-modulated data signals', *Bell Syst. Tech. J.*, **52**, 219–238 (1973)
92. Karnaugh, M. 'Automatic equalizers having minimum adjustment time', *IBM J. Res. Develop.*, **17**, 176–179 (1973)
93. Qureshi, S. U. H. 'Adjustment of the position of the reference tap of an adaptive equalizer', *IEEE Trans. Commun.*, **COM-21**, 1046–1052 (1973)
94. Walzman, T. and Schwartz, M. 'A projected gradient method for automatic equalization in the discrete frequency domain', *IEEE Trans. Commun.*, **COM-21**, 1442–1446 (1973)
95. Taylor, M. G. 'A technique for using a time-multiplexed second order digital filter section for performing adaptive filtering', *IEEE Trans. Commun.*, **COM-22**, 326–330 (1974)

96. Godard, D. 'Channel equalization using a Kalman filter for fast data transmission', *IBM J. Res. Develop.*, **18**, 267–273 (1974)
97. Cho, Y. S. 'Mean square error equalization using manually adjusted equalizers', *Bell Syst. Tech. J.*, **53**, 847–865 (1974)
98. Lee, T. S. and Cunningham, D. R. 'Kalman filter equalizer for QPSK digital communications channel', *Conf. Rec. IEEE Int. Conf. Communications*, pp. 25A.1–25A.5 (1974)
99. Fleischer, P. E. 'Active adjustable loss and delay equalizers', *IEEE Trans. Commun.*, **COM-22**, 951–955 (1974)
100. Kleibanov, S. B., Privalskii, V. B. and Time, I. V. 'Kalman filter for equalization of digital communications channel', *Autom. Remote Control*, **35**, No. 7, Pt. 1, 1097–1102 (1974)
101. Eriksson, L. E. 'Transmitter and receiver filters for digital PAM using transversal filters with few taps', *IEEE Trans. Commun.*, **COM-22**, 1215–1225 (1974)
102. Allen, J. B. and Mazo, J. E. 'A decision-free equalization scheme for minimum phase channels', *IEEE Trans. Commun.*, **COM-22**, 1732–1733 (1974)
103. Kosovych, O. S. and Pickholtz, R. L. 'Automatic equalization using successive overrelaxation iterative technique', *IEEE Trans. Inform. Theory*, **IT-21**, 51–58 (1975)
104. Mueller, K. H. 'A new fast-converging mean square algorithm for adaptive equalizer with partial response signalling', *Bell Syst. Tech. J.*, **54**, 143–153 (1975)
105. Mueller, K. H. and Spaulding, D. A. 'Cyclic equalization—A new rapidly converging equalization technique for synchronous data communication', *Bell Syst. Tech. J.*, **54**, 369–406 (1975)
106. Luvision, A. and Pirani, G. 'A method to compute optimal gains in recursive linear filtering applications', *IEEE Trans. Commun.*, **COM-23**, 399–400 (1975)
107. Fitch, S. M. and Kurz, L. 'Recursive equalization in data transmission— A design procedure and performance evaluation', *IEEE Trans. Commun.*, **COM-23**, 546–550 (1975)
108. Gorog, E. 'A new approach to time-doman equalization', *IBM J. Res. Develop.*, **9**, 228–232 (1965)
109. Boyd, R. T. and Monds, F. C. 'Adaptive equaliser for multipath channels', *Electronics Letters*, **6**, 556–558 (1970)
110. Boyd, R. T. and Monds, F. C. 'Equaliser for digital communication', *Electronics Letters*, **7**, 58–60 (1971)
111. Monsen, P. 'Feedback equalization for fading dispersive channels', *IEEE Trans. Inform. Theory*, **IT-17**, 56–64 (1971)
112. Tomlinson, M. 'New automatic equalizer employing modulo arithmetic', *Electronics Letters*, **7**, 138–139 (1971)
113. Taylor, D. P. 'Nonlinear feedback equaliser employing a soft limiter', *Electronics Letters*, **7**, 265–267 (1971)
114. George, D. A., Bowen, R. R. and Storey, J. R. 'An adaptive decision feedback equalizer', *IEEE Trans. Commun. Technol.*, **COM-19**, 281–293 (1971)
115. Costello, P. J. and Patrick, E. A. 'Unsupervised estimation of signals with intersymbol interference', *IEEE Trans. Inform. Theory*, **IT-17**, 620–622 (1971)
116. Ungerboeck, G. 'Nonlinear equalization of binary signals in Gaussian noise', *IEEE Trans. Commun. Technol.*, **COM-19**, 1128–1137 (1971)

117. Bershad, N. J. and Vena, P. A. 'Eliminating intersymbol interference—a state space approach', *IEEE Trans. Inform. Theory*, **IT-18**, 275–281 (1972)
118. Brownlie, J. D. 'Effects of decision-errors on adaptive equalization', *IERE Conf. Proc.*, No. 23, 221–240 (1972)
119. Price, R. 'Nonlinearly feedback equalized PAM vs. capacity for noisy filter channels', *Conf. Rec. IEEE Int. Conf. Communications*, pp. 22.12–22.17 (1972)
120. Harashima, H. and Miyakawa, H. 'Matched-transmission technique for channels with intersymbol interference', *IEEE Trans. Commun.*, **COM-20**, 774–780 (1972)
121. Westcott, R. J. 'An experimental adaptively equalised modem for data transmission over the switched telephone network', *Radio Electron. Eng.*, **42**, 499–507 (1972)
122. Monsen, P. 'Digital transmission performance on fading dispersive diversity channels', *IEEE Trans. Commun.*, **COM-21**, 33–39 (1973)
123. Clark, A. P. 'Design technique for nonlinear equalisers', *Proc. IEE*, **120**, 329–333 (1973)
124. Mark, J. W. 'A note on the modified Kalman filter for channel equalization', *Proc. IEEE*, **61**, 481–483 (1973)
125. Mark, J. W. and Budihardjo, P. S. 'Performance of jointly optimized prefilter-equalizer receivers', *IEEE Trans. Commun.*, **COM-21**, 941–945 (1973)
126. Kobayashi, H. and Tang, D. T. 'A decision-feedback receiver for channels with strong intersymbol interference', *IBM J. Res. Develop.*, **17**, 413–419 (1973)
127. Taylor, D. P. 'The estimate feeedback equalizer: A suboptimum nonlinear receiver', *IEEE Trans. Commun.*, **COM-21**, 979–990 (1973)
128. Salz, J. 'Optimum mean-square decision feedback equalization', *Bell Syst. Tech. J.*, **52**, 1341–1373 (1973)
129. Messerschmitt, D. G. 'A geometric theory of intersymbol interference. Part 1: Zero-forcing and decision feedback equalization', *Bell Syst. Tech. J.*, **52**, 1483–1519 (1973)
130. Falconer, D. D. and Foschini, G. J. 'Theory of minimum mean square error QAM systems employing decision feedback equalization', *Bell Syst. Tech. J.*, **52**, 1821–1849 (1973)
131. Salazar, A. C. 'Design of transmitter and receiver filters for decision feedback equalization', *Bell Syst. Tech. J.*, **53**, 503–523 (1974)
132. Koeth, H. and Schollmeier, G. 'An adaptive equalizer for partial response signals with improved convergence properties', *IEEE Trans. Commun.*, **COM-22**, 884–885 (1974)
133. Duttweiler, D. L., Mazo, J. E. and Messerschmitt, D. G. 'An upper bound on the error probability in decision feedback equalization', *IEEE Trans. Inform. Theory*, **IT-20**, 490–497 (1974)
134. Monsen, P. 'Adaptive equalization of the slow fading channel', *IEEE Trans. Commun.*, **COM-22**, 1064–1075 (1974)
135. Clark, A. P. and Tint, U. S. 'Linear and non-linear transversal equalizers for baseband channels', *Radio Electron. Eng.*, **45**, 271–283 (1975)
136. Dwight, H. B. *Tables of Integrals and other Mathematical Data*, fourth edition, pp. 1–18, Macmillan, New York (1961)
137. Swokowski, E. *Fundamentals of College Algebra*, 2nd ed., Prindle, Weber and Schmidt, Boston, Mass. (1971)
138. Paley, H. and Weichsel, P. M. *Elements of Abstract and Linear Algebra*, Holt, Rinehart, and Winston, New York (1972)

139. Berlekamp, E. R. *Algebraic Coding Theory*, McGraw-Hill, New York, (1968)
140. Thomas, J. B. *An Introduction to Statistical Communication Theory*, Wiley, New York (1969)
141. Clark, A. P. *Principles of Digital Data Transmission*, Pentech Press, London (1976)

5 Adaptive detection processes

5.1 INTRODUCTION

It is shown in Chapter 4 that a data-transmission system using the optimum nonlinear equalizer at the receiver sometimes achieves a considerable advantage in tolerance to additive white Gaussian noise over the corresponding data-transmission system using a linear equalizer, and its performance is never inferior. However, it can be seen from the method of operation of the nonlinear equalizer that, except in the particular case where the channel introduces pure phase distortion, when the nonlinear equalizer degenerates into a pure phase equalizer and provides the best available detection process for the received signal, only a *portion* of a received signal-element is used in the detection of that element. Each element value s_i is detected from just the first nonzero component of that element, at the output of the linear filter, the remaining part of the element being eliminated in the nonlinear filter. Clearly, if the *whole* of a received signal-element could be used effectively in its detection, an even better tolerance to additive noise should be obtained. This implies that *all* the sample values at the output of the linear filter that are dependent upon s_i should be used in the detection of s_i and not just the first of these. Furthermore, even the received sample values not directly dependent on s_i carry information that can be used to assist in the detection of s_i, so that the more sample values, at the output of the linear filter, that are used in the detection of s_i, the better the tolerance of the system to noise.

Again, a useful reduction in equipment complexity could be achieved in the optimum nonlinear equalizer, if the linear filter could be omitted. This is partly because the linear filter often requires more taps than the nonlinear filter, if optimum equalization is to be achieved, and partly because, with no linear filter, the tap gains of the nonlinear filter are given directly by the sampled impulse-response of the channel, which can be determined both simply and accurately from the received signal, whether or not the noise is Gaussian or stationary. When a linear filter is used in a nonlinear equalizer, the tap gains of both the linear and nonlinear filters are no longer simply related to the sampled impulse-response of the channel, and must be determined through the technique of cross correlation described in Section 4.5.

The technique of adaptive detection to be described here is a development of a pure nonlinear equalizer (Section 4.4.1). Thus it is based on an arrangement in which, after the detection of a received signal-element, all components of that element in the received sampled signal are removed by subtraction, ready for the detection of the next signal element. Each received signal-element is now detected from *several* of the received sample values and *no* linear filter is used.

Any number of different detection processes can be used here, so that it is of interest to determine the detection process that achieves the best compromise between performance and equipment complexity, in other words, the most cost-effective system. To do this it is necessary first to have a clear understanding of what is the best available detection process, regardless of equipment complexity.

In recent years various studies have been made of the best available detection process for a received message that comprises a serial stream of signal elements.[5-36] It is clear from these studies that there are at least two different commonly accepted definitions of what is the best detection process.[9,10]

The detection process that minimizes the probability of error in the detection of an individual received *signal-element* and hence minimizes the average error rate in the detected element values, when a message comprising a stream of signal elements is received in the presence of additive noise, is the process that evaluates the *a posteriori* probability (Section 3.2.8) of each of the possible values of a received element, given the *whole* of the received signal, and then accepts the element value having the largest *a posteriori* probability as the detected value.[1,9,10] The process is repeated for each element. The detection process that minimizes the probability of error in the detection of the received *message*, that is, in the detection of the whole stream of signal elements, is the process that evaluates the *a posteriori* probability of each possible received message and accepts as the detected message that having the highest *a posteriori* probability of being correct (Section 3.2.9).[6] The difference between the two detection processes is that the latter minimizes the probability of there being *one or more* element errors in the detected message, without distinguishing between the *number* of element errors.[9,10]

With the reception of a long message, either detection process is far too complex to be implemented as such. The best that can be done in practice is to limit the number of signal elements involved in a detection process to a relatively small group, so that many detection processes are involved in the detection of the complete message. Under these conditions and at high signal/noise ratios, with additive white Gaussian noise, even if the average element-error rate is 1 in 10^3 and there are 100 elements in a group, there are normally only one or two element-errors in a detected group. Thus the detection process, that minimizes the probability of error (one or more element-errors) in the detection of a group of elements, will for practical purposes also minimize the average probability of error in the detection of an individual element and hence minimize the average element-error rate. This applies even when the basic detection process introduces error-extension effects, so that errors occur in bursts and there are either no errors in a group or usually quite a number. Since the detection process that minimizes the probability of error (one or more element-errors) is appreciably less complex to implement than the other,[1,12,17] it will here be considered as the optimum detection process. It is also the optimum detection process described in Section 3.2.9.

5.2   BASIC ASSUMPTIONS

Consider the synchronous serial binary data-transmission system shown in Figure 5.1. The signal at the input to the baseband channel is a sequence of regularly spaced impulses the ith of which, $s_i\delta(t-iT)$, occurs at time $t = iT$ and has the value

$$s_i = \pm k \tag{5.2.1}$$

where k is a positive constant. As before, the $\{s_i\}$ are statistically independent and equally likely to have either binary value. The baseband channel is as described in Section 4.2. Thus the average transmitted energy per signal element at the input to the transmission path is k^2, and the noise components $\{w_i\}$, in the received samples $\{r_i\}$ at the output of the sampler in the receiver, are statistically

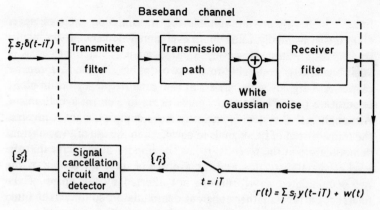

Fig. 5.1 *Data-transmission system*

independent Gaussian random variables with zero mean and variance $\sigma^2 = \frac{1}{2}N_0$. $\frac{1}{2}N_0$ is the two-sided power spectral density of the additive white Gaussian noise at the input to the receiver filter. A high signal/noise ratio is assumed.

The impulse response of the baseband channel in Figure 5.1 is $y(t)$, and the sampled impulse-response of the channel is given by the $(g+1)$-component row vector

$$V = y_0 \; y_1 \ldots y_g \qquad (5.2.2)$$

where $y_i = y(iT)$ and $y_0 \neq 0$, the delay in transmission, other than that involved in the time dispersion of the signal, being neglected as before. Of course, $y_i = 0$ for $i < 0$ and $i > g$. Thus the received sample at the time instant $t = iT$ is

$$r_i = \sum_{j=0}^{g} s_{i-j} y_j + w_i \qquad (5.2.3)$$

and this is fed to the signal cancellation circuit and detector, in Figure 5.1.

It is assumed that the receiver has prior knowledge of V and of the two possible values of s_i. The detector may operate on the $\{r_i\}$ in any one of a number of different ways to give the detected values of the $\{s_i\}$, which are designated the $\{s_i'\}$.

The arrangement considered here includes the case where a vestigial sideband suppressed carrier AM signal is transmitted over a voice-frequency channel, such as a telephone circuit or HF radio link.[37] Such systems are suitable for transmission rates of up to 4800 bits/s, normally using multilevel signal-elements at the higher transmission rates. However, at transmission rates of 4800 or 9600 bits/s, it is preferable to use two double sideband suppressed carrier AM signals with carriers at the same frequency but in phase quadrature (at 90°), the two signals being in element synchronism. Although such a system is basically more complex, since it involves the transmission of two signals in parallel and the use of two coherent demodulators at the receiver, it has the important property that the level of signal distortion and the average of the total power of the two demodulated waveforms are not affected by the phase of the reference carrier in either coherent demodulator, so long as the two reference carriers remain in phase quadrature. This means that in a detection process which suitably corrects or compensates for the signal distortion introduced in transmission, it is no longer necessary to hold the phase of the reference carrier in a coherent demodulator at any particular angle relative to the carrier phase of the corresponding data signal, as must be done in a vestigial-sideband system. Thus there is no need for the transmission of a pilot carrier with the data signal, nor therefore for the receiver to extract the pilot carrier from the received signal. This leads to a significant reduction in equipment complexity and avoids the serious problems of phase jitter sometimes encountered in vestigial-sideband systems. These, of course, normally require the transmission of a pilot carrier to achieve the required carrier phase synchronization.

In a data-transmission system employing two double sideband suppressed carrier AM signals in phase quadrature, variations in the relative phases of the received signal carrier and the demodulator reference-carriers can be corrected for by means of a simple phase-locked loop or predictor that operates on the demodulated and sampled signal, no phase or frequency correction being here applied to the local oscillator itself. A description of the technique is however beyond the scope of the present analysis. Furthermore, in a system of this type, the $\{r_i\}$, $\{s_i\}$, $\{w_i\}$ and $\{y_i\}$ are complex numbers, but, apart from this, the basic principles of operation and the relative performances of the different techniques to be studied are not significantly different from those of the corresponding vestigial-sideband system. Since the additional conceptual difficulties introduced through the use of complex numbers are not compensated for by any deeper insight into the basic mechanisms of the different techniques, only *real* signals will now be considered, as previously assumed.

5.3 DETECTION PROCESSES USING DECISION-DIRECTED SIGNAL CANCELLATION

5.3.1 *Decision-directed cancellation of intersymbol interference*

Instead of detecting s_i from the corresponding sample value x_i, obtained from the received sample r_i after the removal of *all* intersymbol interference in a pure nonlinear equalizer (Section 4.4.1), s_i is now detected from n sample values $\{r'_j\}$, which are obtained from the corresponding n received samples $\{r_j\}$, for $j = i, i+1, \ldots, i+n-1$, through the removal of the intersymbol interference from the *preceding* signal elements. The removal of the intersymbol interference is achieved in a signal cancellation circuit that is a development of the pure nonlinear equalizer. The arrangement is as shown in Figure 5.2. The signal cancellation circuit here comprises the buffer store and multiplier.

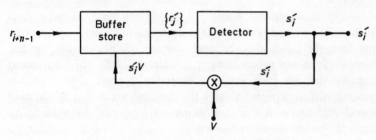

Fig. 5.2 Signal cancellation circuit and detector

The multiplier multiplies each component of V by the detected element value s'_i, to give, at the corresponding $g+1$ input terminals to the buffer store, the $g+1$ components of the detected signal-vector $s'_i V$.

The buffer store holds the n sample values used for the detection of s_i. These are the corresponding n samples $\{r_j\}$, for $j = i, i+1, \ldots, i+n-1$, modified by the cancellation (removal) of all components of the previously detected signal-elements, assuming for the moment that they have been correctly detected and that $n > g$.

Figure 5.2 shows the signals present just after the detection of

s_i and just before the cancellation of the components of the corresponding signal-element from the n sample values $\{r'_j\}$ that have just been used in the detection process. It is assumed here, for convenience, that the last received sample r_{i+n-1}, which is in fact also the nth of the $\{r'_j\}$ used for the detection of s_i, is held at the input until the receipt of the next sample r_{i+n}, at time $t = (i+n)T$. Immediately after the detection of s_i, its detected value is used to generate the $(g+1)$-component vector

$$s'_i V = s'_i y_0 \ \ s'_i y_1 \ldots s'_i y_g \qquad (5.3.1)$$

With the correct detection of s_i, $s'_i = s_i$ and the vector $s'_i V$ contains the $g+1$ components of the ith received signal-element that are present in the corresponding samples $\{r'_j\}$ held in the buffer store.

The $g+1$ components of $s'_i V$ are now removed by subtraction from the corresponding storage elements in the buffer store. This is an arrangement of decision-directed signal cancellation in which *all* components of $s'_i V$ are removed *simultaneously* from the corresponding received samples, rather than successively, at the appropriate time instants, as in a nonlinear feedback transversal equalizer (Figure 4.19).

The modified sample values $\{r'_j\}$ held in the buffer store are now each shifted by one place, so that the second of these, r'_{i+1}, becomes the first, the third, r'_{i+2}, becomes the second, and so on, the last sample of the new group being r_{i+n}, which is the first of the received samples not yet involved in a detection process. The detection process is then repeated to give the detected value s'_{i+1} of the next signal-element, which is then cancelled (removed by subtraction from the buffer store), and so on.

If $n < g+1$, the buffer store holds $g+1$ sample values instead of n, and the detection process operates on only the first n of these. Ideally $n > g+1$ and this is assumed during the following discussion unless otherwise stated.

In a detection process, *all* the $\{s_j\}$, that are involved in the sample values held in the buffer store, are detected simultaneously in an iterative process. However, only the detected value s'_i of the *first* of the $\{s_j\}$ is accepted as correct. The components of the corresponding signal-element, which is the vector $s_i V$, are now known and are cancelled.

If the n sample values held in the buffer store are renamed so that they are given by the n components of the row vector

$$R' = r'_1 \ r'_2 \ldots r'_n \tag{5.3.2}$$

and if the element values s_{i-1} to s_{i-g} have been correctly detected and the corresponding signal components cancelled (removed from the buffer store), then immediately preceding the detection of s_i at time $t = (i+n-1)T$,

$$R' = \sum_{j=1}^{n} s_{i+j-1}(\overbrace{0 \ldots 0}^{j-1} \ y_0 \ y_1 \ldots y_{n-j})$$
$$+ (w_i \ w_{i+1} \ldots w_{i+n-1}) \tag{5.3.3}$$

where $y_i = 0$ for $i > g$, and the contents of a pair of brackets represent an n-component row vector.

It can be seen that, with the correct detection and cancellation of the preceding g signal-elements, the vector R' contains the set of n sample values that would have been received, starting at time $t = iT$, if the ith signal-element $s_i V$ had been the first signal-element received. Thus, once a signal element has been correctly cancelled, the received signal is as though that signal element had never been transmitted.

To simplify the nomenclature in Equation 5.3.3, rename s_i as s_1, s_{i+1} as s_2, and so on, and let S be the n-component row vector

$$S = s_1 \ s_2 \ldots s_n \tag{5.3.4}$$

Similarly, rename w_i as w_1, w_{i+1} as w_2, and so on, and let W be the n-component row vector

$$W = w_1 \ w_2 \ldots w_n \tag{5.3.5}$$

Let Y be the $n \times n$ matrix whose ith row is given by the n-component row vector

$$Y_i = \overbrace{0 \ldots 0}^{i-1} \ y_0 \ y_1 \ldots y_{n-i} \tag{5.3.6}$$

where again $y_i = 0$ for $i > g$. Then, in the absence of noise, the n-component vector held in the buffer store for the detection of s_1 is

$$\sum_{i=1}^{n} s_i Y_i = SY \tag{5.3.7}$$

and in the presence of noise this vector becomes

$$R' = SY + W \tag{5.3.8}$$

where SY and W are the signal and noise vectors, respectively. s_1 is now detected from R'.

It is important to note that the processes of *detection* and *signal cancellation* are two quite separate operations. The detection of an element is the *decision* as to its value s_1, and the cancellation of an element is the *removal* of the components of the element from the received signal. In order to be able to *cancel* the detected signal-element $s_1 Y_1$, the detector must know both s_1 and Y_1, so that it has a complete knowledge of the sample values of this element. However, it can *detect* s_1 with surprisingly little prior knowledge of s_1 and Y_1. If it knows Y_1, it requires no prior knowledge of $|s_1|$, and if it knows $|s_1|$, it requires only limited prior knowledge of Y_1. This will be clarified in the following sections.

5.3.2 System 1A

Since $y_0 \neq 0$, Y is a nonsingular upper-triangular matrix and both vectors S and SY have 2^n possible values which are equally likely and are known at the receiver. Since the $\{w_i\}$ are statistically independent Gaussian random variables with zero mean and fixed variance, the detection process that minimizes the probability of error in the detection of S from R' selects the possible value of the vector S for which $|R' - SY|$ has the minimum value. This is shown in Section 3.2.9. $|R' - SY|$ is the distance between the vectors R' and SY in the n-dimensional Euclidean vector space containing these vectors. At high signal/noise ratios, the detection process also minimizes the probability of error in the detection of s_1.[1]

The detection process operates as follows. The receiver generates the n-component vector X, such that X has one of the 2^n possible values of the vector S. The value of x_1, the first component of X, is fed into a store. The receiver now forms the vector XY and evaluates $|R' - XY|^2$, which is the square of the distance between R' and XY. This value is fed into another store. The receiver then sets the vector X to another of the possible values of S and evaluates $|R' - XY|^2$, as before. It then compares the new value of $|R' - XY|^2$ with the stored value. If the new value is smaller than the stored value, the receiver replaces the stored values of both x_1 and $|R' - XY|^2$ by the new values, and continues with the third of the possible values of S. If, on the other hand, the new value of

$|R' - XY|^2$ is equal to or greater than the corresponding stored value, the receiver takes no action and continues with the third of the possible values of S. The process continues in this way until all 2^n possible values of S have been tested. At the end of the process, the stored value of x_1 is the value corresponding to the smallest $|R' - XY|$ and is therefore taken as the required detected value of s_1.

With the correct detection of s_1, the receiver knows the corresponding signal-element vector $s_1 Y_1$ and removes this vector by subtraction from R' to give $R' - s_1 Y_1$. The resultant sample values are moved one place to the left to give the new vector R', whose extreme right-hand component is the first of the received samples not yet involved in a detection process.

The detection process for s_1 from R', just described, is known here as System 1A.

To start the process of detection and signal cancellation, a known sequence of at least g $\{s_i\}$ can be transmitted and the components of the corresponding received signal-elements can then be cancelled automatically in the buffer store, without the detection of the corresponding $\{s_i\}$. The process of detection and signal cancellation now proceeds as previously described.

If a signal element is incorrectly detected, the cancellation of this element doubles instead of removes its intersymbol interference in the following elements, thus greatly increasing the probability of error in their detection. The arrangement therefore suffers from error extension effects, and errors in the detected element values tend to occur in bursts. Tests by computer simulation have shown that, with System 1A, the error rate may be increased by typically up to about four times due to the wrong cancellation of signal elements. With inferior detection processes the error rate may be increased by typically up to about fifteen times. However, in the presence of additive white Gaussian noise, at high signal/noise ratios giving error rates of 1 in 10^5 or 1 in 10^6, an increase of up to fifteen times in the error rate corresponds to a reduction in tolerance to Gaussian noise of only up to a little over 1 dB, which is not very serious. Of course, at low signal/noise ratios or with non-Gaussian noise, the error extension effects may introduce a significant reduction in tolerance to noise.

Since the system recovers automatically from errors in detection, there is in fact no need to transmit a sequence of known element-values at the start of transmission, correct operation now being achieved after the reception of a few signal elements. Furthermore, the system also recovers automatically after the temporary loss of the received signal or after a burst of high-level noise, just so long as there is a recovery of the correct time-synchronization (phase of the sampling instants) at the receiver and, at the same time, no large change in the sampled impulse-response of the channel.

If n, the number of samples used in a detection process, is increased so that the detection of each signal element involves the *whole* of the remaining part of the message, then the detection process minimizes the probability of error (that is, one or more element errors) in the detection of the whole message. This detection process is in fact now equivalent to that described in Section 3.2.9, and no other detection process can give a lower error probability. At high signal/noise ratios, the detection process also approximately minimizes the probability of error in the detection of s_1 and therefore in the detection of each signal-element.[1] Clearly, as n becomes very large, the probability of error in the detection of a received signal-element in System 1A, given a high signal/noise ratio and the correct detection of the preceding g elements, approaches the minimum value obtainable with any detection process of any type.[5-36]

Again, when $n < g + 1$, it is clear that the receiver is not using the whole of its available prior knowledge of the received signal, since it now makes no use of the last $g - n + 1$ components of the signal element whose value is being determined in a detection process. Clearly, the tolerance of the system to noise cannot now be optimum but must decrease with n, reaching its minimum value when $n = 1$.

The accurate theoretical analysis of the probability of error in the detection of s_1 by System 1A, even given the correct detection and cancellation of the preceding g elements, becomes very difficult when n is large and there is appreciable signal distortion. However, an approximate upper bound to the probability of error in the detection of s_1, at high signal/noise ratios and with small values of n, is readily obtained as follows. Under the assumed conditions, practically all errors in the detection of s_1 occur when the vector S has the value S_b and is incorrectly detected as S_c, where S_c differs from S_b in its first component and, subject to this constraint, minimizes the distance $|S_b Y - S_c Y|$ between the vectors $S_b Y$ and $S_c Y$, the minimum being that for *all* possible vectors S_b and S_c. Clearly there are two or more vectors $\{S_b\}$ and any one of these may be associated with one or more vectors $\{S_c\}$ such that $|S_b Y - S_c Y|$ has the given minimum value. Obviously, the set of vectors $\{S_b\}$ *includes* the set of vectors $\{S_c\}$.

In the detection process of System 1A, the decision boundary separating the vectors $S_b Y$ and $S_c Y$ is the $(n-1)$-dimensional hyperplane that perpendicularly bisects the line joining the two vectors in the n-dimensional Euclidean vector space. When $S = S_b$ and R' lies on the side of the decision boundary closer to $S_c Y$, s_1 is incorrectly detected. Under these conditions, the orthogonal projection of the noise vector W on to the line joining $S_b Y$ to $S_c Y$ lies in the direction from $S_b Y$ towards $S_c Y$ and has a magnitude greater than

$$d = \tfrac{1}{2}|S_b Y - S_c Y| \tag{5.3.9}$$

It is shown in Section 3.2.7 that the value of the noise component (orthogonal projection of W) along *any* direction in the vector space is a Gaussian random variable with zero mean and variance σ^2. Thus the probability that S_b is incorrectly detected as S_c is

$$\int_d^\infty \frac{1}{\sqrt{(2\pi\sigma^2)}} \exp\left(-\frac{w^2}{2\sigma^2}\right) dw = Q\left(\frac{d}{\sigma}\right) \tag{5.3.10}$$

It can now be seen from the discussion in Section 3.2.9 that the average probability of error in the detection of s_1 can be taken to be $ef Q(d/\sigma)$, where e and f are positive quantities defined as follows. e is the fraction of the 2^n different possible vectors $\{SY\}$, that are known here as the vectors $\{S_b Y\}$ and have one or more of the other possible vectors $\{SY\}$, known here as $\{S_c Y\}$, differing in the first component of S and at the given minimum distance from the first vector $S_b Y$. f gives the increase in error probability due to the average number of vectors $\{S_c Y\}$ associated with each vector $S_b Y$. f can never be greater than this average number and is normally somewhat smaller. Clearly, $e<1$ and $f>1$, and in general $ef<1$.

Since a change in the probability of error by a factor of two or three times, at high signal/noise ratios with additive Gaussian noise, corresponds to a change of only a fraction of 1 dB in the signal/noise ratio, it can be seen that an approximate upper bound to the probability of error in the detection of s_1 is

$$P_e = Q\left(\frac{d}{\sigma}\right) \tag{5.3.11}$$

where $2d$ is the minimum distance between two signal vectors $\{SY\}$

differing in the first components of the vectors $\{S\}$. When $n \not> 10$, d can readily be evaluated by computer.

For values of n up to 5 or 6, Equation 5.3.11 generally gives quite an accurate estimate of the signal/noise ratio for a given probability of error in the detection of s_1, assuming the correct detection of the preceding g signal-elements. The discrepancy in the estimate is normally less than 1 dB. However, for the larger values of n, the estimated signal/noise ratio may be up to a few dB too high. This is because, for certain values of the sampled impulse-response of the channel, there are only *two* possible vectors $\{S\}$, differing in their first components, for which the corresponding vectors $\{SY\}$ are at the minimum distance. Furthermore, there is also now a much greater distance between any other pairs of vectors $\{SY\}$ for which the vectors $\{S\}$ differ in their first components. When these conditions hold, they usually apply for all values of n. Thus there is now a probability of 2^{1-n} that S has one of the two values that maximize the probability of error in the detection of s_1, and, for the larger values of n, the chance of S having one of these two values in any given detection process becomes negligible.

In general, computer simulation is the only truly reliable method of determining the tolerance to additive white Gaussian noise of System 1A, for any particular channel and value of n. It takes account not only of the effect just described but also of error-extension effects which have been neglected in the present discussion.

System 1A has the serious weakness that it requires 2^n sequential operations to determine the vector SY at the minimum distance from R' and hence to detect the value of s_1. This number becomes excessive when $n > 10$. It is necessary therefore to consider other detection processes that do not involve either complex equipment or an excessive number of sequential operations, but whose tolerance to noise can be made to approach close to that of System 1A.

5.3.3 Example

It is interesting now to compare the tolerance to additive white Gaussian noise of System 1A with that of the optimum nonlinear equalizer, for the case where the baseband channel in Figure 5.1 has the z-transform

$$Y(z) = 1 + 2.5z^{-1} + z^{-2} \tag{5.3.12}$$

Assume that $n = 3$, so that three sample values are used in the detection of the element-value s_1 of a received signal-element $s_1 Y_1$. Assume also that the preceding two elements are correctly cancelled, so that their intersymbol interference has been eliminated. The received vector R' used in the detection of s_1 is now

$$R' = SY + W \qquad (5.3.13)$$

where R', S and W are 3-component row vectors, given by Equations 5.3.2, 5.3.4 and 5.3.5, respectively. $s_i = \pm k$ for each i, and the components $\{w_i\}$ of the noise vector W are statistically independent Gaussian random variables with zero mean and variance σ^2. Y is the 3×3 matrix

$$Y = \begin{bmatrix} 1 & 2.5 & 1 \\ 0 & 1 & 2.5 \\ 0 & 0 & 1 \end{bmatrix} \qquad (5.3.14)$$

The possible values of the 3-component vector SY are given by the following eight vectors:

$$\begin{array}{llll}
k & (1 & 3.5 & 4.5) \\
k & (1 & 3.5 & 2.5) \\
k & (1 & 1.5 & -0.5) \\
k & (1 & 1.5 & -2.5) \\
k & (-1 & -1.5 & 2.5) \\
k & (-1 & -1.5 & 0.5) \\
k & (-1 & -3.5 & -2.5) \\
k & (-1 & -3.5 & -4.5)
\end{array}$$

An error is caused in the detection of s_1 if one of the upper four of these vectors is detected as one of the lower four, or vice versa.

The decision boundaries in the detection process are the 16 planes that perpendicularly bisect the 16 lines joining the upper four vectors to the lower four in the 3-dimensional Euclidean vector space containing these vectors. The distance from any vector to a decision boundary is *half* the distance between the two vectors separated by this boundary. Thus the minimum distance to a decision boundary, in the detection of s_1, is *half* the minimum distance between the upper four and lower four vectors, in the set of eight vectors $\{SY\}$ listed above.

It can be seen that the *minimum* distance from a vector SY in one group of four to a vector SY in the other group of four is

$$(2^2 + 3^2 + 1^2)^{\frac{1}{2}}k = 3.742k \qquad (5.3.15)$$

so that the minimum distance to a decision boundary in the detection of s_1 is

$$d = 1.871k \qquad (5.3.16)$$

The value of the orthogonal projection of the noise vector W on to *any* direction in the 3-dimensional vector-space is a Gaussian random variable with zero mean and variance σ^2.

Thus the probability of an error in an element detection process, assuming the correct detection of the preceding two signal elements, is approximately

$$P_e = Q\left(\frac{d}{\sigma}\right) = Q\left(\frac{1.871k}{\sigma}\right) \qquad (5.3.17)$$

The previously considered arrangement which gives the best tolerance to additive white Gaussian noise, when $Y(z)$ satisfies Equation 5.3.12, is the optimum nonlinear equalizer. From Equation 4.4.119, the probability of error in the detection of a signal element, assuming the correct detection of the preceding two elements, is here

$$P_e = Q\left(\frac{2k}{\sigma}\right) \qquad (5.3.18)$$

Thus the nonlinear equalizer has an advantage in tolerance to additive white Gaussian noise of 0.6 dB over System 1A. On the other hand, if n is increased from 3 to 6 in System 1A, the latter gains a small advantage over the nonlinear equalizer. The error-extension effects, which have so far been neglected, are smaller in System 1A than in the optimum nonlinear equalizer, so long as $n > 3$.

5.3.4   System 2A

This detection process minimizes the probability of error in the detection of S from R', when the receiver has prior knowledge of the $n \times n$ matrix Y but no knowledge of the $\{|s_i|\}$ or σ^2, assuming a binary polar transmitted signal, as before. It is shown in Section 3.4 that the detection process involves first the maximum-likelihood estimation of S from R', using a process of exact linear equalization to give the vector X whose ith component is x_i, and then the detection of the $\{s_i\}$ from the signs of the corresponding $\{x_i\}$. From Equation 3.3.56, the maximum-likelihood estimate of S from R' is the n-component row vector

$$X = R'Y^T(YY^T)^{-1} = R'Y^{-1} \qquad (5.3.19)$$

since Y is here an $n \times n$ nonsingular matrix. Thus the n-component vector R' at the output of the buffer store in Figure 5.2 is fed to the n input terminals of the $n \times n$ linear network Y^{-1} to give at its n output terminals the n-component vector

$$X = (SY+W)Y^{-1} = S+U \qquad (5.3.20)$$

where

$$U = WY^{-1} \qquad (5.3.21)$$

Each component u_i of the noise vector U is a Gaussian random variable with zero mean and a variance which is not normally equal to σ^2 and which may vary from one component to another.

Equation 5.3.20 shows that the network Y^{-1} performs a process of exact linear equalization on the input vector R', such that

$$x_i = s_i + u_i \qquad (5.3.22)$$

for $i = 1, 2, \ldots, n$, and all intersymbol interference is eliminated at the output of the network. The correct detection and hence cancellation of the g signal-elements preceding $s_1 Y_1$ is, of course, assumed here.

Since only the first component s_1 of S is in fact required to be detected, the detection process is completed by detecting s_1 as k or $-k$, depending upon whether x_1 is positive or negative, respectively.

Now Y is an upper-triangular square matrix and so also is Y^{-1}. Furthermore, the component in the first row and first column of Y is y_0, so that the component in the first row and first column of Y^{-1} is $1/y_0$. From Equation 5.3.19,

$$x_1 = \frac{r'_1}{y_0} \qquad (5.3.23)$$

since the first column of Y^{-1} is all zeros except for the first component $1/y_0$. But from Equation 5.3.13,

$$r'_1 = s_1 y_0 + w_1 \qquad (5.3.24)$$

so that

$$\frac{r'_1}{y_0} = s_1 + \frac{w_1}{y_0} \qquad (5.3.25)$$

It follows that the detection of s_1 from x_1, at the output of the linear network, is exactly equivalent to the detection of s_1 from r'_1, using the following decision rule. If $r'_1/y_0 > 0$, s_1 is detected as k, and if $r'_1/y_0 < 0$, s_1 is detected as $-k$. But this is exactly equivalent to the detection process of System 1A when only a single sample value is used ($n = 1$), since now s_1 is detected as k or $-k$ depending upon whether r'_1 is closer to ky_0 or $-ky_0$, respectively. Thus, when $n = 1$, System 1A becomes the same as System 2A and is the optimum detection process for the case where the receiver has no prior knowledge of the $\{|s_i|\}$ or σ^2. Again, as can be seen from Section 4.4.1, System 2A is an arrangement of pure nonlinear equalization that achieves the exact equalization of the channel.

Clearly, System 2A is best implemented by detecting s_1 as k or $-k$, depending upon whether r'_1/y_0 is positive or negative, respectively.

It can be seen from Equation 5.3.24 or 5.3.25 that an error occurs in the detection of s_1 from r'_1, when w_1 has a magnitude greater than $k|y_0|$ and the opposite sign to $s_1 y_0$. As before, w_1 is a Gaussian random variable with zero mean and variance σ^2. Thus the probability of error in the detection of a signal element by System 2A, following the correct cancellation of the preceding g elements, is

$$\int_{k|y_0|}^{\infty} \frac{1}{\sqrt{(2\pi\sigma^2)}} \exp\left(-\frac{w^2}{2\sigma^2}\right) dw = Q\left(\frac{k|y_0|}{\sigma}\right) \qquad (5.3.26)$$

Due to error extension effects, the average probability of error may be typically up to fifteen times the value given by Equation 5.3.26, but as this corresponds to a reduction in tolerance to additive Gaussian noise of only up to a little over 1 dB, it can for our purposes be neglected. Thus the average error probability can be taken to be

$$P_e = Q\left(\frac{k|y_0|}{\sigma}\right) \qquad (5.3.27)$$

When the first nonzero sample value, y_0, of the sampled impulse-response of the channel is small, System 2A has a correspondingly low tolerance to noise. When n is large and some of the roots of the channel z-transform lie outside the unit circle in the z plane, a better estimate of s_1 than that given by x_1 in Equation 5.3.23 is often obtained by feeding the sequence of sample values given by R' into the appropriate linear equalizer for the channel. The decision-directed cancellation of intersymbol interference, as implemented in System 2A, no longer in fact serves any very useful purpose and can therefore be omitted. On the other hand, when all the roots of the channel z-transform lie inside the unit circle, System 2A has a tolerance to additive white Gaussian noise that is at least as good as or better than that of any equalizer involving either a linear filter or a combination of linear and nonlinear filters (of the type described in Chapter 4), so that no advantage would now be gained by modifying System 2A in the manner just considered.

System 2A is important for two reasons. Firstly, it is the *least* effective detection process for s_1 from R' that is of any real practical value, any useful detection process having a tolerance to noise somewhere between that of System 1A and that of System 2A. Secondly, System 2A can be implemented by means of an iterative process, which can then be modified to improve its tolerance to noise much of the way towards that of System 1A, but without either a serious increase in the equipment complexity or else an excessive number of sequential operations in a detection process. Thus System 2A provides the key to a truly cost-effective detection process.

5.3.5 The point Gauss-Seidel iterative process

The linear estimate X of the vector S, obtained from R' by means of the $n \times n$ network Y^{-1} (Equation 5.3.19), can alternatively be obtained by means of the point Gauss-Seidel iterative method.[2-4] It is implemented as in Figure 5.3 and operates as follows.

The n-component row vector

$$X = x_1 \ x_2 \ldots x_n \qquad (5.3.28)$$

is initially set to zero, and the n-component vector XY is subtracted from R' to give the vector $R' - XY$. The square marked Y^T in Figure 5.3 is an $n \times n$ linear network that performs the linear transformation Y^T on the input vector $R' - XY$. Thus the n input terminals of this network hold the n components of the vector

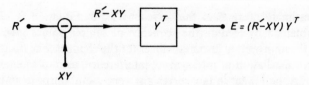

Fig. 5.3 Linear estimation of S from R' by means of an iterative process

$R' - XY$, and the n output terminals hold the n components of the vector

$$E = e_1 \; e_2 \dots e_n \tag{5.3.29}$$

where

$$E = (R' - XY)Y^T \tag{5.3.30}$$

Clearly,

$$e_i = (R' - XY)Y_i^T = R'Y_i^T - XYY_i^T$$
$$= R'Y_i^T - \sum_{j=1}^n x_j Y_j Y_i^T \tag{5.3.31}$$

where Y_i is the n-component row vector formed by the ith row of the $n \times n$ matrix Y and is given by Equation 5.3.6.

When $X = 0$ and $R' \neq 0$, $E \neq 0$. The point Gauss-Seidel iterative process now operates as follows. x_1 is adjusted to set e_1 to zero. x_2 is then adjusted to set e_2 to zero, and so on sequentially to x_n. Since the n vectors $\{Y_i\}$ are not normally all orthogonal, it can be seen from Equation 5.3.31 that a change in x_j changes some or all $\{e_i\}$. The cycle of adjustments in the $\{x_i\}$ is now repeated several times, in the same order as before, until eventually $E \simeq 0$, when

$$(R' - XY)Y^T \simeq 0 \tag{5.3.32}$$

so that

$$X \simeq R'Y^T(YY^T)^{-1} = R'Y^{-1} \tag{5.3.33}$$

This is of course the same vector as that obtained with the linear network Y^{-1} in Equation 5.3.19.

The iterative process always converges to give the vector X in Equation 5.3.33,[2,4] and it can be implemented very simply.[13] In the final stage of the detection process, s_1 is detected from the sign of x_1, as before. Unfortunately, when there is severe amplitude distortion in the received signal, the number of cycles of the iterative process required for convergence to a sufficient degree of accuracy in X (that is, $E \simeq 0$) can become quite excessive, even with relatively small values of n, say $n = 8$. However, it has been shown that the basic process can be modified in various ways to reduce considerably the number of cycles required, and to give different arrangements, which, when $n>8$, require far fewer sequential operations in a detection process than does System 1A. In fact, System 2A, employing the Gauss-Seidel iterative process, forms the basis of a whole class of simple and effective detection processes,[4, 13, 18, 23] and for this reason the iterative process itself is of considerable interest.

It is clear from Equation 5.3.33 that the iterative process *inverts* the matrix Y. There are, of course, many other iterative processes and further techniques for inverting a matrix, but these generally involve either more complex equipment or more sequential operations.[2–4, 13] Alternatively, they cannot be developed both simply and effectively, in the same way as can the point Gauss-Seidel process, to improve the performance of the system.[23] Other simple techniques require special conditions to be satisfied by the matrix Y, in addition to its assumed properties, and they are not therefore suitable for general applications.[3, 4]

It is interesting now to consider the point Gauss-Seidel iterative process in a little more detail. It can be seen from Equation 5.3.31 that in the general case, where x_i is adjusted to set e_i to zero, the change in x_i is

$$\Delta x_i = e_i(Y_i Y_i^T)^{-1} = \frac{e_i}{|Y_i|^2} \qquad (5.3.34)$$

where e_i is the value of the ith component of E immediately before the change in x_i, and $|Y_i|$ is the length of the vector Y_i. Thus, since Y_i is known, Δx_i can be determined directly from e_i. Furthermore, the iterative process can be modified, without affecting the correct operation or convergence of the process, by making the change in x_i

$$\Delta x_i = c \frac{e_i}{|Y_i|^2} \qquad (5.3.35)$$

where c is a real constant such that $0<c<2$. This is known as over-relaxation,[2, 4] and Equation 5.3.34 is a special case of Equation 5.3.35.

The conditions that must be satisfied to ensure the convergence of the iterative process can be demonstrated quite simply by considering the effect of the change in x_i (Equation 5.3.35) on the vector $R'-XY$, in the n-dimensional vector space containing R' and XY. Figure 5.4 shows the two-dimensional subspace containing both Y_i and $R'-XY$, when a change Δx_i is made in x_i. A is the vector

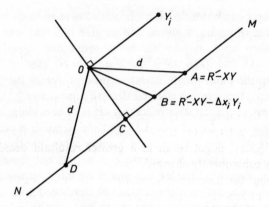

Fig. 5.4 Convergence of the iterative process

$R' - XY$, and B is the resultant vector $R' - XY - \Delta x_i Y_i$ obtained after the change in x_i. O is the origin of the vector space. Clearly, B must lie on the line MN that contains the point A and is parallel to the line OY_i.

If the iterative process converges, it adjusts the vector X in steps until eventually $XY \simeq R'$, when $R' - XY \simeq 0$ and the point A approximately coincides with the origin O. Since R' can (in principle) lie anywhere in the n-dimensional vector space, as can be seen from Equation 5.3.13, a necessary (but not sufficient) condition to ensure the convergence of the process is that XY can also lie anywhere in the vector space. This means that the n vectors $\{Y_i\}$ must span the vector space, so that the $\{Y_i\}$ must be *linearly independent*. When this applies, a sufficient condition to ensure convergence is that the change in each of the n $\{x_i\}$, in every cycle of the iterative process, is such as to *reduce* the distance d from $R' - XY$ to the origin by a quantity that is not vanishingly small. d is, of course, the length of the line OA in Figure 5.4. Under these conditions, B is always *closer* to the origin than A, which means that B may lie anywhere between (but not including) the points A and D on the line MN in Figure 5.4, since both these points are at a distance d from the origin. The iterative process will now always converge.

Suppose next that Δx_i is selected so that B is at the *minimum* distance from the origin. It can be seen from Figure 5.4 that this occurs when B coincides with the point C, which is such that OC is orthogonal (at right angles) to OY_i. Now

$$(R' - XY - \Delta x_i Y_i)Y_i^T = 0 \qquad (5.3.36)$$

so that, from Equation 5.3.31,

$$(R' - XY)Y_i^T - \Delta x_i Y_i Y_i^T = e_i - \Delta x_i |Y_i|^2 = 0 \qquad (5.3.37)$$

Thus Equation 5.3.34 is satisfied and x_i is adjusted to set e_i to zero. This does not necessarily give the most *rapid* rate of convergence, because of the interaction between the changes in the different $\{x_i\}$. For instance, a larger (or smaller) change in x_i than that given by Equation 5.3.34, might result in a greater resultant decrease in d when x_j is subsequently changed.

To prevent the distance d from XY to the origin being increased as a result of any step in the iterative process, it is necessary that B does not lie on the line AM or DN, in Figure 5.4. It is evident, however, that $AC = CD$, which means that the change in x_i needed to carry the vector B from the point A to the point D is *twice* that needed to carry it from the point A to the point C, and is therefore $\Delta x_i = 2e_i/|Y_i|^2$. Thus to prevent B from lying on the line DN, it is necessary that, when $\Delta x_i e_i > 0$

$$|\Delta x_i| < 2 \frac{|e_i|}{|Y_i|^2} \qquad (5.3.38)$$

Again, to prevent B from lying on the line AM, it is necessary that Δx_i has the same sign as e_i. Both of these conditions are satisfied if the constant c in Equation 5.3.35 is such that $0 < c < 2$.

It now follows that, if the $\{Y_i\}$ are linearly independent and $0 < c < 2$ in Equation 5.3.35, the iterative process converges to the required vector X, such that $XY \simeq R'$. The most rapid rate of convergence is normally obtained when $1 \leqslant c \leqslant 1.5$, and the least rapid rate when $c \to 0$ or $c \to 2$. Furthermore, there is a progressive reduction in $|R' - XY|$ as the iterative process proceeds, regardless of whether the $\{x_i\}$ are changed in the same order in successive cycles, or of whether c varies from one cycle to another or even over individual cycles, so long as $0 < c < 2$. Clearly, the process can be modified in many different ways while still maintaining correct convergence.

It can be shown by a more rigorous theoretical analysis and has been confirmed with many practical tests that the iterative process converges to give the required vector X in Equation 5.3.33, so long as $0 < c < 2$ and the matrix YY^T is real, symmetric and positive definite.[2,4,13] This applies regardless of whether or not the n vectors

$\{Y_i\}$ satisfy Equation 5.3.6. When the $\{Y_i\}$ satisfy Equation 5.3.6, as is assumed here, the correct convergence of the iterative process is ensured, so long as $0 < c < 2$ and $y_0 \neq 0$. This is because the n vectors $\{Y_i\}$, which are real, must now also be linearly independent, so that YY^T is real, symmetric and positive definite.

The use of a different value of c (for which $0 < c < 2$) in a given iterative process, does not affect the value of X obtained at the end of the process, but it changes the rate of convergence of the iterative process, so that the required vector X is obtained after a different number of cycles of the process. The end of the iterative process is recognized when, for each i, $|e_i| < \varepsilon$, where ε is a suitably small positive constant. This, of course, ensures that $E \simeq 0$.

If the linear network Y^T is omitted from Figure 5.3, and the iterative process is implemented by adjusting x_i to set a corresponding component of $R' - XY$ to zero, the iterative process no longer necessarily converges, unless some rather special conditions are satisfied by the sampled impulse-response of the channel.[4]

. In the practical implementation of the iterative process,[13] the $n \times n$ network Y^T is replaced by the $n \times 1$ network Y_i^T. Immediately prior to the adjustment of x_i, the network Y_i^T forms the signal $e_i = (R' - XY)Y_i^T$ at its output terminal. Following the resultant change in x_i, the network is changed to Y_{i+1}^T, by shifting, one place to the right, the stored values $\{y_j\}$ that formed the vector Y_i and hence the network Y_i^T. The network Y_{i+1}^T forms the next output signal e_{i+1}, and so on.[13] It follows that at any instant in time, when the network Y^T has become Y_i^T, it gives the same output signal as a linear feedforward transversal filter whose tap gains are given by the respective nonzero components of Y_i. The linear filter has m taps where $1 \leqslant m \leqslant g + 1$, and the signals stored in the filter are the appropriate m components of the vector $R' - XY$. This suggests that the practical implementation of the network Y^T resembles that of an m-tap linear feedforward transversal filter.[13] When the signals are handled as binary-coded numbers, a single multiplier is used to provide, in turn, each of the different multipliers, either for the network Y_i^T or for the corresponding transversal filter. The output signal e_i from the network Y_i^T is obtained by adding the m products formed by the multiplier. This summation process is the same operation as that carried out in the adder of the corresponding transversal filter, and can be implemented simply and rapidly.

It can be seen from Equation 5.3.35 that a single multiplication is required to derive Δx_i from e_i. The new vector $R' - XY$, in Figure 5.3, is now formed by multiplying the m nonzero components of Y_i by Δx_i, to give the n-component vector $\Delta x_i Y_i$, and then subtracting this vector *directly* from the stored vector $R' - XY$. Thus the vector XY is not itself stored, as suggested by Figure 5.3, the vector X being stored instead and the vector R' being held additionally in a separate store. If preferred, Δx_i may be used to determine the new value of x_i and hence of $x_i Y_i$, to give $R' - XY$. The subtraction circuit in Figure 5.3 is the same as the subtraction circuit in the nonlinear equalizer in Figure 4.18, except that it operates on n signals instead of one. The operation of subtraction can be implemented simply and rapidly.

An *individual operation* of the iterative process is defined to be the whole of the procedure involved in changing the value of x_i. Thus one cycle of the iterative process involves n sequential operations, and each of these involves between 3 and $2g+3$ multiplications.

A reduction in the complexity of an individual operation can be achieved by operating directly on the n-component output vector from the network Y^T. Thus, at the start of an iterative process, the vector $R' Y^T$ is determined and also the matrix YY^T. Following each change Δx_i in x_i, determined as before, the n-component vector $\Delta x_i Y_i Y^T$ is now formed by multiplying the ith row of the matrix YY^T by Δx_i, and $\Delta x_i Y_i Y^T$ is subtracted from the vector $R' Y^T - XYY^T$ at the output of the network Y^T. The ith component of $R' Y^T - XYY^T$ is, of course, e_i. If preferred, Δx_i may be used to determine the new value of x_i and hence of $x_i Y_i Y^T$, to give $R' Y^T - XYY^T$.

An *individual operation* of the iterative detection process of System 1A is defined to be the generation of the vector $R' - XY$, where X now has one of the 2^n possible values of S, and the measurement of $|R' - XY|^2$, which is the square of the distance between the vectors R' and XY. The generation of the vector $R' - XY$, from a new vector X, involves at least as many multiplications as are involved in forming the vector $\Delta x_i Y_i$ in System 2A, and often several more. Of course, when n is small, the 2^n possible values of XY may be held in store. The measurement of $|R' - XY|^2$ involves n multiplications in forming the squares of the n components of $R' - XY$, the sum of these components being of course $|R' - XY|^2$. An individual operation of System 1A therefore usually involves more multiplications than does an individual operation of System 2A, and it involves a similar number of additions, subtractions and other simpler functions. However, since the vector X in System 1A has one of the possible values of S, the multiplications involved in generating the vector $R' - XY$ usually involve simpler binary-coded numbers for the components of X than for the $\{\Delta x_i\}$ in System 2A, and they are therefore generally performed more rapidly. It is evident therefore that the individual operations of the Systems 1A and 2A can be considered to be equivalent, both in the time required and in the equipment complexity involved.

Finally, it appears that the equipment complexity involved in the iterative process of Figure 5.3 is of the same order as the equipment complexity of the nonlinear equalizer in Figure 4.18, when the signals are handled as binary-coded numbers. The main difference between the two is that the iterative process involves many more sequential operations in processing the received signal than does the equalizer, so that it cannot be used at nearly such high transmission rates.

A useful reduction in the equipment complexity of the iterative process is achieved by quantizing Δx_i, the change in x_i. A fixed quantization such that $\Delta x_i = a$, $-a$ or 0, depending upon whether $e_i \geqslant d$, $e_i \leqslant -d$ or $|e_i| < d$, respectively, where d is a positive constant such that $d \ll k Y_i Y_i^T$, ensures the convergence of the iterative process, so long as the positive constant a is not too large, but unfortunately it requires that $a \ll k$ if the vector X obtained at the end of the iterative process is to satisfy Equation 5.3.19 sufficiently accurately.[4,13] This results in a slow rate of convergence and requires an excessive number of sequential operations in a detection process. A more rapid rate of convergence can be obtained by modifying the process such that, immediately after x_i has been incremented by a, 0 or $-a$ as just described, a positive constant b is added to the *magnitude* of x_i, without changing the *sign* of x_i. A larger value of a can now be used,[4,13] and $b \ll a$. This leads to a more rapid rate of convergence for a given accuracy in the final vector X. However it is necessary now to impose a constraint on the maximum magnitude of each x_i in order to ensure the convergence of the iterative process.[4]

A much better arrangement, proposed by Clements, is as follows. The value of Δx_i satisfying Equation 5.3.35 is normally handled as a binary-coded number comprising a fairly long sequence of 12 or more binary digits. Δx_i is now quantized to the binary number formed by the first four *significant* digits of its binary-coded value satisfying Equation 5.3.35. Thus, at the start of the iterative process, Δx_i can be very large giving a rapid rate of convergence, whereas, at the end of the iterative process, Δx_i can be very small giving an accurate value for X. It has been shown that the arrangement always converges and can be implemented simply.

A particularly cost-effective system has been proposed by Harvey. Each x_i is here expressed as a binary-coded number having just eight digits, so that the evaluation of Δx_i (and hence of x_i) can be carried out by an 8-bit binary multiplier. The arrangement thus effectively quantizes Δx_i. It appears to be as effective as the previous arrangement but is simpler to implement.

The number of cycles required in an iterative process is usually considerably reduced by starting the iterative process, not with $x_i = 0$ for each i, but rather with $x_i = x'_{i+1}$ where x'_{i+1} is the $(i+1)$th component of the vector X at the *end* of the iterative process involved in the detection of the previous signal-element. As before, $x_n = 0$ at the start of the process. x'_{i+1} is a linear estimate of the element value s_i in the current iterative process. Thus the prior knowledge of all but the last of the n $\{s_i\}$, that has been gained during the previous detection process, is used to speed up the current iterative process. This is a most powerful technique that can be exploited to great advantage not only with the basic point Gauss-Seidel process and its simple modifications considered here, but also with the further modifications of the process to be discussed presently.

The effective time required for the accurate convergence of the iterative process can be further considerably reduced by exploiting the fact that, for any given channel sampled impulse response, the number of sequential operations in the iterative process is a function of the signal-vector S. For certain values of S, the number of sequential operations required can be several times the average number. Thus, if a further buffer store is added at the input to the buffer store in Figure 5.2, an iterative detection process is no longer required to terminate before the arrival of the next received sample value, but the iterative process can now continue to convergence, even if this takes considerably longer than the average time available for the process. The longer duration of some of the iterative processes is compensated for by the shorter duration of the others. The maximum transmission rate for a given channel is now determined by the *average* duration of an iterative process rather than by the *maximum* duration, thus increasing the transmission rate by a factor of up to several times. The arrangement was proposed by Clements.

System 2A would not itself be implemented by the point Gauss-Seidel iterative process just described, since it is very much better implemented by the detection of s_1 from the first component of R', as described in Section 5.3.4. The importance of the point Gauss-Seidel process lies in the further developments that can be applied to this process to improve the tolerance to noise of the data-transmission system and at the same time to reduce the number of sequential operations in the process.

5.3.6 System 3A

This is a modification of the point Gauss-Seidel iterative process just described and involves a negligible increase in equipment complexity. It operates as follows. Throughout the iterative process, x_i is incremented by Δx_i as given by Equation 5.3.35 and described for System 2A, but the value of x_i is now constrained so that its magnitude $|x_i|$ satisfies

$$|x_i| \leqslant k \qquad (5.3.39)$$

The constraint overrides and so, if necessary, truncates the change in x_i dictated by Equation 5.3.35. In the presence of noise, X is normally now a nonlinear estimate of S, and E no longer necessarily goes to zero at the end of an iterative process. The iterative process proceeds otherwise as described for System 2A, and it always converges.[4,13] The end of the iterative process is now recognized by a negligible *change* in X over a cycle of the iterative process. At the end of the detection process, s_1 is detected from the sign of x_1, as before.

Not only does System 3A often have a considerably better tolerance to additive white Gaussian noise than System 2A, but it requires far fewer cycles of the iterative process to reach convergence: normally less than 20 when appropriate techniques are used for speeding up the process. For the most rapid rate of convergence, under typical conditions, the constant c in Equation 5.3.35 must now have a value in the range 1.25 to 1.5. It has in fact been shown that this value of c is also the optimum for more general forms of the matrix Y than that considered here.[4] Again, so long as c lies in the range 0 to 2, its value does not affect the value of X obtained at the end of the iterative process.

In System 3A, the point Gauss-Seidel iterative process, implemented as in Figure 5.3, operates so that at the end of the process the vector X is such as to set each component e_i of the output vector E to zero, wherever this is permitted by the constraints applied to the $\{x_i\}$. The remaining $\{e_i\}$ are set to the minimum magnitudes consistent with the constraints.[4] This can be illustrated by the following simple example.

Let the vector in the buffer store of Figure 5.2 be the two-component vector

$$R' = SY + W = s_1 Y_1 + s_2 Y_2 + W \qquad (5.3.40)$$

where

$$Y_1 = 1 \quad 1$$

$$Y_2 = 0 \quad 1$$

and $s_i = \pm 1$. Suppose that

$$R' = 1.5 \quad 1.5$$

The signal vectors are now as shown in Figure 5.5, where 0 is the origin of the two-dimensional vector space containing the vectors. $\pm Y_1$ and $\pm Y_2$ are the two possible values of $s_1 Y_1$ and $s_2 Y_2$, respectively. T_1, T_2, T_3 and T_4 are the four possible vectors $\{SY\}$.

The values of x_1 and x_2, the two components of X, are constrained to lie in the range -1 to 1. The constraint on x_1 prevents XY lying to the right of the line joining T_1 and T_2 or to the left of the line joining T_3 and T_4, and the constraint on x_2 prevents XY lying above the line joining T_3 and T_1 or below the line joining T_4 and T_2.

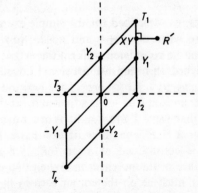

Fig. 5.5   Signal vectors in the two-dimensional vector space

Clearly XY must lie inside or on the quadrilateral formed by the four points T_1, T_2, T_4 and T_3.

Assuming that no overrelaxation is used, so that $c = 1$ in Equation 5.3.35, the iterative process operates as follows. X is first set to zero, so that $XY = 0$ and the output signal-vector in Figure 5.3 is

$$E = (R' - XY)Y^T = 3.0 \quad 1.5 \qquad (5.3.41)$$

x_1 is then set to 1.0, while $x_2 = 0$, so that $E = (1.0, 0.5)$. x_2 is then set to 0.5, while $x_1 = 1.0$, so that $E = (0.5, 0)$. No further change takes place in either x_1 or x_2 in the second or subsequent cycles of the iterative process, so that this has clearly reached convergence at the end of the first cycle. Thus at the end of the iterative process, $X = (1.0, \ 0.5)$ and $XY = (1.0, \ 1.5)$. XY is now as shown in Figure 5.5.

Clearly, x_1 is held at unity by the constraint on its value, so that XY lies on the line joining T_1 and T_2, whereas x_2 has been adjusted to set e_2 to zero, so that

$$e_2 = (R' - XY)Y_2^T = 0 \qquad (5.3.42)$$

This means that the vector $R' - XY$ is orthogonal to Y_2 and hence to the line joining T_1 and T_2. It follows that XY is the *orthogonal projection* of R' on to the nearest side of the quadrilateral whose vertices are the four possible signal-vectors $\{SY\}$. Thus XY is the point on the quadrilateral that is at the *minimum distance* from R'.

If no constraints are applied to the $\{x_i\}$, as in the iterative process of System 2A, then, at the end of the process, $XY = R'$, so that $X = (1.5, 0)$.

The principles just described for the simple example hold also in the general case where $s_i = \pm k$ and $n > 2$. Now, the 2^n possible vectors $\{SY\}$ can be considered to form the vertices of an n-dimensional solid polyhedron in the n-dimensional Euclidean vector space containing these vectors. At the end of the iterative process, each x_i that is not constrained at k or $-k$ is adjusted to set the corresponding e_i to zero, so that $(R' - XY)Y_i^T = 0$. When no x_i is constrained, $E = 0$ and $R' = XY$. R' must now lie inside the solid polyhedron. When one x_i is constrained at k or $-k$, XY must lie on the corresponding face of the polyhedron. When two or more $\{x_i\}$ are constrained, XY must lie on the corner formed by the intersection of the corresponding faces. Since, for each i such that x_i is *not* constrained, $(R' - XY)Y_i^T = 0$, it can be seen that the vector $R' - XY$ must be orthogonal to the $\{Y_i\}$ corresponding to the unconstrained $\{x_i\}$. But the face or corner of the polyhedron that now contains XY must itself be parallel to the subspace spanned by these $\{Y_i\}$ and so must be orthogonal to $R' - XY$. Furthermore, this face or corner of the polyhedron is always that *nearest* to R'. It follows that XY must here be the *orthogonal projection* of R' on to the surface of the polyhedron, so that, from the projection theorem, XY is the point in the solid polyhedron at the *minimum distance* from R'.

These important results may be summarized as follows. Whenever the vector R', from which s_1 is detected, lies inside the solid polyhedron whose vertices are the 2^n possible vectors $\{SY\}$, the vector X obtained by the iterative process of System 3A, satisfies $XY = R'$, just as for System 2A. However, whenever R' lies outside the polyhedron, XY is the orthogonal projection of R' on to the surface of the polyhedron and is therefore the point in the polyhedron that is at the minimum distance from R'. In System 2A, of course, $XY = R'$, regardless of the value of R'. Clearly, the operation of the iterative process is to *minimize* $|R' - XY|$, within the limits of the constraints imposed on X.

5.3.7 System 4A

This is a further development of System 3A. In the first part of a detection process, x_1 is set to k and is held at this value during the whole of the iterative process of System 3A. Thus the $\{x_i\}$, for $i = 2, 3, \ldots, n$, are constrained so that $|x_i| \leqslant k$. At the end of this process, the distance $|R' - XY|$ between the vectors R' and XY is measured. Of course, in the vector X here, $x_1 = k$ and $|x_i| \leqslant k$ for the remaining $\{x_i\}$. In the second part of the detection process, x_1 is set to $-k$ and once more held at this value during the whole of the iterative process of System 3A. $|R' - XY|$ is then measured again. The detected value of s_1 is now taken to be the value of x_1 for which $|R' - XY|$ has the smaller value.

Since the operation of the iterative process is to minimize $|R' - XY|$, within the limits of the constraints imposed on X, it is clear that the smaller of the two values of $|R' - XY|$ gives the value of X that minimizes $|R' - XY|$ subject to the constraints that

$x_1 = \pm k$ and $|x_i| \leqslant k$ for $i = 2, 3, \ldots, n$. The omission of the constraints on the $\{x_i\}$, for $i = 2, 3, \ldots, n$, seriously degrades the performance of the system.

In practice, the receiver measures $|R' - XY|^2$ rather than $|R' - XY|$, since the former is more readily evaluated.

Extensive tests by computer simulation on System 4A suggest that, for any likely value of the sampled impulse-response of the channel and regardless of the value of n, convergence to the required degree of accuracy in the vector X, with x_1 set either to k or $-k$, should normally be obtained in less than *ten* cycles of the iterative process. The vector X at the start of an iterative process here is not set to zero, but instead the ith component of X is set to the value reached by the $(i+1)$th component of X at the *end* of the particular iterative process in the detection of the *previous* signal-element, for which x_1 is set to the value subsequently accepted as the detected value of s_1. The nth component of X is initially set to zero.

In System 4A the possible values of x_1 are constrained to $\pm k$ and the possible values of the remaining $\{x_i\}$ are constrained to the range $-k$ to k. Clearly, the detection process can be modified so that the possible values of the $\{x_i\}$, for $i = 1, 2, \ldots, m$ and $1 \leqslant m \leqslant n$, are constrained to $\pm k$, and the possible values of the $\{x_i\}$, for $i = m+1, m+2, \ldots, n$, are constrained to the range $-k$ to k. The detection process now operates as follows. The $\{x_i\}$, for $i = 1, 2, \ldots, m$, are first all set to k and held at this value while the iterative process of System 3A is applied to the $\{x_i\}$, for $i = m+1, m+2, \ldots, n$. At the end of the iterative process, the distance $|R' - XY|$ between the vectors R' and XY is measured. The $\{x_i\}$, for $i = 1, 2, \ldots, m$, are now set to another of their possible combinations (each $x_i = \pm k$) and are then held at these values while the iterative process of System 3A is applied to the remaining $\{x_i\}$ (each $|x_i| \leqslant k$). At the end of this process, $|R' - XY|$ is again measured. The detection process continues in this way until all 2^m combinations of the first m $\{x_i\}$ have been used. The detected value of s_1 is now taken to be the value of x_1 for which $|R' - XY|$ has the *smallest* value. It can be seen that if $m = 1$, the detection process becomes that of System 4A, and if $m = n$, it becomes that of System 1A. Unfortunately, when $m > 1$, this is not a very useful system because of the large number of sequential operations involved in a detection process. It is of interest, however, because it clarifies the relationship that exists between Systems 1A and 4A.

System 4A achieves a tolerance to additive white Gaussian noise typically within about 1.5 dB of that of the optimum detection process, System 1A. It does not involve either complex equipment or an excessive number of sequential operations.

5.3.8 *Relationship between the Systems 1A, 2A, 3A and 4A*

There is an interesting and important relationship between the different systems just described. Each system can be considered as an arrangement which first forms the n-component vector X, which is an *estimate* of the signal-vector S and need not, of course, have one of the possible values of S. In each system, X is such that the n-component vector XY is at the minimum distance from R', subject to the particular constraints that are applied to the vector X. s_1 is then detected from the sign of x_1. In System 1A, X is constrained to just the 2^n possible values of S. In System 2A, there are no constraints on X. In System 3A, X is constrained so that $|x_i| \leqslant k$ for each i. In System 4A, X is constrained so that $x_1 = \pm k$ and $|x_i| \leqslant k$ for $i = 2, 3, \ldots, n$.

The constraints applied to X in any detection process represent the prior knowledge of the possible values of S that is used in the process. As far as the receiver is concerned in any of the detection processes considered here, X is equally likely to take on any of its permitted values, so that the receiver makes no use of any further prior knowledge, beyond the fact that S cannot take on any of the prohibited values of X. Thus the range of the permitted values of X represents the lack of prior knowledge of S, the greater the range of values, the smaller the prior knowledge. Obviously, if the receiver has prior knowledge that S cannot have a certain value, then the consideration of this as a possible value in the detection process must necessarily reduce the use of the available prior knowledge of S and so, in general, reduce the tolerance of the system to noise, so long as no other changes are at the same time introduced into the detection process.

If X is the *maximum-likelihood* estimate of S, given the vector R', X is the value of S_a that maximizes $p(R'|S_a)$. S_a is here a possible value of S, as far as the receiver is concerned, so that S_a may have any permitted value of X and is therefore not necessarily confined to the actual possible values of S. (This matter is considered further in Section 5.3.3.). $p(R'|S_a)$ is the conditional probability density of the random vector with sample value R', at the given value R' and given that $S = S_a$. If $S = S_a$, the noise vector in R' becomes

$$U = R' - S_a Y \qquad (5.3.43)$$

Thus X is the value of S_a that maximizes

$$p(R'|S_a) = p(R' - S_a Y|S_a) = p(U|S_a) = p(U) \qquad (5.3.44)$$

which is the probability density of the random vector with sample value U, at the given value U. U is of course given by the values of R' and S_a in Equation 5.3.43, and need not therefore be the *actual* received noise vector W.

Since the components of U are sample values of statistically independent Gaussian random variables with zero mean and variance σ^2, U is equally likely to lie in any direction in the n-dimensional Euclidean vector space containing the vectors R' and $S_a Y$. Furthermore, the value of the orthogonal projection of U on to any direction in the vector space is a sample value of a Gaussian random variable with zero mean and variance σ^2, as is shown in Section 3.2.7. Thus the value of $p(U)$ always increases as $|U|$ decreases, where $|U|$ is the length of the noise vector. It follows from Equation 5.3.44 that $p(R'|S_a)$ is maximum at the value of S_a for which $|U|$ is minimum, so that the maximum-likelihood estimate of S, given by $S_a = X$, corresponds to the noise vector U having the *minimum length*. But in each of the Systems 1A–4A, the vector X determined by the iterative process corresponds to the noise vector U having the smallest possible length, subject to the given constraints on X. Thus the vector X is in every case the *maximum-likelihood* estimate of S, *subject to these constraints*.

Clearly, the smaller the permitted range of values of X (just so long as these include all the 2^n possible values of S), the better should be the estimate of S given by X. Furthermore, if the permitted range of values of X is reduced, with particular emphasis on the permitted range of values of x_1, the lower should be the probability of error in the detection of s_1. This does not always hold but is nevertheless a good general rule. Thus the detection processes, listed in the order of their tolerances to additive white Gaussian noise and starting with the best system, should be Systems 1A, 4A, 3A and 2A.

5.3.9 System 5A

Perhaps the most attractive approach to the practical realization of the optimum detection process is to employ the basic technique of System 1A, in which the n-component vector X in the iterative process is only permitted to take on the possible values of S, but now limiting the number of different values of S tested to a small fraction of the total number 2^n.

An effective method of achieving this, known here as System 5A, has been developed by Harvey. The detection process starts by adjusting the vector X as follows. x_i is set to s'_{i+1}, for $i = 2, 3, \ldots,$ $n-1$, where s'_{i+1} is the detected value of s_{i+1} in the detection of the previous signal element, s_{i+1} being of course renamed s_i in the current detection process. x_1 is then set to k, and x_n is determined as the value $\pm k$ for which $|R' - XY|^2$ has the smaller value. The resultant vector X together with the corresponding value of $|R' - XY|^2$ are now stored. x_{n-1} is then changed to the other of its two possible values and $|R' - XY|^2$ is measured, x_{n-1} being then returned to its previous value. This is repeated for $x_{n-2}, x_{n-3}, \ldots, x_2$ in turn. If the *smallest* value of $|R' - XY|^2$ obtained as a result of these changes is less than the stored value of $|R' - XY|^2$, the corresponding x_i is changed to the other of its two possible values to give a new stored vector X. The smallest value of $|R' - XY|^2$ now replaces the stored value of $|R' - XY|^2$, and the whole process is repeated, starting this time with x_n. If the stored value of $|R' - XY|^2$ is not changed as a result of the process just described, X is left unchanged and the process is not repeated. The process is, in any case, not repeated a second time. The values of the $\{x_i\}$, for $i = 2, 3, \ldots, n$, are next changed in *pairs* over all possible combinations of the $\{x_i\}$ taken two at a time, and the corresponding values of $|R' - XY|^2$ measured, the values of the two $\{x_i\}$ being restored after each change. The smallest value of $|R' - XY|^2$ obtained as a result of these changes is then compared with the stored value of $|R' - XY|^2$. If it is smaller than this value, the corresponding two $\{x_i\}$ are changed to their other possible values, giving a new stored vector X and a new stored value of $|R' - XY|^2$.

x_1 is now changed to $-k$ and the $\{x_i\}$, for $i = 2, 3, \ldots, n-1$, are set to their *original* values, used at the start of the previous operation. x_n is then determined as the value $\pm k$ for which $|R' - XY|^2$ has the smaller value, and $|R' - XY|^2$ together with the associated vector X are stored. The whole of the previous operation is now repeated, with x_1 this time held at $-k$, to give a second stored value of $|R' - XY|^2$ and of the associated vector X.

Finally, the detected vector S' is taken as the vector X corresponding to the smaller of the two stored values of $|R' - XY|^2$. Only the detected value of s_1 is accepted, this being the value of the first component of S'.

Various simplifications of the above detection process have been tested but it has not been found possible to achieve a useful reduction in the number of sequential operations without at the same time a significant reduction in tolerance to noise.

The tolerance to additive white Gaussian noise of System 5A is typically no more than 0.8 dB below that of System 1A. System 5A is more complex than System 4A but, when $n < 12$, it requires fewer sequential operations in a detection process.

5.3.10 Comparison of Systems 1A–5A

The tolerances to additive white Gaussian noise of the Systems 1A–5A, together with that of the optimum nonlinear equalizer (Section 4.4.6), have been tested by computer simulation for different channels. Eight sample values are used in each detection process of the Systems 1A–5A, so that R' is an 8-component vector. The data-transmission system assumed in these tests is as shown in Figure 5.1 and described in Section 5.2. The results of the tests are

Table 5.1 Reduction in tolerance to additive white Gaussian noise, measured in dB, when the given channel replaces one introducing no distortion or attenuation

Channel	Sampled impulse-response of channel	System 1A	System 2A	System 3A	System 4A	System 5A	Optimum nonlinear equalizer
1	$2^{-\frac{1}{2}}(1,1,0,0,0)$	1.3	3.4	2.0	2.2	1.7	3.4
2	$1.5^{-\frac{1}{2}}(1,0.5,0.5,0,0)$	1.0	2.3	1.6	1.5	0.9	2.3
3	$1.5^{-\frac{1}{2}}(0.5,1,0.5,0,0)$	2.7	8.9	5.5	4.2	3.4	8.9
4	$1.5^{-\frac{1}{2}}(0.5,1,-0.5,0,0)$	0.0	8.9	0.0	0.0	0.0	0.2
5	$2^{-\frac{1}{2}}(0.235,0.667,1,0.667,0.235)$	6.2	17.8	9.4	7.3	7.0	11.9
6	$2^{-\frac{1}{2}}(-0.235,0.667,1,0.667,-0.235)$	1.6	17.9	2.4	2.5	2.3	4.3

given in Table 5.1. This quotes the number of dB reduction in tolerance to additive white Gaussian noise, at an element error rate of 4 in 10^3, when the given channel replaces one that introduces no distortion or attenuation. All detection processes have the same tolerance to noise in the presence of no distortion or attenuation. The sampled impulse-response of each channel is given by a vector of unit length, so that no signal attenuation (or gain) is introduced by any channel. The 95% confidence limits of the quoted results are about ± 0.4 dB.

It can be seen from Table 5.1 that, at the given error rate, the relative tolerances to additive white Gaussian noise of the Systems 1A–4A are in the same order as that predicted theoretically, the nonlinear equalizer having a tolerance to noise between that of System 3A and that of System 2A. Systems 1A, 3A and 4A are in every case better than the nonlinear equalizer, with sometimes a considerable advantage in tolerance to noise.

It is interesting to observe that Channel 4 introduces mainly phase distortion, whereas Channels 3, 5 and 6 introduce pure amplitude distortion. Furthermore, all roots of the z transforms of the Channels 1 and 3 lie on the unit circle in the z plane, and all the roots of Channel 2 lie inside the unit circle. Thus, in the case of Channels 1, 2 and 3, the optimum nonlinear equalizer becomes a pure nonlinear equalizer and therefore equivalent to System 2A. Channels 1 and 3 cannot, of course, be equalized linearly.

Further tests with Systems 1A and 4A over Channel 5 have shown that the tolerance to noise increases rapidly with n, the number of sample values used in a detection process, as n increases from 1 to 4, giving a reduction in tolerance to additive white Gaussian noise of 7.7 dB for both systems, when $n = 4$ and a channel introducing no signal distortion or attenuation is replaced by Channel 5. The tolerances to noise of the systems then increase much more slowly with n until, when $n = 8$, they are as given in Table 5.1. This variation with n of the tolerance to noise of each system is to be expected, because, when $n = 4$, most of the energy of the signal-element $s_1 Y_1$ is involved in the detection process.

Finally, computer simulation tests carried out by Harvey on Systems 1A–4A, operating over Channel 5 (Table 5.1) and using 8 components for R', have shown that the variation of error rate with signal/noise ratio for each of the different systems is as given in Figure 5.6. It can be seen here that, as the error rate decreases, so

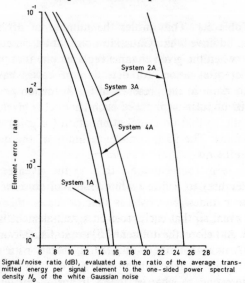

Signal/noise ratio (dB), evaluated as the ratio of the average transmitted energy per signal element to the one-sided power spectral
density N_0 of the white Gaussian noise

Fig. 5.6   Error rates of Systems 1A–4A at different signal/noise ratios

the signal/noise ratio required by System 3A increases more rapidly
than the signal/noise ratio required by any of the other three systems.
The reason for this effect is explained in Section 5.4.6, where it is
shown that, at sufficiently high signal/noise ratios, the performance
of System 3A approaches that of System 2A. The performance of
System 4A, however, improves slightly relative to the performances
of the other systems, as the error rate decreases.

System 5A has also been tested at different signal/noise ratios and
its performance closely follows that of System 4A, being in fact a
little better at the lower signal/noise ratios and slightly inferior at
the higher signal/noise ratios.

For small values of n ($n < 8$) and with binary signals, the preferred
detection process is System 1A. It achieves the best tolerance to
noise, it is simple to implement and does not now involve an
excessive number of sequential operations. For larger values of
n ($n > 8$) and with binary signals, the preferred detection processes
are Systems 4A and 5A.

Many other detection processes have been developed and tested
and some of these have small advantages over the better of the
Systems 1A–5A, either in performance or in equipment complexity.
It seems more than likely that further processes will be developed,
with significant advantages over Systems 4A and 5A. The ideal

system is, of course, one which has the same performance as System 1A, but requires only a few sequential operations in a detection process and is, at the same time, simple to implement. The particular detection processes described here have been selected as those which demonstrate most clearly the *basic* principles and techniques involved in such processes.

5.3.11 Detection of multilevel signal elements

Consider now the case where s_i is equally likely to have any one of the m different values given by $(2l-m+1)k$ for $l = 0, 1, \ldots, m-1$, where m is even, so that each received signal-element is an m-level polar signal. As before, the different $\{s_i\}$ are statistically independent.

System 1A now requires m^n sequential operations in a detection process for s_1 and the arrangement becomes even less attractive than it is for a binary signal, when n is large. In System 3A, the constraint on X throughout a detection process is such that

$$|x_i| \leqslant (m-1)k \qquad (5.3.45)$$

for each i. It is clear that the greater the value of m, the less effect do the constraints have on the detection process, since many of the n $\{s_i\}$ involved in the detection of s_1 are likely to have values other than $\pm(m-1)k$. Thus as m increases, the tolerance to additive white Gaussian noise of System 3A approaches that of System 2A. Finally, System 4A now involves an appreciably greater number of sequential operations in the detection of s_1 than does System 3A, since the iterative process must here be repeated for each of the m possible values of s_1.

It follows that, apart from System 2A which is not significantly affected by the value of m but has in any case a poor performance, the detection processes now either have a reduced tolerance to noise or else they require an excessive number of sequential operations.

Various techniques have been studied for improving the tolerance to noise and at the same time reducing the number of sequential operations of the more effective detection processes. The essential feature of some of these techniques is to reduce the detection process to that of a binary signal. A particular example is described in Section 5.4.10, but unfortunately this results in an unacceptably low tolerance to noise in the present application.

Fortunately, System 4A can readily be modified for use with multilevel signal-elements, as follows. x_1 is set, in turn, to two *selected* possible values of s_1, and these are taken as the two possible values of s_2 nearest to the component x_2 of the vector X for which $|R' - XY|$ has the smaller value, in the detection of the *previous* received signal-element. The constraints applied to the $\{x_i\}$, for $i = 2, 3, \ldots, n$, are $|x_i| \leqslant (m-1)k$. The iterative process is further speeded up by setting the vector X at the *start* of the process so that its ith component, for $i = 2, 3, \ldots, n-1$, is equal to the $(i+1)$th component of the other vector X, for which $|R' - XY|$ has the smaller value at the end of the previous detection process. The nth component of X is set to zero.

With 4-level signals, the tolerance to Gaussian noise of this detection process is usually within about 1.5 dB of that obtained with System 1A, and no significant penalty is paid here in tolerance to noise through setting x_1 to only *two* of the four possible values of s_1. However, a 4-level signal typically requires about four times as many sequential operations as the corresponding binary signal. Special techniques must be used with higher-level signals in order to restrict the number of sequential operations to an acceptable value.

It is desirable now to clarify an important relationship that exists between polar baseband signals having different numbers of element values (numbers of levels). If it is required to double the transmission (information) rate of a data-transmission system, this may be achieved by doubling the information content of each signal element (using more signal levels) or by doubling the signal element rate, or, of course, by appropriately increasing both the number of levels and the signal element rate. Assume that the signal elements are statistically independent and equally likely to have any of the possible element values, and assume a high signal/noise ratio. Suppose also that the transmission path has a very wide bandwidth and introduces no signal distortion, regardless of the signal element rate within the range considered. If now a binary signal is replaced by a 4-level signal, at the same element rate and average power level, this doubles the transmission rate at the expense of a reduction of 7 dB in tolerance to additive white Gaussian noise.[37] On the other hand, if the element rate of the binary signal is doubled, while maintaining the same average power level, this doubles the transmission rate at the expense of a reduction of only 3 dB in tolerance to additive white Gaussian noise.[37] Again, if a 4-level signal is replaced by a 16-level signal, at the same element rate and average power level, this doubles the transmission rate at the expense of a reduction of about 12 dB in tolerance to additive white Gaussian noise, compared to a reduction of only 3 dB if the element rate of the 4-level signal is doubled. Similarly, if a 16-level signal is replaced by a 256-level signal, at the same element rate and average power level, this doubles the transmission rate at the expense of a reduction of about 24 dB in tolerance to additive white Gaussian noise, compared to a reduction of only 3 dB if the element rate of the 16-level signal is doubled.[37] Of course, an increase in the number of levels does not change the signal bandwidth, whereas a doubling of the signal element rate doubles the bandwidth used. This means that when the transmission path is bandlimited to a frequency band not much wider than that of the transmitted signal, as is normally the case in practice, any increase in the signal element rate results in a greater level of amplitude distortion in the received signal, with a consequent reduction in tolerance to additive white Gaussian noise, even with the very best detection process. Furthermore, the noise samples are now usually correlated, so that a noise-whitening network (Section 3.5) is needed if the best tolerance to noise is to be achieved by the detection processes considered here. However, an adequate approximation to the ideal noise-whitening network is likely to be achieved relatively simply, since the tolerance to noise is often not seriously degraded even with the complete omission of the noise-whitening network.[16,18] The available evidence suggests that, when the element rate is doubled, the reduction in tolerance to additive white Gaussian noise of a near optimum detection process is more than 3 dB but not often as much as several times this value. A greater reduction usually occurs if there is a greater level of amplitude distortion in the original signal. It appears therefore that, except possibly in the presence of very severe amplitude distortion, a smaller reduction in tolerance to additive white Gaussian noise is normally experienced by a near optimum detection process, when the signal element rate is doubled, than is experienced when the information content per element is doubled, particularly when starting with a signal having four or more levels. It follows that, as a general rule, it is best to operate at the highest element rate that can conveniently be handled and therefore with the lowest number of signal levels, bearing in mind that an increase in the element rate correspondingly increases the number of components in the sampled impulse-response of the channel and at the same time reduces the time available for a detection process. Whenever possible, binary signal-elements should be used, these having the additional advantage that a near optimum detection process can here be implemented relatively simply.

When the transmission path in the data-transmission system of Figure 5.1 contains a telephone circuit or HF radio link, the impulse response of the resultant baseband channel may have a duration of up to several milliseconds. At a transmission rate of 9600 bits/s, the preferred signal over the telephone circuit or HF radio link is a 4-level PM signal formed by two binary double sideband suppressed carrier AM signals in phase quadrature, the two signals being in element synchronism. The sampled impulse-response of the resultant baseband channel may now have up to about 50 non-zero samples, and the values of these samples are complex numbers (Section 5.2). Unfortunately, there are only some 200 μs available for the detection of a received 4-level signal element, which strictly limits the number of samples of the received signal that may be used in a detection process. System 4A is here the preferred detection process and it can operate with up to about 16 samples of the received signal. Under these conditions, the transmitter and receiver filters in Figure 5.1 should be designed to concentrate as much as possible of the energy of the impulse response in the early part of this response, which is the part involved in a detection process. Of course, following the detection of a received signal-element, *all* components of the element must be cancelled and not just the particular components involved in the detection of the element.

Since the signal element rate over the resultant baseband channel in Figure 5.1 now exceeds the Nyquist rate, the bandwidth of the transmitter and receiver filters should be appropriately adjusted to a value appreciably narrower than that assumed in Figure 5.1, which means that there is considerable correlation between neighbouring noise samples at the receiver. A reasonable approximation to the ideal noise whitening network (Section 3.5) should therefore be used in the detection process of System 4A, to avoid the possibility of a serious degradation in performance. If no noise-whitening network is to be used, it may be better to transmit two 4-level double sideband suppressed carrier AM signals in phase quadrature and to use the appropriately modified arrangement of System 4A at the receiver. The ideal receiver filter does not now introduce any correlation between the different noise samples.

5.3.12   The Viterbi algorithm

An alternative approach towards the optimum detection of the received $\{s_i\}$ in the data-transmission system of Figure 5.1, that has recently received some attention in the published literature, uses the Viterbi algorithm.[17,20] The technique differs fundamentally from the systems studied here in that the detection process operates directly on the individual samples $\{r_i\}$ of the received signal, by comparing these with their different possible values in the absence of noise, for the different possible $\{s_i\}$ and the given sampled impulse-response of the channel. It uses the fact that the $\{r_i\}$ are formed by the addition of the statistically independent Gaussian noise samples $\{w_i\}$ to the sequence formed by the convolution of the $\{s_i\}$ and $\{y_i\}$. The process determines which of the different possible sequences has the maximum likelihood, given the received sequence of the $\{r_i\}$, and it accepts the corresponding $\{s_i\}$ as the detected element values.

As before, the transmitted $\{s_i\}$ are assumed to be statistically independent m-level polar signals, s_i being equally likely to have any of the m possible values, and the noise samples $\{w_i\}$ are statistically independent Gaussian random variables with zero mean and variance σ^2.

We shall not here consider the detailed implementation of the Viterbi algorithm but just sufficient of the basic principles of the system to form an overall picture of its operation. The emphasis is on *what* the system does rather than on *how* it does it.

The ith received sample value in the data-transmission system of Figure 5.1, at time $t = iT$, is

$$r_i = q_i + w_i \qquad (5.3.46)$$

where

$$q_i = \sum_{l=0}^{g} s_{i-l}\, y_l \qquad (5.3.47)$$

Assuming that transmission started at time $t = T$, $s_i = 0$ for $i \leqslant 0$, and the resultant received vector, at time $t = iT$, is

$$R_i = r_1\ r_2 \ldots r_i \qquad (5.3.48)$$

The corresponding received signal-vector is

$$Q_i = q_1\ q_2 \cdots q_i \qquad (5.3.49)$$

and the corresponding received noise-vector is

$$W_i = w_1 \ w_2 \dots w_i \qquad (5.3.50)$$

so that

$$R_i = Q_i + W_i \qquad (5.3.51)$$

With an m-level signal and with no particular relationships assumed between the $g+1$ $\{y_i\}$ or between these and the possible values of s_i, q_i in general has m^{g+1} different possible values, when $i > g$. Clearly, $q_1 = s_1 y_0$, $q_2 = s_1 y_1 + s_2 y_0$, and so on.

It is shown in Section 5.3.8 that the maximum-likelihood estimate of the signal-vector S from the vector R' is the estimate that corresponds to the noise-vector having the *smallest length*, subject to any constraints that may be applied to the estimate of S. We are here concerned with estimating Q_i from R_i, subject to the constraint that the estimate of Q_i must have a *possible value* of Q_i. The estimate of Q_i is the i-component vector

$$X_i = x_1 \ x_2 \dots x_i \qquad (5.3.52)$$

Let

$$R_i = X_i + U_i \qquad (5.3.53)$$

where

$$U_i = u_1 \ u_2 \dots u_i \qquad (5.3.54)$$

so that U_i is the estimate of W_i that corresponds to the estimate X_i of Q_i.

The maximum-likelihood vector X_i, given that $x_i = p_i$, where p_i has one of the m^{g+1} possible values of q_i (assuming, for convenience, that $i > g$), is the vector X_i such that $|U_i|$ has the minimum value, subject to the constraint that $x_i = p_i$. $|U_i|$ is the length of the vector U_i in Equation 5.3.53.

When $x_{i+1} = p_{i+1}$, where p_{i+1} has one of the m^{g+1} possible values of q_{i+1}, there are m different possible values of x_i. For any one of these values of x_i, say $x_i = p_i$, there is the corresponding maximum-likelihood vector X_i, given that $x_i = p_i$. Thus, associated with $x_{i+1} = p_{i+1}$, there are m maximum-likelihood vectors $\{X_i\}$, each for the corresponding given value of x_i.

The maximum-likelihood vector X_{i+1}, given that $x_{i+1} = p_{i+1}$, is the vector corresponding to the minimum value of $|U_{i+1}|^2$, subject to the constraint that $x_{i+1} = p_{i+1}$. But for *any* vector X_{i+1}, the squared length of the corresponding vector U_{i+1} (Equation 5.3.53 with i replaced by $i+1$) is

$$|U_{i+1}|^2 = u_1^2 + u_2^2 + \ldots + u_i^2 + u_{i+1}^2$$

$$= |U_i|^2 + u_{i+1}^2 \tag{5.3.55}$$

where U_i is the i-component vector formed by the first i components of U_{i+1}, and

$$u_{i+1} = r_{i+1} - x_{i+1} \tag{5.3.56}$$

Clearly, when $x_{i+1} = p_{i+1}$, u_{i+1} is fixed. Thus the maximum-likelihood vector X_{i+1}, *given* that $x_{i+1} = p_{i+1}$, is the vector X_{i+1} whose first i components form the vector X_i, where this is the one of the m possible maximum-likelihood vectors $\{X_i\}$, associated with the given value of x_{i+1}, that gives the *smallest* value of $|U_i|^2$. It follows that the maximum-likelihood vector X_{i+1}, given that $x_{i+1} = p_{i+1}$, can be determined from a knowledge of the appropriate m maximum-likelihood vectors $\{X_i\}$. Thus, if the detector has a knowledge of the maximum-likelihood vector X_i, given that $x_i = p_i$, for each of the m^{g+1} possible values of p_i, it can then readily determine the maximum-likelihood vector X_{i+1}, given that $x_{i+1} = p_{i+1}$, for each of the m^{g+1} possible values of p_{i+1}. Having determined the m^{g+1} maximum-likelihood vectors $\{X_{i+1}\}$, it can then determine the corresponding m^{g+1} maximum-likelihood vectors $\{X_{i+2}\}$, and so on.

Consider the transmitted sequence of element values $s_1, s_2, \ldots, s_j$, giving the corresponding sequence of received samples $r_1, r_2, \ldots, r_{j+g}$, where $r_i = q_i + w_i$. It is assumed that $s_i = 0$ for $i \leqslant 0$ and $i > j$, and also that $j > g + 1$. Just as the number of possible values of q_i (Equation 5.3.47) decreases to m when i decreases to unity, since $q_1 = s_1 y_0$, so also the number of possible values of q_i decreases to m when i reaches $j + g$, since $q_{j+g} = s_j y_g$. If the process just described is carried out on the received sequence, starting on the received sample r_1 and ending on the received sample r_{j+g}, it results in the evaluation of the maximum-likelihood vector X_{j+g}, *given* that $x_{j+g} = p_{j+g}$, for *each* of the m possible values of p_{j+g}. Associated with each of these m vectors is the squared length $|U_{j+g}|^2$ of the corresponding vector U_{j+g}. The *true* maximum-likelihood estimate of the received vector Q_{j+g}, is the maximum-likelihood estimate X_{j+g} with *no* constraints applied to x_{j+g} (other than that it must have a possible value of q_{j+g}). This is simply the one of the m

maximum-likelihood vectors $\{X_{j+g}\}$ that gives the *smallest* value of $|U_{j+g}|^2$. The corresponding value of X_{j+g} is now taken as the *detected* value of Q_{j+g}.

The receiver, in fact, requires to know the sequence of received element values $\{s_i\}$, given by the vector

$$S_j = s_1 \ s_2 \ldots s_j \tag{5.3.57}$$

It can be seen from Equations 5.3.47 and 5.3.49 that S_j is *uniquely* determined by Q_j, and therefore also by Q_{j+g}, regardless of whether or not q_i has as many as m^{g+1} different possible values for any i in the range $g+1$ to j, and therefore regardless of whether or not s_i is uniquely determined by q_i. Thus, from the detected value of Q_{j+g} obtained at the end of the received message, the receiver determines the corresponding possible value of S_j, which is taken as the *detected* value of S_j. This is of course the maximum-likelihood estimate of S_j, subject to the constraint that the estimate can only take on a *possible* value of S_j.

In practice, the vectors $\{X_i\}$, which are the possible values of the vector Q_i, are generated from the possible values of the vector S_i, using a prior knowledge both of the $\{y_l\}$ and of the possible values of the $\{s_l\}$, as can be seen from Equation 5.3.47. Thus the generation of the maximum-likelihood vector X_i, given the received vector R_i, implies also the generation of the corresponding maximum-likelihood estimate of the vector S_i. The receiver therefore searches directly for this estimate of S_i by trying the appropriate set of its possible values. However, to simplify the description, we consider the vectors $\{X_i\}$ rather than the possible values of S_i, bearing in mind the unique one-to-one relationship that exists between the $\{X_i\}$ and the possible values of S_i.

The detection process based on the Viterbi algorithm need not in fact evaluate and store quite as many quantities as is suggested by the preceding analysis. This is because, when $g<i<j$, the m^{g+1} possible values of x_{i+1} can be divided into m^g separate (disjoint) groups of m possible values, and, in the vectors $\{X_{i+1}\}$ whose last components $\{x_{i+1}\}$ have the m values of any one of these groups, the m possible values of x_{i+1} are associated with a given set of m possible values of x_i, such that a vector X_{i+1}, satisfying the above condition, can contain any one of the m possible values of x_i together with any one of the m possible values of x_{i+1} but no other combination of x_i and x_{i+1}. There are m^g separate sets of the m

possible values of x_i. Furthermore, as has already been shown, the vector X_i, formed by the first i components of the maximum-likelihood vector X_{i+1}, for a given value of x_{i+1}, is the one of the m maximum-likelihood vectors $\{X_i\}$, corresponding to the m possible values of x_i, that gives the smallest value of $|U_i|^2$. It follows that the first i components of *each* of the m maximum-likelihood vectors $\{X_{i+1}\}$, corresponding to the m possible values of x_{i+1} in any one of the m^g separate groups, form the *same* vector X_i. Thus the detector need not store the m^{g+1} different maximum-likelihood vectors $\{X_i\}$ corresponding to the m^{g+1} different possible values of x_i. Instead, at the time instant $t = (i+1)T$, the detector holds in store only the m^g different maximum-likelihood vectors $\{X_i\}$, each of whose last component x_i is constrained to have one of the m possible values of a different one of the m^g groups of possible values, and which therefore gives the smallest value of $|U_i|^2$ subject to this constraint. The detector also holds in store the m^g corresponding values $\{|U_i|^2\}$. It then measures u_{i+1}^2 for each of the m^{g+1} possible values of x_{i+1}, using Equation 5.3.56.

Associated with any given value of x_{i+1} is the corresponding maximum-likelihood vector X_i, whose last component is constrained to have one of m given values. The maximum-likelihood vector X_{i+1}, whose last component is constrained to have the given value x_{i+1}, is clearly

$$X_{i+1} = \overbrace{x_1 \; x_2 \ldots x_i}^{x_i} \; x_{i+1} \tag{5.3.58}$$

where X_i is as just described. For every maximum-likelihood vector X_{i+1}, the detector determines the associated value of $|U_{i+1}|^2$, given by Equation 5.3.55. Then, for each group of m possible values of x_{i+1} associated with a different one of the m^g groups of m possible values of x_{i+2} in the possible vectors $\{X_{i+2}\}$, the detector determines the maximum-likelihood vector X_{i+1}, whose last component x_{i+1} is constrained to have one of the m given values and which therefore gives the smallest value of $|U_{i+1}|^2$ subject to this constraint. The m^g maximum-likelihood vectors $\{X_{i+1}\}$ and the associated values $\{|U_{i+1}|^2\}$ now replace the corresponding vectors $\{X_i\}$ and the associated values $\{|U_i|^2\}$, in the appropriate stores. The whole of the procedure just described is repeated for each new received sample r_i.

Obviously, for long messages and therefore for large values of i, the detector cannot, at the start of a selection process, hold in store m^g separate vectors $\{X_i\}$, each with i components. Instead, when the detector has determined the m^g maximum-likelihood vectors $\{X_i\}$, it selects the one of these having the lowest value of $|U_i|^2$, this being therefore the *true* maximum-likelihood estimate of Q_i, and it takes the $(i-n+1)$th component of this vector X_i as the *detected* value of q_{i-n+1}. x_{i-n+1} is then *frozen* at this value, the $\{x_l\}$, for $l = 1, 2, \ldots, i-n$, having been previously frozen. Thus the $\{x_l\}$, for $l = 1, 2, \ldots, i-n+1$, are now *common to all* vectors $\{X_{i+1}\}$ and are stored as the *detected* vector Q'_{i-n+1}. This, in turn, uniquely determines the detected value of S_{i-n+1}. Clearly, the vectors $\{X_i\}$ involved in the procedure for selecting the maximum-likelihood vectors $\{X_{i+1}\}$, have effectively just $n-1$ components. Normally $n > g + 1$. The freezing of x_{i-n+1} can be achieved by omitting from further consideration all vectors $\{X_i\}$ whose values of x_{i-n+1} differ from the frozen value.

It can be seen that the freezing of the values of the $\{x_l\}$, for $l = 1, 2, \ldots, i-n$, is the direct equivalent of the process of decision-directed signal cancellation that is carried out in the Systems 1A–5A. Indeed it is evident that the last n components of the vector X_i which is the *true* maximum-likelihood estimate of Q_i, given that the $\{x_l\}$ for $l = 1, 2, \ldots, i-n$ have been frozen, is the equivalent of the vector XY obtained at the end of the detection process of System 1A (Section 5.3.2), when this operates on the corresponding n received samples $\{r'_l\}$, obtained after the cancellation of the previously detected signal-elements. Thus the detection process based on the Viterbi algorithm is an alternative method of implementing System 1A, and, for a given signal distortion and value of n, it has ideally the same tolerance to additive white Gaussian noise.

For each received m-level signal-element, the Viterbi algorithm involves some m^{g+1} sequential operations, each of which involves just one multiplication (but also some other actions) in determining the detected value of the corresponding s_i. Thus, unlike System 1A, its number of sequential operations per received signal-element is not a function of the effective number of sample values n used in a detection process, $n-1$ here being equal to the delay in the number of sample values received before a firm decision is reached as to the value of any q_i. Clearly, for small values of m and g, relatively few sequential operations are required per received signal-element. Furthermore, the number of sequential operations per signal element can often be reduced considerably below the value of m^{g+1}, by discarding those maximum-likelihood vectors $\{X_i\}$ whose $\{|U_i|^2\}$ exceed the smallest of the m^g values $\{|U_i|^2\}$ by some appropriately large quantity. These vectors are unlikely to form the first part of the true maximum-likelihood vector, finally obtained for the *whole* of the received message, since the value of $|U_{i+l}|^2$ for a maximum-likelihood vector X_{i+l} (given x_{i+l}), whose first i components form one of these vectors, is

$$|U_{i+l}|^2 = |U_i|^2 + u_{i+1}^2 + u_{i+2}^2 + \ldots + u_{i+l}^2 \qquad (5.3.59)$$

where $|U_i|^2$ is the value for the particular vector X_i. Thus $|U_{i+l}|^2$ is unlikely to 'catch up' with the corresponding value for one of the better sequences, as l increases in value.

The disadvantages of the Viterbi algorithm, relative to the Systems 3A–5A, are firstly that it involves a considerable amount of storage in holding the different maximum-likelihood vectors $\{X_i\}$, and secondly that, for large values of m and g, it requires an excessive number of sequential operations per signal element.

The detection process based on the Viterbi algorithm is undoubtedly one of the most important of the recent developments in data transmission, and it is particularly attractive when m and g are both small. For large values of m and g, the process can be modified along the following lines, to give the most promising of all detection processes studied in this book.

The received samples are first fed to a linear feedforward transversal filter that gives a resultant sampled impulse-response for the channel and filter in which the roots (zeros) of the channel z-transform that lie outside the unit circle in the z plane are replaced by their reciprocals, the remaining roots being left unchanged. The linear filter is that used in the optimum nonlinear equalizer and is described in Section 4.4.6. It is a pure phase equalizer (although not for the channel itself) and it does not therefore change the signal/noise ratio nor does it introduce any correlation into its output noise samples (Section 4.3.10). The effect of the linear filter is to minimize the effective number of components in the sampled impulse-response, as far as is possible without equalizing any of the amplitude distortion introduced by the channel and so without causing its output noise samples to be correlated.

The output samples from the linear filter are fed to the Viterbi-algorithm detector. (To avoid the introduction of a further set of symbols, the channel and linear filter are now assumed to replace the baseband channel in Figure 5.1, so that the Equations 5.3.46–5.3.59 apply to the resultant channel). If the new value of g is sufficiently reduced so that m^{g+1} is not excessive, the Viterbi-algorithm detector can be used as it is to operate on the sample values $\{r_i\}$ at the output of the linear filter, the Viterbi-algorithm detector replacing the signal cancellation circuit and detector in Figure 5.1.

J. P. Driscoll has shown that the freezing of x_{i-n+1} at the detected value of q_{i-n+1} may slightly reduce the tolerance to noise relative to the arrangement where no vectors $\{X_i\}$ that have apparently incorrect values of x_{i-n+1} are discarded, so that it is assumed now that the stored vectors $\{X_i\}$ are not altered as a result of the detection of q_{i-n+1}.

When m^{g+1} is excessive, the Viterbi-algorithm detector must be modified to limit to an acceptable value the number of vectors $\{X_i\}$ that are involved in the detection of q_{i-n+1}. To simplify the description of the modified system, it is best to consider in place of the vector X_i the corresponding i-component row vector D_i, whose lth component d_l, for $l = 1, 2, \ldots, i$, has the possible value of s_l that is assumed in X_i. D_i is uniquely determined by X_i, and vice versa, and the detection of q_{i-n+1} corresponds to the detection of s_{i-n+1}. In the systems considered here the detector selects only a small number of vectors $\{D_i\}$ for use in the detection of s_{i-n+1}. One of the problems involved in such a selection is that all the selected vectors may have the *same* value of d_{i-l} for some $l = 0$, 1, 2, \ldots, so that s_{i-l} is now effectively detected as d_{i-l} after only the corresponding small delay, with the resultant reduction in tolerance to noise. The following arrangement avoids the possibility of this happening.

In the detection of s_{i-n+1} from R_i, the receiver holds in store m^2h n-component vectors $\{E_i\}$, where h is a suitable positive integer less than n, and

$$E_i = d_{i-n+1}\ d_{i-n+2} \ldots d_i$$

If $n < g+1$, n is replaced in E_i by $g+1$. Each vector E_i is associated with the corresponding i-component vector X_i and therefore with the corresponding $|U_i|^2$, which is stored together with E_i. s_{i-n+1} is now detected as the value of d_{i-n+1} in the vector E_i associated with the smallest $|U_i|^2$. The receiver then selects, for each of the

m possible values of d_{i-h+1}, the vector E_i with the given value of d_{i-h+1} and with the smallest $|U_i|^2$. The process is repeated, in turn, for $d_{i-h+2}, d_{i-h+3}, \ldots, d_i$, a vector once selected not being available for selection a second time. The selection of the $\{E_i\}$ in the order described, rather than for the m values of each of $d_i, d_{i-1}, \ldots, d_{i-h+1}$, in that order, was suggested by J. D. Harvey. It appears that the maximum number of mh vectors is always selected here.

When the next received sample r_{i+1}, at the output of the linear filter, is received, each of the mh selected vectors $\{E_i\}$ forms a common part of m different vectors $\{D_{i+1}\}$. Each of the resultant m^2h vectors $\{D_{i+1}\}$ gives the corresponding vector E_{i+1} and has the associated value of

$$|U_{i+1}|^2 = |U_i|^2 + \left(r_{i+1} - \sum_{l=0}^{g} d_{i-l+1} y_l \right)^2$$

which is determined. The m^2h vectors $\{E_{i+1}\}$ together with the associated $\{|U_{i+1}|^2\}$ are stored. s_{i-n+2} is now detected as the value of d_{i-n+2} in the vector E_{i+1} associated with the smallest $|U_{i+1}|^2$. The receiver then selects, for each of the m possible values of d_{i-h+2}, the corresponding vector E_{i+1} with the smallest $|U_{i+1}|^2$. The process is repeated, in turn, for $d_{i-h+3}, d_{i-h+4}, \ldots, d_{i+1}$, as previously described, to give a set of mh selected vectors $\{E_{i+1}\}$ which are held in store together with the associated $\{|U_{i+1}|^2\}$, the remaining $\{E_{i+1}\}$ and $\{|U_{i+1}|^2\}$ being discarded. The detection process continues in this way. It is highly flexible and can give a near optimum tolerance to noise.

When the output samples $\{r_i\}$ from the linear filter are fed to System 3A or 4A, in place of the Viterbi-algorithm detector, the resultant arrangement not only minimizes the number of samples that must be used in a detection process to achieve a good tolerance to noise, but it also minimizes the number of sequential operations required to reach convergence in the iterative process. It is interesting to observe that when System 2A is used here, the arrangement becomes the optimum nonlinear equalizer. Since the Systems 1A, 3A, 4A and 5A, when suitably modified for use with multilevel signals, achieve a tolerance to additive white Gaussian noise that is at least as good as or better than that of System 2A, it is evident that when used with the linear filter they must be at least as good as or better than the optimum nonlinear equalizer, for all channels. This clearly also applies to the Viterbi-algorithm detector, when used with the linear filter. The linear filter can be held adaptively adjusted for a slowly time varying channel by using an appropriately modified version of the arrangement described for the adaptive nonlinear equalizer in Section 4.5, the detected element values now being obtained from the following detection process. The sampled impulse-response of the channel and linear filter is given here by the tap gains of the nonlinear filter, which is not otherwise involved in the detection process.

5.3.13 Estimation of the sampled impulse-response of the channel

When transmitting data over a time-varying channel and using one of the detection processes previously described, it is necessary to estimate the sampled impulse-response of the channel from the received data signal, after the detection of each s_i, so that the detector can be held correctly adjusted for the channel. The estimate of the channel response makes use of the fact that the signal vector SY is obtained from the *convolution* of the n-component signal vector S in Equation 5.3.4 and the n-component vector

$$V = y_0 \ y_1 \ldots y_{n-1} \qquad (5.3.60)$$

which represents the sampled impulse-response of the channel but now taken to n components, where $n > g$, instead of to $g + 1$ components as in Equation 5.2.2. Clearly, $V = Y_1$, where Y_1 is the n-component row vector given by the first row of the matrix Y, whose ith row is given by the n-component row vector

$$Y_i = \overbrace{0 \ldots 0}^{i-1} \ y_0 \ y_1 \ldots y_{n-i} \qquad (5.3.61)$$

where $y_i = 0$ for $i > g$. SY in fact contains only the first n components of the convolution of S and V, which has $2n - 1$ components.

Let Z be the $n \times n$ matrix whose ith row is given by the n-component row vector

$$Z_i = \overbrace{0 \ldots 0}^{i-1} \ s_1 \ s_2 \ldots s_{n-i+1} \qquad (5.3.62)$$

Clearly, $Z_1 = S$. Assume furthermore, as before, that the $\{s_i\}$ are polar signals having an even number of possible values, so that

$s_i \neq 0$. Thus the $\{Z_i\}$ are linearly independent and Z is a nonsingular upper-triangular matrix.

It can now be seen that

$$VZ = SY \qquad (5.3.63)$$

But, from Equation 5.3.8,

$$R' = SY + W \qquad (5.3.64)$$

so that

$$R' = VZ + W \qquad (5.3.65)$$

After the correct *detection* of S from R', the receiver knows S and therefore also the matrix Z. The receiver can now use its knowledge of Z to estimate V from R'.

Just as the maximum-likelihood estimate of S from R', when the receiver has prior knowledge of Y, is the n-component vector

$$X = R'Y^{-1} \qquad (5.3.66)$$

as can be seen from Section 3.3.3, so the maximum-likelihood estimate of V from R', when the receiver has prior knowledge of Z, is the n-component vector

$$F = R'Z^{-1} \qquad (5.3.67)$$

where

$$F = f_0 \, f_1 \ldots f_{n-1} \qquad (5.3.68)$$

and Z^{-1} is an $n \times n$ upper-triangular matrix whose components along the main diagonal all have the value s_1^{-1}.

If the vector R' is fed to the n input terminals of the $n \times n$ linear network Z^{-1}, the n output terminals of this network hold the vector F. From Equations 5.3.65 and 5.3.67,

$$F = (VZ + W)Z^{-1} = V + U \qquad (5.3.69)$$

where

$$U = WZ^{-1} \qquad (5.3.70)$$

Thus F is an unbiased linear estimate of V, such that

$$f_i = y_i + u_{i+1} \qquad (5.3.71)$$

where u_{i+1} is the $(i+1)$th component of the n-component vector U and is a Gaussian random variable with zero mean.

Considerable equipment complexity would obviously be involved in setting up the $n \times n$ network Z^{-1} after each detection process, but the transformation Z^{-1} can fortunately be implemented very much more simply by means of the feedback shift register shown in Figure 5.7.

Each square marked T in Figure 5.7 represents a storage element that holds the corresponding sample value r'_i in analogue or binary-coded form. Each circle represents a multiplier that multiplies its input signal by the quantity marked in the circle. Each time the storage elements are triggered, the stored values are shifted one place to the right. Before the estimation of V, all $\{s_i\}$ in Figure 5.7 are set to zero and the vector R' is fed into the shift register, to the position shown. The $\{s_i\}$ are now set to their *detected* values (these being assumed to be correct), and the shift register is triggered $n-1$ times to give the n-component output vector F.

From Equation 5.3.67,

$$R' = FZ \tag{5.3.72}$$

so that R' contains the first n terms of the sequence formed by the *convolution* of F and S. The feedback shift register in Figure 5.7 performs a process of *deconvolution* in that it removes from R' all components of S to leave just the vector F.

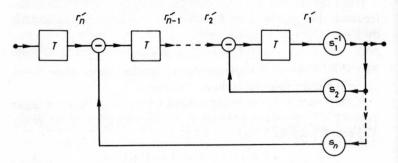

Fig. 5.7   Estimation of the sampled impulse-response of the channel from R'

In the n-component row vector R', the first two and last components are as follows:

$$r'_1 = f_0 s_1$$
$$r'_2 = f_0 s_2 + f_1 s_1 \tag{5.3.73}$$
$$r'_n = f_0 s_n + f_1 s_{n-1} + \ldots + f_{n-1} s_1$$

The multiplier s_1^{-1} in Figure 5.7 operates on r'_1 to give f_0 as the first component of the output vector. The multipliers $s_2, s_3, \ldots, s_n$ then form the terms $f_0 s_2, f_0 s_3, \ldots, f_0 s_n$ which are subtracted from the corresponding points along the shift register, removing the corresponding terms from the signals at these points, as can be seen from Equations 5.3.73. The shift register is then triggered to move the stored signals one place to the right, and the multiplier s_1^{-1} operates on its new input signal $r_2' - f_0 s_2 = f_1 s_1$, to give f_1 as the second component of the output vector. The multipliers $s_2, s_3, \ldots,$ s_{n-1} and associated subtraction circuits then remove all terms dependent on f_1 from the corresponding $n-2$ stored signals, as can be seen from Equations 5.3.73, and the shift register is triggered to give f_2 at the output of the multiplier s_1^{-1}. The multipliers $s_2, s_3, \ldots, s_{n-2}$ and associated subtraction circuits then remove all terms dependent on f_2 from the corresponding $n-3$ stored signals, and the shift register is triggered to give f_3 at the output, and so on. Clearly, the $(i+1)$th component of the output vector is just f_i and the feedback shift register converts FZ to F, thus performing the linear transformation Z^{-1}. The correct detection of S is, of course, assumed here.

When the receiver has prior knowledge that $y_i = 0$ for $i > g$, the feedback shift register may be simplified somewhat by replacing n by $g+1$, wherever it occurs in Figure 5.7. The vector F now has only $g+1$ components and is obtained by triggering the shift register g times.

From Equation 5.3.66,

$$R' = XY \qquad (5.3.74)$$

Thus, if each multiplier in Figure 5.7 is changed through the replacement of s_i by y_{i-1}, to give the arrangement in Figure 5.8, the

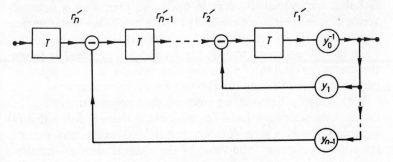

Fig. 5.8   Estimation of S from R'

feedback shift register now removes from R' all components of V, to leave just the vector X, and it therefore performs the linear transformation Y^{-1}. Unfortunately it has not so far been found possible to modify the feedback shift register in a simple manner so that this performs the same *nonlinear* transformation as that of the iterative process of System 3A, but in only the one sequence of operations corresponding to the triggering of the shift register $n-1$ times.

The estimation of V from R' relies here on the correct detection of S from R'. Since the detection of the last few components of S is not likely to be nearly as reliable as that of the first component, a better arrangement is as follows.

For each received signal-element, the vector V is estimated immediately after the detection of s_1, but the estimation process uses the vector R' corresponding to the $(n-1)$th previously detected element, together with the corresponding vector S. Each component of this vector S has been detected as s_1 in the appropriate detection process. The vector R' used in the detection of any signal element is, of course, the corresponding set of n sample values of the received signal, obtained after the removal of all intersymbol interference from the preceding elements. The receiver must therefore hold in a store the last n vectors $\{R'\}$, or alternatively it must generate the appropriate vector R' from the received samples $\{r_i\}$.

At high signal/noise ratios, the detected value of s_1 will normally be correct. When a received signal-element is incorrectly detected, n successive detected vectors $\{S'\}$, used for the estimation of V, contain the incorrect value for one of their components $\{s_i'\}$ and the n estimates of V corresponding to these $\{S'\}$ are therefore likely to have significant errors. The situation is further aggravated by the fact that errors tend to occur in bursts, so that when a detected vector S' is in error it is quite likely to have more than one incorrect component. Of course, even with the correct detection of S, any noise in R' leads to an error in the estimate of V. Thus, to ensure the correct operation of the estimation process, it is necessary to prevent the use of estimates of V that have serious errors.

With a slowly time-varying channel, the receiver can exploit the fact that the sampled impulse-response of the channel does not vary significantly over a time duration appreciably longer than that of the impulse response. The value of the channel sampled-impulse-response that is used in a detection process is the stored estimate F_s

of the sampled impulse-response V. Each new estimate F of the sampled impulse-response is first checked by measuring the distance $|F - F_s|$ between the vectors F and F_s in the n-dimensional Euclidean vector space containing these vectors. Whenever

$$|F - F_s| \geqslant c \qquad (5.3.75)$$

where c is a suitable positive constant such that $c < |F_s|$, the new estimate F is rejected, leaving the stored estimate at F_s. Whenever,

$$|F - F_s| < c \qquad (5.3.76)$$

the stored estimate F_s is updated to $bF + (1-b)F_s$, where b is a small positive constant, usually in the range 0.0001 to 0.1. The use of a small value for b provides an averaging or integrating action and thereby reduces the effect of noise on F_s. Clearly, the smaller the value of b, the greater the tolerance of the system to noise, but at the same time, the slower the greatest rate of variation of V that can be followed reasonably accurately by the stored estimate F_s. If $b = 1$, the new estimate F is itself used as the stored estimate for the next detection process. The system will now tolerate rapid variations in the sampled impulse-response of the channel, so long as there is negligible noise. However, the system has a very low tolerance to noise and may in fact become unstable in the presence of any significant noise. It is not therefore a practical arrangement. The preferred value of b always gives a compromise between the noise level and the rate of variation of the channel response that are tolerated by the system.

One of the problems in the estimation of the sampled impulse-response of the channel is that the mean-square noise level increases steadily along the length of the vector F, and where F has many components, say 50, the noise level can become quite large towards the end of the vector. Fortunately, it often happens that, when V has many non-zero components, only the first few of these have large magnitudes, the remainder being relatively small. Use can be made of this property by constraining the magnitudes of all but the first few components of F to an appropriate maximum level. Use can also be made of the fact that each individual component of F should only differ slightly from the corresponding component of F_s, by not permitting more than a given small discrepancy between corresponding components. These techniques can greatly reduce the effects of noise on F_s.

With a slowly time varying channel, the cancellation of the detected signal-elements is not normally exact (even with the correct detection of all signal elements). This is because the delay involved in the determination of F_s means that V has in the meantime changed a little, so that, even in the absence of noise, F_s usually differs slightly from its correct value V. Thus the vector R' (Equation 5.3.64) contains, in addition to the signal vector SY and the noise vector W, an error vector whose magnitude (length) is small compared with that of SY. Furthermore, the error vector has zero mean and is uncorrelated with SY, so that the error vector and SY are statistically orthogonal (as is also the case with W and SY). It follows that the averaging process used in the generation of F_s reduces the effects of the error vector (and noise vector) on F_s.

An important assumption that has been made so far is that the first component y_0 of the sampled impulse-response of the channel is nonzero. Over a time-varying channel it is of course quite possible for y_0 to become very small or even zero. Under these conditions, the detection process of System 2A fails, unless it is implemented by means of the point Gauss-Seidel iterative process.[18] In this case, if y_j is the first nonzero component of V, the first j components of R' are automatically ignored and the iterative process operates on just the last $n-j$ components of R'. In fact, the only effect of y_0 becoming zero, on any of the iterative processes of Systems 1A–5A, is to reduce automatically the effective value of n, as just mentioned, and hence to reduce the number of signal elements involved in a detection process. Unlike the practical implementation of System 2A, the iterative detection processes, together with the technique used for estimating V, assume only that $y_i = 0$ for $i < 0$ and $i > g$. None of the latter processes is therefore critically dependent on the precise position of the first nonzero component of V, so long as it is reasonably placed within the permitted range.

At the start of a transmission over a time-varying baseband channel (Figure 5.1) it is necessary to obtain an estimate of the sampled impulse-response of the channel, *before* the cancellation of intersymbol interference in a received sequence of signal elements of *known* element values that precedes the normal process of detection and signal cancellation. This can be achieved in various ways, a simple method being to send a sequence of unit impulses at regular intervals of $(g+1)T$ seconds to the input of the baseband channel, no other signals being transmitted during this time. The $g+1$

samples at intervals of T seconds, at the receiver, that correspond to each transmitted impulse, form a linear estimate of the sampled impulse-response of the channel. The average of these linear estimates is now taken to be the sampled impulse-response of the channel at the start of transmission.

The estimation of V from R', using the detected value of S, relies on the fact that S is the vector of a digital signal that has one of a *finite* number of different possible values, accurately known at the receiver. Thus, so long as S is correctly detected, its value is accurately known. If S were now the vector of a signal with a continuous range (and therefore an infinite number) of possible values, the receiver could not gain a knowledge of the *exact* value of S from a detection process on R'. The best it could do would be to *estimate* S, given R' and the stored estimate of V. If the receiver were now to use this estimate of S to estimate V, it would simply obtain the original estimate of V that was used to estimate S. This is, of course, because R' is now formed by the first n components of the convolution of the *estimates* of S and V. The estimate of S, obtained from R' and the estimate of V, cannot therefore itself be used with R' to give a *better* estimate of V.

An important feature of the technique described here for estimating the sampled impulse-response of the channel is that it does not rely on the absence of correlation between the received element values, so that it may be used satisfactorily with *any* sequence of element values. In the absence of noise, S is correctly detected and the subsequent estimate of V is exactly equal to V, regardless of the value of S. Furthermore, the technique enables a good estimate of the sampled impulse-response of the channel to be obtained from far fewer samples of the received data signal than would be possible with the corresponding arrangement of cross correlation, similar to the techniques used for adjusting the tap gains of linear and nonlinear equalizers. Thus the arrangement of adaptive detection and decision-directed signal cancellation tolerates a more rapidly time-varying channel, at a given signal/noise ratio, than do either linear or nonlinear equalizers.

5.3.14 Assessment of the basic technique

The technique of adaptive detection and decision-directed signal cancellation studied here can give a near optimum tolerance to additive white Gaussian noise, without excessive equipment complexity and at transmission rates of typically up to 10 000 bits/s over time-invariant or slowly time-varying channels. It can often achieve a considerable advantage in tolerance to additive noise over the corresponding linear or nonlinear equalizer and is much better suited to adaptive operation over time-varying channels. Furthermore, it does not involve any very significant increase in equipment complexity, when converted from operation over a time-invariant channel to operation over the corresponding time-varying channel. It is therefore ideally suited for applications of data transmission over voice-frequency channels such as telephone circuits and HF radio links. However, at transmission rates above 100 000 bits/s, there is insufficient time for the use of the more effective iterative detection processes, so that the optimum nonlinear equalizer is now probably the most cost-effective arrangement, particularly where the channel is time-invariant and known, so that adaptive operation is not necessary.

The basic technique employed in any of the iterative detection processes studied here is to break the detection process down into a number of identical or very similar operations that are performed sequentially by a simple piece of equipment. In this way, a sophisticated and powerful detection process can be implemented very simply, although more time is now required for the completion of the detection process.

5.4 DETECTION OF SEPARATE GROUPS OF SIGNAL ELEMENTS

5.4.1 *Introduction*

The technique of adaptive detection and decision-directed signal cancellation, described in Section 5.3, although normally capable of giving a good performance, has certain weaknesses that can sometimes cause problems.

It has been assumed throughout the analysis here that the receiver has knowledge of the correct sampling instants. Various techniques are available for determining the sampling instants. They may be determined from the received signal, either by means of a special timing signal transmitted with the data signal, or else from the received data signal itself.[37] However, with a time-varying channel that sometimes introduces severe signal distortion, it may not always be simple to ensure an accurate estimate of the correct sampling instants from the received signal.

An adaptive equalizer or detector normally requires a training signal, or at least a special synchronizing signal, to be sent at the start of transmission for the initial adjustment of the receiver. If now there is a break in transmission, during which time there is a large change in the sampled impulse-response of the channel, the training or special synchronizing signal must be retransmitted before correct operation can be restored. It would obviously be desirable that correct operation should be resumed automatically.

Again, for certain types of signal distortion, the combination of the sampled impulse-response of the channel and some particular sequence of transmitted element values can result in the complete loss of the received signal over the duration of the sequence. Correct operation of the data-transmission system clearly cannot be maintained over the prolonged transmission of such a sequence, regardless of the detection process used. For instance, a channel with the z-transform $1 + z^{-1}$ gives no output signal for the sequence of element values where $s_i = -s_{i-1}$ for all $\{i\}$, and thus prevents the prolonged transmission of this sequence. To avoid this difficulty it is necessary to scramble the transmitted signal to ensure that the $\{s_i\}$ are, for practical purposes, uncorrelated. The prolonged transmission of any particular simple sequence is now very unlikely. The scrambling of the transmitted signal is also necessary, both for nonlinear equalizers and for detection processes using decision-directed signal cancellation, to prevent an unfortunate combination of the sampled impulse-response of the channel and some repetitive sequence of transmitted element values from causing serious error-extension effects. It is sometimes desirable, however, that the data-transmission system should be able to handle *any* possible sequence of transmitted element values without the need for scrambling.

The technique to be studied here ensures that the necessary timing signals can readily be obtained at the receiver from the received data signal itself, however severe the signal distortion, and it avoids the need for scrambling the data signal, under all conditions.

The data-transmission system is as shown in Figure 5.9. The essential feature of this system is that the transmitted data signal

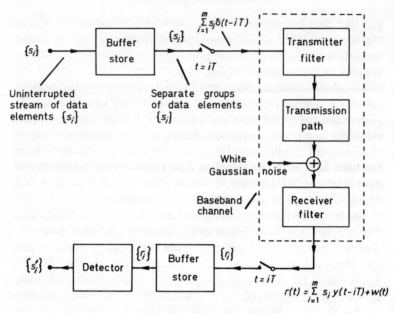

Fig. 5.9   Data-transmission system

at the input to the baseband channel is no longer a continuous
stream of regularly spaced signal-elements (impulses), but instead
the signal elements are arranged in separate groups, with sufficient
gaps (time guard bands) between adjacent groups to ensure that the
groups do not overlap each other at the receiver. Thus there is no
intersymbol interference between different groups at the receiver.
Each received group of signal elements is detected in a separate
process.

The signal elements at the input and output of the buffer store in
the transmitter (Figure 5.9) normally have rectangular waveforms.
The buffer store contains two stores, each with m storage elements.
At any instant, one of the two stores is filled with the corresponding
m successive element values of the incoming data stream, and the
other store is receiving the incoming data stream at the rate of one
element value every τ seconds. When one of the two stores has been
filled, its storage elements are sampled, in turn, once every T seconds,
where

$$T = \frac{m}{m+g}\tau \qquad (5.4.1)$$

g is the smallest positive integer such that the duration of the impulse response $y(t)$ of the baseband channel is always less than $(g+1)T$ seconds. Each output signal s_i from the buffer store is sampled at the appropriate time instant $t = iT$ and is fed to the baseband channel in the form of the corresponding impulse $s_i\delta(t-iT)$. When all m storage elements have been sampled, the next g impulses, fed to the baseband channel at intervals of T seconds, are all set to zero, so that no signal is transmitted during this period of gT seconds. By the end of this time, the second of the two stores has been filled with the incoming data element values, so that this store is now sampled while the other receives the incoming data, and so on. Clearly, after each group of m signal-elements, at the input to the baseband channel, there is a time gap of gT seconds. Thus, if $m = 4$ and $g = 2$, the signal could appear as in Figure 5.10.

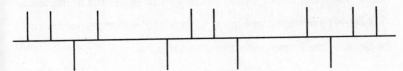

Fig. 5.10   Typical transmitted signal

The baseband channel has all the properties assumed in Sections 5.2 and 5.3, so that the sampled impulse-response of the channel is given by the $(g+1)$-component row vector

$$V = y_0 \; y_1 \ldots y_g \qquad (5.4.2)$$

where $y_i = y(iT)$. Clearly, the effect of the baseband channel is to spread out each group of signal elements so that it extends over possibly the whole of the following time-gap of gT seconds (but no further) in addition to its original time duration of mT seconds. The correct operation of the system relies on the fact that the impulse-response $y(t)$ of a practical channel has effectively a finite duration and does not for practical purposes extend to infinity.

It can be seen that the transmitted group of m signal-elements $\sum_{i=1}^{m} s_i\delta(t-iT)$ in Figure 5.9 arrive as the waveform $\sum_{i=1}^{m} s_i y(t-iT)$ at the receiver input, to give the received waveform

$$r(t) = \sum_{i=1}^{m} s_i y(t-iT) + w(t) \qquad (5.4.3)$$

where $w(t)$ is the received Gaussian noise waveform having the same properties as before. The m signal-elements are spread out over the $m+g$ received samples $\{r_i\}$, for $i = 1, 2, \ldots, m+g$, where $r_i = r(iT)$. These samples are independent of the other received groups of elements and are used for the detection of the m element values $\{s_i\}$ in Equation 5.4.3. Similarly, each of the other received groups of m elements is detected from the corresponding $m+g$ received samples that depend only on the group. For convenience, let

$$n = m+g \qquad (5.4.4)$$

As before, it is assumed that $s_i = \pm k$, the $\{s_i\}$ being statistically independent and equally likely to have either binary value.

To analyse the performance of the system it is necessary to consider only a single group of signal elements. Suppose therefore that at the input to the baseband channel there are the m impulses $\sum_{i=1}^{m} s_i \delta(t - iT)$ as shown in Figure 5.9. Let the m element values $\{s_i\}$ be given by the m-component row vector

$$S = s_1 \ s_2 \ldots s_m \qquad (5.4.5)$$

The corresponding n received samples are given by the n-component row vector

$$R = r_1 \ r_2 \ldots r_n \qquad (5.4.6)$$

and are fed to the buffer store in the receiver of Figure 5.9. The n noise components in the received vector R are given by the n-component row vector

$$W = w_1 \ w_2 \ldots w_n \qquad (5.4.7)$$

where $w_i = w(iT)$. As before, the noise components $\{w_i\}$ are statistically independent Gaussian random variables with zero mean and variance σ^2.

Let Y be the $m \times n$ matrix whose ith row is given by the n-component row vector

$$Y_i = \overbrace{0 \ldots 0}^{i-1} y_0 \ y_1 \ldots y_g \ 0 \ldots 0 \qquad (5.4.8)$$

The ith individual received signal-element of the group of m is given by the n-component row vector $s_i Y_i$, so that

$$\sum_{i=1}^{m} s_i Y_i = SY \qquad (5.4.9)$$

and, from Equation 5.4.3,

$$R = SY + W \qquad (5.4.10)$$

It may readily be shown that the $\{Y_i\}$ are linearly independent, so that Y has rank m. Y is here an $m \times n$ convolution matrix, such that SY is the convolution of the vectors S and V. The important differences between Equations 5.4.9 and 5.3.7 are firstly that, in Equation 5.3.7, S is an n-component vector and Y is an $n \times n$ matrix. Secondly, in Equation 5.4.9, the length of the vector Y_i is

$$|Y_i| = |V| \qquad (5.4.11)$$

for $i = 1, 2, \ldots, m$, where V is given by Equation 5.4.2 and is, of course, the sampled impulse-response of the channel. Equation 5.4.11 does *not* hold for all $\{i\}$ in Equation 5.3.7.

The buffer store at the receiver contains two separate stores, each with n storage elements. While one store holds the n received sample values $\{r_i\}$ for a detection process, the other store is receiving the next n sample values in preparation for the next detection process. Thus a period of nT seconds is available for the detection of the m received signal-elements of a group.

The detector detects the m $\{s_i\}$ from the n $\{r_i\}$, that is, it detects the vector S from the received vector R. The detector is assumed to have prior knowledge of the sampled impulse-response of the channel, V, and therefore of the matrix Y. In the case of a time-varying channel, the change in Y is assumed to be negligible over the duration of a received group of m signal-elements and sufficiently slow to enable Y to be correctly estimated from the received signal.

It can be seen from Equations 5.4.8 and 5.4.9 that, except for the trivial case where $V = 0$, $SY \neq 0$. Thus regardless of either the signal distortion introduced by the channel or the sequence of the binary element values $\{s_i\}$, there *cannot* be a loss of any received group of m signal-elements.

The main weakness of the arrangement is the fact that for a given transmitted information rate, the signal-element rate over a group of elements is greater than the element rate where these are transmitted continuously (without inserted time gaps). This increases the intersymbol interference between the signal elements of a group and so tends to reduce the tolerance of the system to noise. Obviously the ratio m/g of the duration of a group of elements to the duration of the time gap between adjacent transmitted groups must be kept as large as possible and preferably such that $m > 2g$. On the other hand, the greater the value of m, the greater the complexity of the system, so that m must be no larger than necessary to achieve a satisfactory performance. In a typical application, $8 \leqslant m \leqslant 32$ and $\frac{1}{4}m \leqslant g \leqslant m$.

5.4.2 System 1B

The detector here has prior knowledge both of Y and of the $\{|s_i|\}$, so that it knows the 2^m possible values of SY. The detector now selects the possible value of S for which $|R - SY|$ has the minimum value, where $|R - SY|$ is of course the distance between the vectors R and SY in the n-dimensional Euclidean vector space containing these vectors. The m element values $\{s_i\}$ are here detected in a single detection process, and all m detected values $\{s_i'\}$ are accepted. It is shown in Sections 3.2.8 and 3.2.9 that this detection process minimizes the probability of error in the detection of S from R, under the assumed conditions. At high signal/noise ratios it also minimizes the probability of error in the detection of any given s_i.[1]

The weakness of the detection process is that it involves 2^m sequential operations, which becomes excessive when $m > 10$ and the transmission rate approaches 10 000 bits/s. The important property of System 1B is that, for the given received signal, it achieves the best available tolerance to additive white Gaussian noise, so that no other detection process can give a lower probability of error in the detection of S from R.

It can be seen from Section 5.3.2 that the probability of error in the detection of S from R, in System 1B, may be estimated by evaluating the minimum distance $2d$ between two of the possible vectors $\{SY\}$ and taking the error probability to be $Q(d/\sigma)$. This is in fact an approximate upper bound to the correct error probability. At high signal/noise ratios and for $m < 10$, the tolerance to additive white Gaussian noise given by $Q(d/\sigma)$ is usually within about 1 dB of its correct value.[16,23]

5.4.3 Example

It is interesting now to evaluate the tolerance to additive white

Gaussian noise of System 1B for the case where the baseband channel in Figure 5.9 has the z-transform

$$Y(z) = 1 + 2.5z^{-1} + z^{-2} \qquad (5.4.12)$$

Suppose that the signal elements are transmitted in groups of three, with adjacent groups separated by two elements set to zero. Thus $m = 3$, $g = 2$ and $n = 5$. As before, $s_i = \pm k$.

S is a 3-component row vector and the matrix Y is the 3×5 matrix

$$Y = \begin{bmatrix} 1 & 2.5 & 1 & 0 & 0 \\ 0 & 1 & 2.5 & 1 & 0 \\ 0 & 0 & 1 & 2.5 & 1 \end{bmatrix} \qquad (5.4.13)$$

S and SY each have 8 possible values, the 8 possible values of SY being as follows:

$$
\begin{array}{llrrrr}
k & (\ 1 & 3.5 & 4.5 & 3.5 & 1) \\
k & (\ 1 & 3.5 & 2.5 & -1.5 & -1) \\
k & (\ 1 & 1.5 & -0.5 & 1.5 & 1) \\
k & (\ 1 & 1.5 & -2.5 & -3.5 & -1) \\
k & (-1 & -1.5 & 2.5 & 3.5 & 1) \\
k & (-1 & -1.5 & 0.5 & -1.5 & -1) \\
k & (-1 & -3.5 & -2.5 & 1.5 & 1) \\
k & (-1 & -3.5 & -4.5 & -3.5 & -1)
\end{array}
$$

R is a 5-component row vector given by

$$R = SY + W \qquad (5.4.14)$$

where the components $\{w_i\}$ of the noise-vector W are statistically independent Gaussian random variables with zero mean and variance σ^2.

In a detection process on the three signal elements of a group, one or more of the elements are wrongly detected if S is wrongly detected. Clearly, the average error probability in the detection of an s_i (the average being taken over the three $\{s_i\}$) cannot exceed the error probability in the detection of S and it cannot be less than a third of this error probability. Thus no significant inaccuracy is introduced by taking the average probability of error in the detection of s_i as being equal to the error probability in the detection of S, so long as the signal/noise ratio is high.

In the detection process considered here the decision boundaries are hyperplanes (4-dimensional planes) which perpendicularly bisect the lines joining the different possible vectors $\{SY\}$ in the 5-dimensional vector space containing R. The distance from any vector to a decision boundary is *half* the distance between the two vectors separated by this decision boundary. Thus the *minimum* distance from a vector to a decision boundary, taken over all vectors and their associated decision boundaries, is *half* the minimum distance between two of the possible vectors $\{SY\}$ in the vector space.

It can be seen from the 8 possible values of SY that the minimum distance between two vectors is

$$k(2^2 + 3^2 + 3^2 + 2^2)^{\frac{1}{2}} = 5.099k \qquad (5.4.15)$$

so that the minimum distance to a decision boundary is

$$d = 2.550k \qquad (5.4.16)$$

The value of the projection of the noise-vector W on to any direction in the 5-dimensional vector-space is a Gaussian random variable with zero mean and variance σ^2. Thus the probability of one or more element errors in a detection process is approximately

$$P_e = \int_d^\infty \frac{1}{\sqrt{(2\pi\sigma^2)}} \exp\left(-\frac{w^2}{2\sigma^2}\right) dw$$

$$= Q\left(\frac{d}{\sigma}\right) = Q\left(\frac{2.55k}{\sigma}\right) \qquad (5.4.17)$$

so that the average probability of error in the detection of a received signal-element, at high signal/noise ratios, is also approximately $Q(2.55k/\sigma)$.

The tolerance to additive white Gaussian noise of System 1B can now be compared with the tolerances of some of the other detection processes previously studied, for the particular channel whose z transform is $1 + 2.5z^{-1} + z^{-2}$, and using the appropriate equations for P_e that have been derived for these systems. The results of the comparison are given in Table 5.2, which shows the relative noise levels, for a given transmitted signal level and a given error-probability P_e of less than 10^{-4}. All systems are here compared with the linear equalizer whose tolerance to noise is taken as 0 dB, and the results quoted are only approximate. The information rate in System 1B is, of course, only $\frac{3}{5}$ of that in the other systems.

Table 5.2 Relative tolerances to additive white Gaussian noise of the different systems for a channel with z-transform $1+2.5z^{-1}+z^{-2}$

System	Relative tolerance to noise (dB)
Linear equalizer (Section 4.3.8)	0
Pure nonlinear equalizer (Section 4.4.2)	−1.3
Linear and nonlinear equalization of complementary factors of the channel response (Section 4.4.5)	3.4
Optimum nonlinear equalizer (Section 4.4.7)	4.7
System 1A (Section 5.3.3)	4.1
System 1B (Section 5.4.3)	6.8

Different values of the channel z-transform give different relative tolerances to noise for the various systems, so that the quoted values of the relative tolerances to noise in Table 5.2 apply only to the particular channel z-transform assumed here. The channel is nevertheless quite typical of practical channels that introduce appreciable amplitude distortion.

5.4.4 System 2B

This detection process minimizes the probability of error in the detection of S from R, when the detector has prior knowledge of the $m \times n$ matrix Y but no knowledge of the $\{|s_i|\}$ or σ^2. A binary polar signal is assumed, as before.

It is shown in Section 3.4 that this detection process determines the maximum-likelihood estimate of S from R, using a process of exact linear equalization, to give the m-component row vector

$$X = RY^T(YY^T)^{-1} \tag{5.4.18}$$

whose ith component is x_i. The $\{s_i\}$ are then detected from the signs of the corresponding $\{x_i\}$.

The principles of the maximum-likelihood estimation of S from R are considered in detail in Section 3.3. Clearly, if the n components of the received vector R are fed to the n input terminals of the $n \times m$ linear network $Y^T(YY^T)^{-1}$, then the signals at the m output terminals are the m components $\{x_i\}$ of the vector X.

Since the m vectors $\{Y_i\}$ are linearly independent, they span an m-dimensional subspace of the n-dimensional Euclidean vector space containing the vectors R, SY and W. But, from Equation 5.4.9, SY is a linear combination of the m $\{Y_i\}$, so that SY lies in the m-dimensional subspace for all values of S. The vectors R and W, however, do not normally lie in the subspace. The process of linear equalization, performed by the network $Y^T(YY^T)^{-1}$, determines the m-component vector X such that XY is the orthogonal projection of R on to the subspace. XY is therefore the point in the subspace

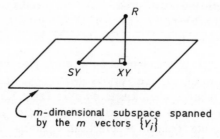

Fig. 5.11   Relationship between the vectors R, SY and XY

at the minimum distance from R. This corresponds to the noise-vector W of minimum length, given the vector R and the matrix Y, and so corresponds to the most likely value of the noise vector under the given conditions, as can be seen from Section 5.3.8. The relationship between the vectors R, SY and XY is illustrated in Figure 5.11, where of course $R-SY = W$. Again, from Equation 3.3.59, the ith component of X is

$$x_i = s_i + u_i \tag{5.4.19}$$

where u_i is a zero-mean Gaussian random variable. Thus X is an unbiased estimate of S and is obtained from R by a process of exact linear equalization.

It is shown in Section 3.4 that the probability of error in the detection of s_i from the sign of x_i is $Q(d_i/\sigma)$, where

$$d_i = k(Y_i(I - L_i^T(L_i L_i^T)^{-1}L_i)Y_i^T)^{\frac{1}{2}} \qquad (5.4.20)$$

L_i is the $(m-1) \times n$ matrix obtained from Y by deleting the ith row, and I is an $n \times n$ identity matrix.

In practice, the linear transformation $Y^T(YY^T)^{-1}$ is best performed by means of the point Gauss-Seidel iterative process,[13] as described in Section 5.3.5 and illustrated in Figure 5.3. R' is now replaced by R, X becomes an m-component vector and Y becomes an $m \times n$ matrix, but the arrangement is otherwise the same as System 2A, described in Section 5.3.5. Thus X is initially set to zero and its components are then adjusted, in turn and cyclically, the change in x_i being given by Δx_i in Equation 5.3.35, where preferably $1 \leqslant c \leqslant 1.5$. At the end of the iterative process, X satisfies Equation 5.4.18, where now $Y^T(YY^T)^{-1} \neq Y^{-1}$. Again, the practical implementation of the iterative process can be greatly simplified[13] by using the linear network Y_i^T instead of Y^T in Figure 5.3. Convergence is normally reached in less than 100 cycles of the iterative process.

The detection of s_1 from x_1 is *not* now equivalent to the detection of s_1 from r_1 and often gives a much better tolerance to noise. This is because the linear network $Y^T(YY^T)^{-1}$ exploits the prior knowledge of the fact that SY lies in the m-dimensional subspace spanned by the m $\{Y_i\}$, this prior knowledge not being used when s_1 is detected from r_1. Again, if a conventional linear feedforward transversal equalizer is used with the received signal here, a lower tolerance to noise is obtained. This is because a conventional equalizer makes no use of the prior knowledge of the fact that between adjacent groups of signal elements there are g elements set to zero. The transversal equalizer treats *all* received elements the same, including those set to zero and forming the gaps between adjacent groups. Furthermore, for the *given* sampled impulse-response of the channel, in Equation 5.4.2, the network $Y^T(YY^T)^{-1}$ gives *exact* linear equalization of the received signal, whereas a linear transversal equalizer achieves only the approximate equalization of the received signal. In practice, of course, y_i is not exactly equal to zero for $i < 0$ and $i > g$ (the delay in transmission being neglected here), so that the equalization of the received signal is not in fact exact when the network $Y^T(YY^T)^{-1}$ is used.

System 2B is the least effective detection process for S from R that is of any real practical value, any useful detection process having a tolerance to noise somewhere between that of System 2B and that of System 1B. Again, the Gauss-Seidel iterative process, used to implement System 2B, can be modified to improve its tolerance to noise much of the way towards that of System 1B, without either a serious increase in equipment complexity or else an excessive number of sequential operations in a detection process. The resultant detection processes will now be studied.

5.4.5 System 3B

This is a modification of the point Gauss-Seidel iterative process as applied to System 2B, and involves a negligible increase in equipment complexity. It operates in exactly the same way as System 3A, described in Section 5.3.6. Thus X is initially set to zero and its components are then adjusted, in turn and cyclically, the change in x_i being given by Δx_i in Equation 5.3.35, where $0 < c < 2$ (but preferably $1.25 \leqslant c \leqslant 1.5$)[4] and x_i is now constrained so that $|x_i| \leqslant k$, the constraint overriding and so if necessary truncating the change dictated by Equation 5.3.35.

As before, the operation of the iterative process is such as to adjust the vector X to minimize $|R - XY|$ within the limits of the constraints imposed on X. Thus, as can be seen from Section 5.3.6, XY is the orthogonal projection of R on to the m-dimensional solid polyhedron whose 2^m vertices are formed by the 2^m possible vectors $\{SY\}$. When the orthogonal projection of R on to the m-dimensional subspace spanned by the m $\{Y_i\}$ lies inside the polyhedron, XY is equal to this orthogonal projection, which is the point in the polyhedron at the minimum distance from R. When the orthogonal projection of R on to the subspace lies outside the polyhedron, XY is the point in the polyhedron at the minimum distance from the orthogonal projection of R on to the subspace, and it is therefore again at the minimum distance from R itself. Clearly, System 3B uses more of the available prior knowledge of the received signal than does System 2B, but not as much as System 1B.

Not only does System 3B often have a considerably better tolerance to additive white Gaussian noise than System 2B, at the lower signal/noise ratios, but it requires fewer cycles of the iterative process to reach convergence.

System 3B has the important advantage over the corresponding detection process System 3A, that in one iterative detection process of System 3B the detected values of all m $\{s_i\}$ are accepted whereas in System 3A only the detected value of s_1 is accepted. On the other hand, at the start of the iterative process in System 3B, the vector X must be set to zero, since the receiver has no prior knowledge of the values of the $\{s_i\}$. At the start of an iterative detection process of System 3A, the vector X can be set to value that is usually quite close to its final value, since the receiver here normally has considerable prior knowledge of the $\{s_i\}$. This reduces the number of sequential operations in the detection process, so that, although System 3B usually requires many fewer sequential operations per received signal-element than does System 3A, the improvement is not as great as might be expected. All the other techniques previously described, for speeding up the iterative processes of Systems 2A and 3A may, of course, also be applied to the iterative processes of Systems 2B and 3B, as well as to the further developments of these systems.

5.4.6 Example

The relationship between the Systems 1B, 2B and 3B can be illustrated by the following simple example. Suppose that

$$R = SY + W \qquad (5.4.21)$$

where

$$SY = s_1 Y_1 + s_2 Y_2 \qquad (5.4.22)$$

As before, $s_1 = \pm k$, $s_2 = \pm k$, and $|Y_1| = |Y_2|$.

Clearly, SY has four possible values that lie in the two-dimensional subspace spanned by Y_1 and Y_2. The four possible values of SY are shown by the points T_1, T_2, T_3 and T_4 in Figure 5.12, where the points K, L, M and N are the vectors kY_1, kY_2, $-kY_1$ and $-kY_2$, respectively.

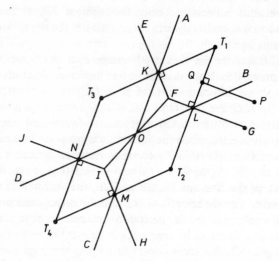

Fig. 5.12   Decision boundaries in the two-dimensional subspace

Suppose that P is the orthogonal projection of R on to the two-dimensional subspace, so that P is the vector XY obtained at the end of the iterative process of System 2B. Let Q be the orthogonal projection of P on to the nearest side of the quadrilateral $T_1T_2T_4T_3$, so that Q is the vector XY obtained at the end of the iterative process of System 3B. Clearly, Q is the orthogonal projection of P on to the line T_1T_2 in Figure 5.12. If P lies inside the quadrilateral $T_1T_2T_4T_3$, $Q = P$.

It is shown in Section 3.4.4 that System 1B is equivalent to the arrangement that divides the two-dimensional subspace into the four decision regions bounded by *EFG*, *GFIH*, *HIJ* and *JIFE*, where *EF*, *FG*, *HI* and *IJ* are the perpendicular bisectors of T_3T_1, T_1T_2, T_2T_4 and T_4T_3, respectively. The arrangement forms the vector P and then selects the vector T_i that lies in the same decision region. The corresponding possible value of the vector S is the detected vector.

System 2B is equivalent to the arrangement that divides the two-dimensional subspace into the four decision regions bounded by *AOB*, *BOC*, *COD* and *DOA*, where *AC* and *BD* are the lines defining the one-dimensional subspaces spanned by Y_1 and Y_2, respectively. As before, the arrangement forms the vector P and then selects the vector T_i that lies in the same decision region, the corresponding possible value of the vector S being the detected vector.

It can now be seen that System 3B is equivalent to the arrangement that operates as System 2B, having the same decision boundaries as System 2B but using the vector Q instead of the vector P. Alternatively, System 3B is equivalent to the arrangement that divides the two-dimensional subspace into the four decision regions bounded by *EKOLG*, *GLOMH*, *HMONJ* and *JNOKE*, and uses the vector P to give the detected vector S', as before. The interesting thing about this is that the decision boundaries are a *compromise* between those for System 1B and those for System 2B, where each system uses the vector P.

For each of the Systems 1B, 2B and 3B, the vector P is the sum of the received signal-vector $T_i = SY$ and the orthogonal projection of the noise-vector W on to the two-dimensional subspace spanned by Y_1 and Y_2. This can be seen from Figure 5.11, bearing in mind that $W = R - SY$. An error occurs in the detection of S when the projected noise vector carries P on to the other side of a decision boundary with respect to the received vector T_i. The projected noise-vector is equally likely to lie along any direction in the subspace. Furthermore, the probability of its magnitude exceeding a given positive value depends only on this value and on the noise variance σ^2, and decreases rapidly as the value increases. It can be seen in Figure 5.12 that, along any direction from any one of the four $\{T_i\}$, the distance to the nearest decision boundary is in general greater for System 1B than it is for System 2B, and it is never smaller. Thus, for a given signal/noise ratio, System 1B has a lower probability of error than System 2B, and System 3B has a probability of error somewhere between that of System 1B and that of System 2B.

It can be seen from Figure 5.12 that whenever P lies *outside* the quadrilateral $T_1T_2T_4T_3$, the effective decision boundaries in the detection process of System 3B are *EK*, *GL*, *HM* and *JN*, which form part of the decision boundaries of System 1B. Thus the tolerance to additive white Gaussian noise is now the same as that of System 1B. Similarly, when P lies *inside* the quadrilateral $T_1T_2T_4T_3$ (so that $P = Q$), the effective decision boundaries in the detection process of System 3B are *KO*, *LO*, *MO* and *NO*, which form part of the decision boundaries of System 2B. Thus the tolerance to additive white Gaussian noise is now the same as that of System 2B.

Since the orthogonal projection of the noise vector W on to the two-dimensional subspace in Figure 5.12 is equally likely to lie along *any* direction in the subspace, it can be seen that when the received signal-vector is say T_1 in Figure 5.12, the probability that P lies *outside* the quadrilateral $T_1T_2T_4T_3$ is $(360 - \theta)/360$, where $\theta°$ is the acute angle formed by T_3T_1 and T_2T_1. Similarly, when the received signal-vector is T_i, in the general case of an m-dimensional subspace, the probability that P lies *outside* the solid polyhedron whose vertices are the 2^m points $\{T_j\}$, is the fraction of the total solid (m-dimensional) angle at the vertex T_i that lies outside the polyhedron. As m increases for a given sampled impulse-response of the channel, so there is an increase in the average value of the fraction of the total solid angle at a vertex that lies outside the polyhedron. This effect can be observed by considering the decision boundaries in one-, two- and three-dimensional subspaces. Thus the probability that P lies outside the polyhedron increases with m. Clearly, for large values of m, P nearly always lies outside the polyhedron, and so, as m increases, the tolerance of System 3B to additive white Gaussian noise approaches that of System 1B. An alternative explanation of this effect is that, as m increases, so, in the detection process of System 3B, there is an increase in the *fraction* of the decision boundary, surrounding any signal-vector T_i, that is common to (coincident with) the decision boundary of System 1B, and hence the more closely does the performance of System 3B approach that of System 1B, the approximation improving as the signal/noise ratio *decreases*. At sufficiently high signal/noise ratios, however, the tolerance to additive white Gaussian noise is still essentially determined by the *minimum* distance from T_i to the decision boundary surrounding it. Thus, as the signal/noise ratio increases for any given value of m, so the performance of System 3B approaches that of System 2B. Of course, the larger the value of m, the higher the signal/noise ratio at which the tolerances of the two systems to additive white Gaussian noise become essentially the same.

The simple example, that has just been used to demonstrate the relationship between Systems 1B, 2B and 3B, may also be applied to the Systems 1A, 2A and 3A to show exactly the same relationship between these three systems. The only difference is that now the vector space has only two dimensions and is spanned by Y_1 and Y_2, so that $P = R'$. Also $|Y_1| \neq |Y_2|$ which leads to rather less symmetry in the decision boundaries.

5.4.7   System 4B

It is clear from the basic structure of the matrix Y that in the received signal-vector SY (Equation 5.4.9) there is on average less intersymbol interference in $s_1 Y_1$ and $s_m Y_m$ than in the remaining $\{s_i Y_i\}$, so that, with System 3B, there is usually a lower probability of error in the detection of s_1 or s_m than in the detection of any of the remaining $\{s_i\}$. System 4B exploits this fact by using an arrangement of decision-directed cancellation of intersymbol interference, operating as described in Section 5.3.1. System 4B is a modification of System 3B, in which the detection process of System 3B is carried out exactly as described in Section 5.4.5, but only the detected value of s_1 is accepted. Assuming the correct detection of s_1, the components of $s_1 Y_1$ are all known. These components are now removed by subtraction from R to give $R - s_1 Y_1$, and x_1 is set and held at zero throughout the rest of the process. The detection process of System 3B is then repeated on the vector $R - s_1 Y_1$, to give the detected value of s_2. With the correct detection of s_2, the components of $s_2 Y_2$ are all known. These components are now removed by subtraction from $R - s_1 Y_1$ to give $R - s_1 Y_1 - s_2 Y_2$, and x_2 is set and held at zero throughout the rest of the process. The detection process continues in this way. The arrangement was proposed by Ghani.

Obviously, if any signal element is incorrectly detected, its intersymbol interference in the following elements instead of being removed is doubled, so that there are now likely to be further errors in the detection of the remaining elements. It can be seen that in the detection of s_i, given the correct detection of s_1 to s_{i-1}, the signal-element $s_i Y_i$ is the *first* of the signal elements remaining uncancelled, so that the probability of error in the detection of s_i, given the correct detection of s_1 to s_{i-1}, is on average no greater (and often less) than the probability of error in the detection of s_1. Thus, at high signal/noise ratios, the probability of error in the detection of S from R by System 4B can be taken to be equal to the probability of error in the detection of s_1 from R.

The process can be speeded up through the following modification, proposed by Clements. The receiver accepts the detected values of both s_1 and s_m at the end of the first iterative process, then subtracts both $s_1 Y_1$ and $s_m Y_m$ from R, and so on. Furthermore, in the second and subsequent iterative processes, the vector X at the start of the process is set so that its components, corresponding to the $\{s_i\}$ still to be detected, have the values they had at the end of the preceding iterative process, the remaining $\{x_i\}$ being held at zero. Since far fewer cycles of the iterative process of System 3B are usually needed for the satisfactory detection of s_1 and s_m than are needed for the detection of the remaining $\{s_i\}$, the number of sequential operations involved in a detection process of System 4B is considerably less than for System 3B.

At error rates above 1 in 10^3, System 4B achieves a tolerance to additive white Gaussian noise typically within about 3 dB of that of System 1B. It does not involve either complex equipment or an excessive number of sequential operations.

5.4.8　System 5B

This is a modification of System 4B that operates exactly as the System 4A described in Section 5.3.7. Thus in the detection of s_1, x_1 is first set to k and held at this value during the whole of the iterative process of System 3B. At the end of this process, the distance $|R - XY|$ is measured. x_1 is then set and held at $-k$, and the process repeated to give a second measure of $|R - XY|$. The detected value of s_1 is now taken to be the value of x_1 for which $|R - XY|$ has the smaller value. Assuming the correct detection of s_1, the components of $s_1 Y_1$ are all known. These components are then removed by subtraction from R to give $R - s_1 Y_1$, and x_1 is set and held at zero throughout the rest of the process. The iterative process of System 3B is now repeated with x_2 held first at k and then at $-k$, to give the two corresponding values of $|R - s_1 Y_1 - XY|$, and s_2 is detected as the value of x_2 for which $|R - s_1 Y_1 - XY|$ has the smaller value. With the correct detection of s_2, the components of $s_2 Y_2$ are all known. These components are then removed by subtraction from $R - s_1 Y_1$ to give $R - s_1 Y_1 - s_2 Y_2$, and x_2 is set and held at zero throughout the rest of the process. The detection process continues in this way.

System 5B makes more use of the available prior knowledge of the received signal than does System 4B, for the reasons given in Section 5.3.8, bearing in mind that System 5B corresponds to System 4A, whereas System 4B corresponds to System 3A.

System 5B achieves a tolerance to additive white Gaussian noise typically within about 2 dB of that of System 1B. It does not involve either complex equipment or an excessive number of sequential operations.

5.4.9   Comparison of Systems 1B–5B

From the theoretical considerations in the preceding sections it can be seen that the detection processes, listed in the order of their tolerances to additive white Gaussian noise and starting with the best system, should be Systems 1B, 5B, 4B, 3B and 2B.

The tolerances to additive white Gaussian noise of the Systems 1B–5B have been tested by computer simulation for different channels, with groups of eight signal-elements and a channel sampled-impulse-response extending over up to five consecutive samples, so that $m = 8$ and $g = 4$. The data-transmission system assumed in these tests is as shown in Figure 5.9 and described in

Table 5.3 Reduction in tolerance to additive white Gaussian noise, measured in dB, when the given channel replaces one introducing no distortion or attenuation

Channel	Sampled impulse-response of channel	System 1B	System 2B	System 3B	System 4B	System 5B
1	$2^{-\frac{1}{2}}(1,1,0,0,0)$	1.2	6.0	2.3	1.6	1.6
2	$1.5^{-\frac{1}{2}}(1,0.5,0.5,0,0)$	0.5	3.3	1.6	1.5	1.5
3	$1.5^{-\frac{1}{2}}(0.5,1,0.5,0,0)$	2.4	13.7	7.1	3.4	3.4
4	$1.5^{-\frac{1}{2}}(0.5,1,-0.5,0,0)$	0.0	0.3	0.3	0.0	0.1
5	$2^{-\frac{1}{2}}(0.235,0.667,1,0.667,0.235)$	4.6	17.6	14.0	7.3	6.5
6	$2^{-\frac{1}{2}}(-0.235,0.667,1,0.667,-0.235)$	1.2	4.9	2.5	2.0	2.1

Section 5.4.1. The results of the tests are given in Table 5.3. This quotes the number of dB reduction in tolerance to additive white Gaussian noise, at an element error rate of 4 in 10^3, when the given channel replaces one that introduces no distortion or attenuation. All detection processes have the same tolerance to noise in the presence of no distortion or attenuation. The sampled impulse-response of each channel is given by a vector of unit length, so that no signal attenuation (or gain) is introduced by any channel. The

channels here are the same as those in Table 5.1. The 95% confidence limits of the quoted results are about ± 0.4 dB.

It can be seen from Table 5.3 that the relative tolerances to additive white Gaussian noise of the Systems 1B–5B are in the same order as those predicted theoretically. A comparison of Table 5.3 with Table 5.1, however, shows that System 5B gains a smaller advantage over System 4B than the corresponding System 4A gains over System 3A. In fact, System 5B is only marginally better than System 4B. It is, however, somewhat more complex than System 4B and involves more than twice the number of sequential operations in a detection process.

Computer simulation tests carried out by Clements on Systems 1B–5B, operating over Channel 5 (Table 5.3) and using 8 signal-elements in a group, have shown that the variation of error rate with signal/noise ratio for each of the different systems is as given in Figure 5.13. It can be seen here that, as the error rate decreases so

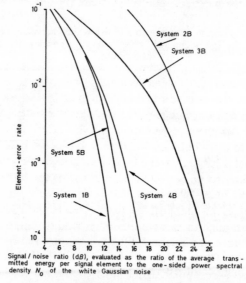

Signal / noise ratio (dB), evaluated as the ratio of the average trans-mitted energy per signal element to the one-sided power spectral density N_0 of the white Gaussian noise

Fig. 5.13   Error rates of Systems 1B–5B at different signal/noise ratios

the signal/noise ratio required by System 3B increases much more rapidly than the signal/noise ratio required by any of the other four systems. At error rates below 1 in 10^4, the performance of System 3B becomes close to that of System 2B, whereas at error rates above 1 in 10, it becomes close to that of System 1B. System 4B exhibits the same effect as System 3B but to a much less marked degree. Its performance only becomes close to that of System 2B at error rates well below 1 in 10^8. Again, the performance of System 5B slowly approaches that of System 1B as the error rate decreases.

Extensive tests by computer simulation on System 4B suggest that the number of sequential operations involved in a detection process is unlikely to exceed the corresponding number involved in 15 cycles of the iterative process of System 3B. This compares with a typical maximum of 20 cycles for the detection process of System 4A. Bearing in mind that m signal-elements are detected in a detection process of System 4B or 5B, compared with only one signal-element in the case of System 4A, the greater number of sequential operations required in a detection process of System 5B, relative to System 4B, is not quite such a serious disadvantage for System 5B.

The analysis of the different systems suggests that System 5B is likely to be the most cost-effective system.

5.4.10 Detection of multilevel signal-elements

Where multilevel signal-elements are transmitted in separate groups, the detection processes, other than System 2B, either have a reduced tolerance to noise or else require an excessive number of sequential operations. An effective technique for the detection of multilevel elements is to carry out an *initial* detection process, based on System 3B, that selects for each s_i the two adjacent possible values that are the most likely to be correct. The detection of the $\{s_i\}$ is then completed by an iterative process that operates *only* on the selected possible values of the $\{s_i\}$, so that it treats the received signal-elements as though these were the corresponding binary elements.[23]

If s_i can have one of the l different values given by $(2i-l+1)k$ for $i = 0, 1, \ldots, l-1$, where l is even and each received signal-element is an l-level polar signal, the two selected values of s_i and s_{i+1}, used in the final detection process, are hk and $(h+2)k$ for s_i, and jk and $(j+2)k$ for s_{i+1}, where h and j are odd integers in the range $-l+1$ to $l-3$. More often than not, h and j are not equal.

The initial detection process is System 3B which gives the m-component vector X, such that $|x_i| \leqslant (l-1)k$ for each i. The two selected values of s_i, for each i, are now the two possible values *nearest* to x_i. If $x_i = (l-1)k$, the two selected values of s_i are $(l-1)k$ and $(l-3)k$. If x_i is *exactly* equal to one of the $l-2$ possible values of s_i that do not include the two extreme values, then this value of s_i together with *either* of the two adjacent possible values are taken as the two selected values.

The final detection process can be one of the Systems 1B, 3B or 4B, operating on the *selected* possible values of the m $\{s_i\}$. The detection processes are otherwise exactly as described for the binary polar signals where $s_i = \pm k$. There would, of course, be no purpose in using System 2B here. In the case of Systems 3B and 4B, the constraints on x_i are now such as to prevent it from taking on a value outside the range of values between the two selected possible values of s_i.

System 5B can be modified for use with multilevel signal-elements in a manner exactly similar to that described for System 4A in Section 5.3.11. System 5B has the important advantage over the other systems here that it does not require a *separate* iterative process for the initial selection of the possible values of the $\{s_i\}$, although all possible values of s_1 must now be considered.

If System 1B is used for the final iterative process here, the resultant tolerance to noise is similar to that if System 1B were used on its own for the complete detection process.[23] Again, the relative tolerances to additive white Gaussian noise of the Systems 1B–4B, when used for the second part of the two-stage detection process just described, are in the same order as those in Table 5.3, where of course the signals are binary coded and the detection processes are used on their own.[23] The most promising complete system is the modified form of the Viterbi-algorithm detector (Section 5.3.12).

Clearly, a near optimum tolerance to additive white Gaussian noise can be obtained with multilevel signal-elements, without either excessively complex equipment or an excessive number of sequential operations in a detection process, so long as m is not too large.

5.4.11 Estimation of the sampled impulse-response of the channel

The sampled impulse-response of the baseband channel in Figure 5.9 is the $(g+1)$-component row vector

$$V = y_0 \ y_1 \dots y_g \qquad (5.4.23)$$

The received n-component signal vector SY is the *convolution* of the vectors V and S, where

$$S = s_1 \ s_2 \ldots s_m \tag{5.4.24}$$

Y is the $m \times n$ convolution matrix whose ith row is given by the n-component row vector

$$Y_i = \overbrace{0 \ldots 0}^{i-1} \ y_0 \ y_1 \ldots y_g \ 0 \ldots 0 \tag{5.4.25}$$

From Equation 5.4.10, the received n-component vector, used for the detection of S, is

$$R = SY + W \tag{5.4.26}$$

Let Z be the $(g+1) \times n$ matrix whose ith row is given by the n-component row vector

$$Z_i = \overbrace{0 \ldots 0}^{i-1} \ s_1 \ s_2 \ldots s_m \ 0 \ldots 0 \tag{5.4.27}$$

Assume furthermore that the $\{s_i\}$ are polar signals having an even number of possible values, so that $s_i \neq 0$. Thus the vectors $\{Z_i\}$ are linearly independent and the matrix Z has rank $g+1$.

It can now be seen that

$$VZ = SY \tag{5.4.28}$$

so that, from Equation 5.4.26,

$$R = VZ + W \tag{5.4.29}$$

After the correct detection of S from R, the receiver knows S and therefore also the matrix Z. The receiver can now use its knowledge of Z to estimate V from R.

Just as the maximum-likelihood estimate of S from R, when the receiver has prior knowledge of Y, is the m-component row vector

$$X = RY^T(YY^T)^{-1} \tag{5.4.30}$$

so the maximum-likelihood estimate of V from R, when the receiver has prior knowledge of Z, is the $(g+1)$-component row vector

$$F = RZ^T(ZZ^T)^{-1} \tag{5.4.31}$$

where

$$F = f_0 \ f_1 \ldots f_g \tag{5.4.32}$$

and $Z^T(ZZ^T)^{-1}$ is an $n \times (g+1)$ matrix of rank $g+1$.

Equation 5.4.31 may alternatively be derived as follows. When the receiver has no prior knowledge of V, then, as far as it is concerned, V may be any real $(g+1)$-component vector F, and VZ is the n-component vector

$$FZ = \sum_{i=0}^{g} f_i Z_{i+1} \tag{5.4.33}$$

Thus FZ is a linear combination of the $g+1$ row vectors $\{Z_i\}$ and FZ must lie in the $(g+1)$-dimensional subspace spanned by the $g+1$ vectors $\{Z_i\}$, in the n-dimensional Euclidean vector space containing the received vector R. As far as the receiver is concerned, before it has received the vector R, FZ is equally likely to lie at any point in the subspace.

Given the received vector R, the most likely value of F, defined as the value that maximizes the conditional probability density $p(R|F)$, is the value of F that minimizes the distance between FZ and R, as can be seen from Section 3.3.3. Thus FZ is the orthogonal projection of R on to the subspace spanned by the $\{Z_i\}$, as shown in Figure 5.14. This corresponds to the noise-vector $W = R - FZ$ of the minimum length, given the vector R and the matrix Z.

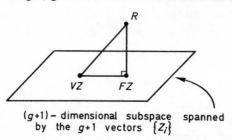

$(g+1)$- dimensional subspace spanned
by the $g+1$ vectors $\{Z_i\}$

Fig. 5.14   Relationship between the vectors R, VZ and FZ

Clearly, the vector $R - FZ$ is now orthogonal to each vector Z_i, so that

$$(R - FZ)Z_i^T = 0 \tag{5.4.34}$$

or

$$(R - FZ)Z^T = 0 \tag{5.4.35}$$

so that F is given by Equation 5.4.31.

Thus, if the vector R is fed to the n input terminals of the $n \times (g+1)$ linear network $Z^T(ZZ^T)^{-1}$, the $g+1$ output terminals of the network hold the vector F. From Equations 5.4.31 and 5.4.29,

$$F = (VZ + W)Z^T(ZZ^T)^{-1} = V + U \qquad (5.4.36)$$

where

$$U = WZ^T(ZZ^T)^{-1} \qquad (5.4.37)$$

Thus F is an unbiased linear estimate of V, such that

$$f_i = y_i + u_{i+1} \qquad (5.4.38)$$

where u_{i+1} is the $(i+1)$th component of the $(g+1)$-component vector U.

Just as the linear transformation $Y^T(YY^T)^{-1}$ is best implemented by the point Gauss-Seidel iterative process, so also is the linear transformation $Z^T(ZZ^T)^{-1}$, the arrangement being as shown in Figure 5.15.

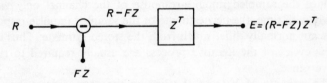

$$R \longrightarrow \ominus \xrightarrow{R-FZ} \boxed{Z^T} \longrightarrow E = (R-FZ)Z^T$$

$$FZ$$

Fig. 5.15   Linear estimation of V from R by means of an iterative process

The vector F is first set to the stored value of the estimate of V, when the $g+1$ output terminals of the linear network Z^T, in Figure 5.15, hold the $g+1$ components of the vector

$$E = e_0 \ e_1 \dots e_g \qquad (5.4.39)$$

where

$$E = (R - FZ)Z^T \qquad (5.4.40)$$

The iterative process now operates as follows. f_0 is adjusted to set e_0 to zero. f_1 is then adjusted to set e_1 to zero, and so on sequentially to f_g. The cycle of adjustments to the $\{f_i\}$ is now repeated several times, in the same order as before, until eventually $E \simeq 0$, when

$$(R - FZ)Z^T \simeq 0 \qquad (5.4.41)$$

so that

$$F \simeq RZ^T(ZZ^T)^{-1} \qquad (5.4.42)$$

and F approximately satisfies Equation 5.4.31.

The iterative process always converges to Equation 5.4.42 without involving an excessive number of sequential operations, and it can be implemented very simply.[13]

When f_i is adjusted to set e_i to zero, the change in f_i is

$$\Delta f_i = e_i (Z_i Z_i^T)^{-1} = \frac{e_i}{|Z_i|^2} \tag{5.4.43}$$

as can be seen from Equation 5.4.40. Thus, since Z_i is known, Δf_i can be determined directly from e_i.

The iterative process always converges[2,4] so long as

$$\Delta f_i = c \frac{e_i}{|Z_i|^2} \tag{5.4.44}$$

where c is a real constant such that $0 < c < 2$. However the most rapid rate of convergence is in general obtained when $1 \leqslant c \leqslant 1.5$.

Since the sampled impulse-response of the channel only varies slowly with time, the new estimate of the sampled impulse-response does not normally differ much from the stored estimate. Thus only a few cycles of the iterative process are usually required to reach convergence.

The estimate of V from R relies on the correct detection of S from R. If S is incorrectly detected, the matrix Z is incorrect, leading to an error in the estimate of V. Of course, even with the correct detection of S, any noise in R usually leads to an error in the estimate of V.

In the detection of S from R, the receiver uses the stored estimate F_s of the channel sampled-impulse-response V. Following the detection of S, the receiver uses the detected value of S to estimate V from R. The estimate F is then checked by measuring the distance $|F - F_s|$ between the vectors F and F_s in the $(g+1)$-dimensional Euclidean vector space containing these vectors. Whenever

$$|F - F_s| \geqslant c \tag{5.4.45}$$

where c is a suitable positive constant such that $c < |F_s|$, the new estimate F is rejected, leaving the stored estimate at F_s. The condition given by Equation 5.4.45 can also be used to indicate a possible error in the detection of S. Whenever

$$|F - F_s| < c \tag{5.4.46}$$

the stored estimate F_s is updated to $bF + (1-b)F_s$ where b is a small positive constant, usually in the range 0.0001 to 0.1. The arrangement operates exactly as that described in Section 5.3.13.

It has been shown by Clements that the correct estimation of the sampled impulse-response of the channel does not here require the transmission of any training signal at the start of transmission. The receiver starts by setting the stored estimate of V to $(1, 0, 0, \ldots)$ and proceeds to the detection of S and estimation of V for each received group of signal elements. If correct operation, indicated by only small variations in the new estimates of V, is not achieved within a certain time, the stored estimate of V is set to $(0, 1, 0, 0, \ldots)$, and so on. As soon as the component 1 in the initial stored estimate of V coincides with the main component of V, correct detection of S and estimation of V is achieved after the reception of only a relatively small number of groups of signal elements, this number depending upon the severity of the signal distortion.

What in fact happens here is that so long as the single nonzero component of the initial stored estimate of V coincides with the main component of V, there are on average more correct than incorrect detected values of the $\{s_i\}$ in the detection of S, and this is sufficient to give a succession of improved estimates of V. During this learning process, a relatively large value of b is used, to permit a rapid rate of variation in the stored estimate of V. One reason why the estimates of V improve is that there are no error extension effects in the detection of successive groups of received signal-elements, the vector R used for a detection or estimation process being unaffected by the correct or incorrect detection of S in the previous detection process.

Whenever the correct operation of the receiver has been disrupted for some reason, such as a prolonged break in transmission or a burst of very high level noise, this is recognized by persistent large variations in the new estimates of V, and the receiver now automatically goes through the learning procedure just described.

One weakness of this procedure is that the estimate finally obtained for V may be the *negative* of its correct value. The effect of this may however readily be overcome by using *differential* coding in the transmitted signal.[37]

In the absence of noise and with the correct stored estimate of V, S is correctly detected and the subsequent estimate of V is exactly equal to V, regardless of the value of S. Furthermore, a good estimate of the sampled impulse-response of the channel is obtained from far fewer samples of the received signal than would be possible with the corresponding arrangement of cross correlation, similar to the techniques used for adjusting the tap gains of linear and non-linear equalizers. Thus the adaptive detection processes, just described, tolerate more rapidly time-varying channels, for a given signal/noise ratio, than do either linear or nonlinear equalizers.

An important feature of the iterative process for estimating the sampled impulse-response of the channel is that when S is detected by means of one of the Systems 2B–5B, the same basic iterative process is used both for the detection of S and for the estimation of V, so that the *same* equipment (with only minor modifications) is used first to detect S and then to estimate V. Thus no significant additional equipment complexity is involved in estimating V.

5.4.12 Assessment of the basic technique

The basic arrangement studied here achieves the greatest advantage over other techniques when two conditions are satisfied. Firstly, there is severe amplitude distortion but g is small, say $g < 3$. Secondly, the data to be transmitted occurs in the form of separate characters (or frames) of fixed length, so that the data is alpha-numeric or otherwise suitably coded. Under these conditions, a negligible penalty in tolerance to noise results from the time gaps between adjacent transmitted groups of elements and no additional equipment complexity is involved in dividing the data into separate groups. Now, severe amplitude distortion with a small value of g is normally obtained over a partial-response channel. Although the additive noise components at the output of such a channel are normally correlated, it has been shown that this does not greatly reduce the tolerance of the system to the noise, relative to the optimum detection process that exploits a prior knowledge of the correlation between the noise samples.[16] Thus the basic arrangement studied here should be ideally suited to the transmission of alpha-numeric data over partial-response channels.

Again, tests carried out by Ghani have indicated that when $m = 2g$ and System 1B, with binary signals, is used, the arrangement usually has a tolerance to additive white Gaussian noise better than or similar to that of the optimum nonlinear equalizer, operating with the corresponding continuous (uninterrupted) signal over the same baseband channel and at the same average transmitted power level and information rate. Thus, even when g is not very much smaller than m, the arrangement compares well with the best channel equalizer. However, the technique of adaptive detection and signal cancellation (Section 5.3), with System 1A operating on only m samples, would now have a significantly better tolerance to noise. (In practice, of course, Systems 4A and 5B would be used in place of Systems 1A and 1B, respectively).

The basic technique considered here has four important properties that are not shared by the arrangement of adaptive detection and signal cancellation, described in Section 5.3.

Firstly, even in the presence of severe signal distortion, the receiver can extract both element (digit) and group (character) timing from the received data signal itself, by using the fact that the envelope of the received signal always decays to near zero at the end of a received group of elements and always rises again at the start of the next group, if the effects of noise are neglected. The receiver, of course, requires prior knowledge of the nominal element rate $1/T$ as well as of m and g, which means that it has a phase-locked oscillator that generates the required element and group timing waveforms at basically the correct frequencies and in the correct relative phase, but requiring to be appropriately phase-synchronized to the received signal. The synchronization is achieved by monitoring the values of the $\{|r_i|\}$, for $i = 1, 2, \ldots, n$, and very slowly adjusting the phase of the group timing waveform (to which the element timing waveform is synchronized) so that the one or more $\{r_i\}$ having the minimum mean magnitude, when averaged over many groups, occur right at the end of the n samples forming one complete group. Ideally, of course, the mean-square value of all the $\{r_i\}$ should be maximized subject to the above constraint. Clearly, there is no need for the transmission of a separate timing signal or for the use of any other technique for extracting the timing signals from the received data signal.

Secondly, the correct estimation of the sampled impulse-response of the channel does not require a training signal at the start of transmission, so that correct operation is automatically restored following a loss of the true estimate without the need for any further training process.

Thirdly, there is no need to scramble the transmitted data since the temporary loss either of the received signal itself or else of the element timing information, due to an unfortunate combination of the channel response and the transmitted element values, cannot now occur.

Finally, there are no error-extension effects affecting the detection of different groups of signal elements.

It may sometimes happen that the sampled impulse-response of the channel has a few large components followed by several very much smaller components. Under these conditions it is possible to use a combination of (or compromise between) the two basic techniques studied in this chapter. Separate groups of signal elements are transmitted here but the time gap between adjacent groups is only just long enough to accommodate the high-level part of the dispersed portion of a received group of elements (which immediately follows the non-dispersed portion), the remainder of the signal overlapping the following group. The detection of a received group of signal elements is carried out as before but of course ignoring the part of the received group that overlaps the following group. This part of the group is then cancelled from the received signal, thus removing the intersymbol interference between the two groups. The arrangement has some of the properties of each of the two basic techniques involved and could be effective in some applications.

REFERENCES

1. Schiff, L. 'On the transmission of information over the Gaussian and related channels', Ph.D. thesis, Polytechnic Institute of Brooklyn (1968)
2. Varga, R. S. *Matrix Iterative Analysis*, Prentice-Hall, New Jersey (1962)
3. Wilde, D. J. *Optimum Seeking Methods*, Prentice-Hall, New Jersey (1964)
4. Clark, A. P. 'The transmission of digitally-coded-speech signals by means of random access discrete address systems', Ph.D. thesis, London University (1969)
5. Jordan, K. L. 'The performance of sequential decoding in conjunction with efficient modulation', *IEEE Trans. Commun. Technol.*, **COM-14**, 283–297 (1966)
6. Chang, R. W. and Hancock, J. C. 'On receiver structures for channels having memory', *IEEE Trans. Inform. Theory*, **IT-12**, 463–468 (1966)
7. Jacobs, I. M. 'Sequential decoding for efficient communication from deep space', *IEEE Trans. Commun. Technol.*, **COM-15**, 492–501 (1967)
8. Gonsalves, R. A. 'Maximum-likelihood receiver for digital data transmission', *IEEE Trans. Commun. Technol.*, **COM-16**, 392–398 (1968)
9. Abend, K., Harley, T. J., Fritchman, B. D. and Gumacos, C. 'On optimum receivers for channels having memory', *IEEE Trans. Inform. Theory*, **IT-14**, 819–820 (1968)
10. Bowen, R. R. 'Bayesian decision procedure for interfering digital signals', *IEEE Trans. Inform. Theory*, **IT-15**, 506–507 (1969)

11. Root, W. L. 'An introduction to the theory of the detection of signals in noise', *Proc. IEEE*, **58**, 610–623 (1970)
12. Abend, K. and Fritchman, B. D. 'Statistical detection for communication channels with intersymbol interference', *Proc. IEEE*, **58**, 779–785 (1970)
13. Clark, A. P. 'Adaptive detection of distorted digital signals', *Radio Electron. Eng.*, **40**, 107–119 (1970)
14. Kobayashi, H. 'Application of probabilistic decoding to digital magnetic recording systems', *IBM J. Res. Develop.*, **15**, 64–74 (1971)
15. Omura, J. K. 'Optimal receiver design for convolutional codes and channels with memory via control theoretical concepts', *Inform. Sci.*, **3**, 243–266 (1971)
16. Clark, A. P. 'A synchronous serial data-transmission system using orthogonal groups of binary signal elements', *IEEE Trans. Commun. Technol.*, **COM-19**, 1101–1110 (1971)
17. Forney, G. D. 'Maximum-likelihood sequence estimation of digital sequences in the presence of intersymbol interference', *IEEE Trans. Inform. Theory*, **IT-18**, 363–378 (1972)
18. Clark, A. P. 'Adaptive detection with intersymbol interference cancellation for distorted digital signals', *IEEE Trans. Commun.*, **COM-20**, 350–361 (1972)
19. Magee, F. R. and Proakis, J. G. 'Adaptive maximum-likelihood sequence estimation for digital signaling in the presence of intersymbol interference', *IEEE Trans. Inform. Theory*, **IT-19**, 120–124 (1973)
20. Forney, G. D. 'The Viterbi algorithm', *Proc. IEEE*, **61**, 268–278 (1973)
21. Frasco, L. A. and Goldfein, H. P. 'Signal design for aeronautical channels', *IEEE Trans. Commun.*, **COM-21**, 534–547 (1973)
22. Qureshi, S. U. H. and Newhall, E. E. 'An adaptive receiver for data transmission over time-dispersive channels', *IEEE Trans. Inform. Theory*, **IT-19**, 448–457 (1973)
23. Clark, A. P. and Ghani, F. 'Detection processes for orthogonal groups of digital signals', *IEEE Trans. Commun.*, **COM-21**, 907–915 (1973)
24. Painter, J. H. and Gupta, S. C. 'Recursive ideal observer detection of known M-ary signals in multiplicative and additive Gaussian noise', *IEEE Trans. Commun.*, **COM-21**, 948–953 (1973)
25. Magee, F. R. and Proakis, J. G. 'An estimate of the upper bound on error probability for maximum-likelihood sequence estimation on channels having a finite-duration pulse response', *IEEE Trans. Inform. Theory*, **IT-19**, 699–702 (1973)
26. Messerschmitt, D. G. 'A geometric theory of intersymbol interference. Part 2: Performance of the maximum likelihood detector', *Bell Syst. Tech. J.*, **52**, 1521–1539 (1973)
27. Falconer, D. D. and Magee, F. R. 'Adaptive channel memory truncation for maximum likelihood sequence estimation', *Bell Syst. Tech. J.*, **52**, 1541–1562 (1973)
28. Pulleyblank, R. W. 'A comparison of receivers designed on the basis of minimum mean-square error and probability of error for channels with intersymbol interference and noise', *IEEE Trans. Commun.*, **COM-21**, 1434–1438 (1973)
29. Ungerboeck, G. 'Adaptive maximum likelihood receiver for carrier-modulated data transmission systems', *IEEE Trans. Commun.*, **COM-22**, 624–636 (1974)
30. Painter, J. H. and Wilson, L. R. 'Simulation results for the decision-directed MAP receiver for M-ary signals in multiplicative and additive Gaussian noise', *IEEE Trans. Commun.*, **COM-22**, 649–660 (1974)

31. Fritchman, B. D., Kanal, L. N. and Womer, J. D. 'Optimum sequential detector performance on intersymbol interference channels', *IEEE Trans. Commun.*, **COM-22**, 788–797 (1974)
32. Fredricsson, S. A. 'Optimum transmitting filter in digital PAM systems with a Viterbi detector', *IEEE Trans. Inform. Theory*, **IT-20**, 479–489 (1974)
33. Cantoni, A. and Kwong, K. 'Further results on the Viterbi algorithm equalizer', *IEEE Trans. Inform. Theory*, **IT-20**, 764–767 (1974)
34. Foschini, J. G. 'Performance bound for maximum-likelihood reception of digital data', *IEEE Trans. Inform. Theory*, **IT-21**, 47–50 (1975)
35. Hayes, J. F. 'The Viterbi algorithm applied to digital data transmission', *Commun. Soc.*, **13**, No. 2, 15–20 (1975)
36. Magee, F. R. 'A comparison of compromise Viterbi algorithm and standard equalization techniques over band-limited channels', *IEEE Trans. Commun.*, **COM-23**, 361–367 (1975)
37. Clark, A. P. *Principles of Digital Data Transmission*, Pentech Press, London (1976)

6 Parallel systems using code-division multiplexing

6.1 INTRODUCTION

In the various data-transmission systems described in Chapters 4 and 5, a serial signal is transmitted. It is now of interest to consider whether any advantage in tolerance to noise can be gained through the use of a parallel signal. It has been shown[23] that an FDM parallel system can be designed both to make an efficient use of bandwidth and to have a good tolerance to amplitude and phase distortions. In this system a number of separate data-signals are transmitted simultaneously and each occupies a different part of the available frequency band. Unfortunately, such an arrangement involves considerable equipment complexity. This is because, at the transmitter, the individual (separate) baseband data signals are first used to form the corresponding modulated-carrier signals, before being multiplexed (added together or combined) and, at the receiver, the received signal is first demultiplexed (separated) into the individual data signals, before these are demodulated to form the corresponding baseband signals. Thus a separate modulator and demodulator is required for each of the parallel signals, resulting in considerable equipment complexity.[23]

An alternative technique is to multiplex the individual baseband signals and to use the resultant baseband signal to form the corresponding modulated-carrier signal for transmission. At the receiver, the received modulated-carrier signal is first demodulated and the resultant baseband signal is then demultiplexed to give the individual baseband data signals which are then detected. The detection process can sometimes operate directly on the original demodulated baseband signal, with no need for a process of demultiplexing. In the arrangements studied here, the multiplexing and demultiplexing operations are carried out on the baseband signals rather than on the corresponding modulated-carrier signals, as just described. Thus only a single process of modulation and demodulation is involved in a one-way link, as for a serial signal.[1-7]

In order to be able to identify the individual baseband signals, so that these can be demultiplexed correctly, it is necessary to *code* the signals. The arrangement is therefore known as a code-division multiplex (CDM) parallel system.[8-22]

6.2 BASIC ASSUMPTIONS

Consider the synchronous parallel binary data-transmission system shown in Figure 6.1. The signal at the input to the transmitter is a sequence of binary element values $\{s_i\}$, where

$$s_i = \pm k \qquad (6.2.1)$$

the $\{s_i\}$ being statistically independent and equally likely to have either binary value.

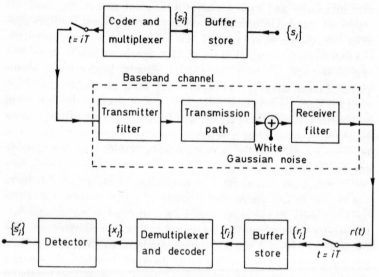

Fig. 6.1 Parallel CDM data-transmission system

The buffer store at the input to the transmitter holds m successive element values $\{s_i\}$. In the coder and multiplexer, the m $\{s_i\}$ are converted into the corresponding m coded signal-elements, each of which has a sequence of components and so can be considered to form a vector. The m coded signal-elements are in element synchronism, such that their corresponding components coincide in time. They are now added together linearly to form a resultant sequence of components. These components are held at the output of the coder and multiplexer, and are sampled, in turn, at the time instants $\{iT\}$, to give the corresponding sequence of impulses, regularly spaced at T seconds, which are fed to the baseband channel.

The transmission path in Figure 6.1 is a bandpass channel together with a linear modulator at the transmitter and a linear demodulator at the receiver, the whole forming a linear baseband channel. The bandpass channel could be a telephone circuit or HF radio link.[23] The transmitter filter, transmission path and receiver filter in Figure 6.1 together form a linear baseband channel, whose impulse response is $y(t)$ and has, for practical purposes, a finite duration of less than $(g+1)T$ seconds. It is assumed that $y(t)$ is either time invariant or varies only slowly with time.

The transmitter and receiver filters are as previously assumed and described in Section 4.2. The transmitter filter is such that an impulse of value (area) k at its input gives a transmitted waveform of energy k^2 at the input to the transmission path, the waveform resulting from any n such impulses, spaced at T seconds, having a total energy of nk^2. The receiver filter is such that the additive white Gaussian noise, with a two-sided power spectral density of $\sigma^2 = \frac{1}{2}N_0$ at its input, gives the Gaussian waveform $w(t)$ at its output. The samples $\{w_i\}$ of this waveform, at the time instants $\{iT\}$, are statistically independent Gaussian random variables with zero mean and variance σ^2.

The sequence of impulses, corresponding to a single group of m signal-elements at the input to the baseband channel in Figure 6.1, is converted into a rounded waveform by the transmitter filter. This waveform is fed to the modulator as the baseband modulating waveform, to give the corresponding modulated-carrier signal at the modulator output. This signal is transmitted over the bandpass channel to the receiver, where it is demodulated and filtered to give a baseband waveform again, the latter being normally different from that at the transmitter due to the effects of noise and distortion. The demodulated baseband waveform is sampled and suitably processed at the receiver to give the detected values of the m signal-elements.

The sampled impulse-response of the baseband channel in Figure 6.1 is given by the $(g+1)$-component row vector

$$V = y_0 \ y_1 \ldots y_g \qquad (6.2.2)$$

where $y_i = y(iT)$. As before, $y_0 \neq 0$, and $y_i = 0$ for $i < 0$ and $i > g$. The channel is assumed to vary at most only slowly with time so that V does not change noticeably over the duration of a received group of m signal-elements.

The received waveform $r(t)$ at the output of the baseband channel in Figure 6.1 is sampled at the time instants $\{iT\}$, for all integers $\{i\}$, the sample value at time $t = iT$ being r_i. The $\{r_i\}$ are fed to the buffer store which contains two separate stores. While one of these holds a set of the received $\{r_i\}$ for a detection process, the other is receiving the next set of $\{r_i\}$ in preparation for the next detection process. A group of m multiplexed signal-elements are detected simultaneously in a single detection process, from the set of $\{r_i\}$ that depend only on these elements and are therefore independent of the preceding and following groups of elements. The receiver is assumed to have prior knowledge both of V and of the codes used for the m signal-elements in a group, and to use this knowledge in the detection of the m element-values $\{s_i\}$ from the received samples $\{r_i\}$. A period of nT seconds is available for the detection process, where

$$n = m + g \qquad (6.2.3)$$

It will be assumed throughout that $m > g$, which is a necessary condition if the data-transmission system is to achieve an acceptably efficient use of the available bandwidth.

Except where otherwise stated, the decoder and demultiplexer in Figure 6.1 together determine from the appropriate set of received $\{r_i\}$ the m estimates $\{x_i\}$ of the m element-values $\{s_i\}$ in a received group of elements. Each x_i is an unbiased estimate of the corresponding s_i, such that

$$x_i = s_i + u_i \qquad (6.2.4)$$

where u_i is a zero-mean Gaussian random variable.

In the detector, each s_i is detected as k or $-k$, depending upon whether $x_i > 0$ or $x_i < 0$, respectively. The detected value of s_i is designated s_i'.

6.3 CDM SYSTEMS

6.3.1 Systems 1C and 1D

In these systems a time gap (time guard band) of gT seconds is inserted between each pair of adjacent groups of m transmitted signal-elements, at the input to the baseband channel, such that the received signal corresponding to any one group of m elements does not overlap the adjacent groups at the output of the baseband channel. Each transmitted group of m elements has just m components, and each received group of elements may be detected from the corresponding n received samples, without intersymbol interference from the other groups. Consider therefore the transmission and detection of just one group of m signal-elements.

The coder and multiplexer in the transmitter of Figure 6.1 accept the sequence of m $\{s_i\}$, for $i = 1, 2, \ldots, m$, and code these to form the m signal-vectors $\{s_i A_i\}$, where A_i is the m-component row vector

$$A_i = a_{i1} \ a_{i2} \ldots a_{im} \qquad (6.3.1)$$

The m vectors $\{s_i A_i\}$ are now added together to give the resultant m-component row vector

$$\sum_{i=1}^{m} s_i A_i = SA \qquad (6.3.2)$$

where S is the m-component row vector

$$S = s_1 \ s_2 \ldots s_m \qquad (6.3.3)$$

and A is the $m \times m$ matrix whose ith row is given by the vector A_i.

The resultant signal fed to the baseband channel is the sequence of m regularly spaced impulses

$$\sum_{i=1}^{m} \sum_{j=1}^{m} s_i a_{ij} \delta(t - jT)$$

assuming for convenience that the first of the m impulses occurs at the time instant $t = T$. The following g impulses, occurring at the time instants $(m+1)T, (m+2)T, \ldots, nT$, are all set to zero, to give a total of n impulses transmitted for the group of m signal-elements. This sequence of impulses is the baseband signal formed by m parallel signal-elements, after these have been coded and multiplexed. The sequence of m impulses

$$\sum_{i=1}^{m} \sum_{j=1}^{m} s_i a_{ij} \delta(t - jT)$$

is more simply represented by the m-component row vector SA.

The next group of m signal-elements, at the input to the baseband channel, is such that the first of its impulses occurs at the time instant $t = (n+1)T$, so that nT seconds are allocated to each group of m elements.

The signal at the output of the baseband channel, corresponding to the m transmitted signal-elements

$$\sum_{i=1}^{m} \sum_{j=1}^{m} s_i a_{ij} \delta(t - jT),$$

is

$$r(t) = \sum_{i=1}^{m} \sum_{j=1}^{m} s_i a_{ij} y(t - jT) + w(t) \qquad (6.3.4)$$

Let Y be the $m \times n$ convolution matrix whose ith row is given by the n-component row vector

$$Y_i = \overbrace{0 \ldots 0}^{i-1} \, y_0 \, y_1 \ldots y_g \, 0 \ldots 0 \qquad (6.3.5)$$

Thus the received signal-vector, that corresponds to the m-component transmitted signal vector SA, is the n-component vector SAY. To this is added the received n-component noise-vector W, and the sum of the two vectors is held in the buffer store at the receiver in Figure 6.1. The resultant set of n received samples $\{r_i\}$ in the store is given by the n-component row vector

$$R = SAY + W \qquad (6.3.6)$$

where the ith components of R and W are r_i and w_i, respectively. There is no intersymbol interference in R from the preceding or following transmitted signal-vectors $\{S\}$. The transmitted vector S is now detected from the corresponding received vector R.

It is assumed that A is a real orthogonal matrix, so that $A^T = A^{-1}$ and the vectors $\{A_i\}$ formed by its rows are orthogonal vectors of unit length.

There are two important cases that are studied here. In the first of these, System 1C, A is an $m \times m$ identity matrix. In the second,

System 1D, A is a Hadamard matrix whose components are multiplied by a suitable positive constant such that the rows of the resultant $m \times m$ matrix are vectors of unit length. Any other $m \times m$ real orthogonal matrix could, if required, be used for A.

A Hadamard matrix is a square matrix of order 2^i, where i is any positive integer.[11] Its components are all ± 1 and such that any two rows or columns of the matrix are orthogonal. The simplest Hadamard matrix is the 2×2 matrix

$$\begin{bmatrix} 1 & 1 \\ 1 & -1 \end{bmatrix}$$

The next simplest Hadamard matrix is the 4×4 matrix

$$\begin{bmatrix} 1 & 1 & 1 & 1 \\ 1 & -1 & 1 & -1 \\ 1 & 1 & -1 & -1 \\ 1 & -1 & -1 & 1 \end{bmatrix}$$

which is formed from the 2×2 Hadamard matrix by replacing each component of the 2×2 matrix by the product of this component and the 2×2 matrix itself. The 8×8 Hadamard matrix is similarly formed by replacing each component of the 2×2 Hadamard matrix by the product of this component and the 4×4 Hadamard matrix, and so on. In the case where the 8×8 Hadamard matrix is used for the matrix A, each component of the Hadamard matrix is multiplied by $8^{-\frac{1}{2}}$ so that the rows of the modified Hadamard matrix are 8-component vectors of unit length.

For any orthogonal matrix A, each of the m signal-elements $\{s_i A_i\}$, whose sum SA is formed by the linear transformation A on the signal vector S, is a vector of length k. Since the $\{s_i A_i\}$ are orthogonal, it follows that the average transmitted energy per signal element is k^2, regardless of the binary values of the transmitted $\{s_i\}$.

The received n-component vector R, given by Equation 6.3.6, is held in the buffer store in Figure 6.1 and is fed to the n input terminals of the demultiplexer and decoder. These operate *linearly* on the vector R to give the m-component row vector

$$X = x_1 \ x_2 \ldots x_m \tag{6.3.7}$$

at their output. X is the *maximum-likelihood estimate* of S. It can be seen from Equation 3.3.56, bearing in mind that the matrix Y is now replaced by the matrix AY, that

$$X = R(AY)^T(AY(AY)^T)^{-1}$$
$$= SAY(AY)^T(AY(AY)^T)^{-1} + W(AY)^T(AY(AY)^T)^{-1}$$
$$= S + U \qquad (6.3.8)$$

where

$$U = W(AY)^T(AY(AY)^T)^{-1} \qquad (6.3.9)$$

as can be seen from Equation 6.3.6. AY is an $m \times n$ matrix of rank m, and $(AY)^T(AY(AY)^T)^{-1}$ is an $n \times m$ matrix of rank m. A prior knowledge of AY is of course used in the derivation of X.

With a time invariant baseband channel, the demultiplexer and decoder could together be implemented by the $n \times m$ linear network $(AY)^T(AY(AY)^T)^{-1}$, such that when the vector R is fed to its n input terminals, the vector X appears at its m output terminals.

With a time-varying channel, the linear transformation $(AY)^T(AY(AY)^T)^{-1}$ is best performed by the point Gauss-Seidel iterative process, implemented as in Figure 6.2. The arrangement

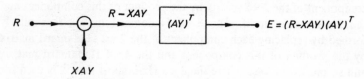

Fig. 6.2   Linear estimation of S from R by means of an iterative process

operates exactly as that described in Section 5.3.5, except that R' is replaced by R and Y by AY. Thus X is initially set to zero and its components are then adjusted, in turn and cyclically, the change in x_i being

$$\Delta x_i = \frac{ce_i}{|A_iY|^2} \qquad (6.3.10)$$

where e_i is the ith component of the m-component output vector E, immediately before the change and in x_i, and $0 < c < 2$. The iterative process always converges, and at the end of the process, $E \simeq 0$, so that

$$(R - XAY)(AY)^T \simeq 0 \qquad (6.3.11)$$

or

$$X \simeq R(AY)^T(AY(AY)^T)^{-1} \qquad (6.3.12)$$

In the detector, each s_i is now detected from the sign of the corresponding x_i, to give the detected vector S.

In the case of System 1C, A is an $m \times m$ identity matrix, so that System 1C is identical to the System 2B described in Section 5.4.4. Thus the serial data-transmission system in which the signal elements are transmitted in separate groups, with time gaps between adjacent transmitted groups, is a special case of the corresponding CDM parallel system.

When the receiver of System 1C or 1D has prior knowledge of the $\{|s_i|\}$ and therefore of the 2^m possible vectors $\{SAY\}$, the detection process corresponding to System 3B can be used here and gives an improved tolerance to noise. The demultiplexer and decoder now combine to give a nonlinear estimate of S, from which S is detected as before. For the minimum probability of error in the detection of S from R, the vector S is detected as the one of its 2^m possible values that minimizes $|R - SAY|$, which is the distance between the vectors R and SAY in the n-dimensional Euclidean vector space containing these vectors. This detection process corresponds to System 1B and it replaces the demultiplexer, decoder and detector in Figure 6.1.

Except where otherwise stated, it will be assumed here that, in the Systems 1C and 1D, S is detected from X in Equation 6.3.12, so that the receiver uses no prior knowledge of the $\{|s_i|\}$ in the detection process.

With a time-varying channel, the receiver needs to estimate the sampled impulse-response of the baseband channel, which is the vector V in Equation 6.2.2. Just as S can be estimated linearly from R, given a prior knowledge of AY, so V can be estimated from R, given prior knowledge of SA. It is clear that SAY is the convolution of the m-component vector SA and the $(g+1)$-component vector V. Thus, if Z is the $(g+1) \times n$ convolution matrix whose ith row is given by the n-component vector

$$Z_i = \overbrace{0 \ldots 0}^{i-1} z_1\, z_2 \ldots z_m\, 0 \ldots 0 \qquad (6.3.13)$$

where z_i is the ith component of SA, then

$$SAY = VZ \qquad (6.3.14)$$

From Equation 6.3.6,

$$R = VZ + W \qquad (6.3.15)$$

and the required linear estimate of V from R is now the $(g+1)$-component vector

$$F = RZ^T(ZZ^T)^{-1} \tag{6.3.16}$$

as can be seen from Equation 5.4.31.

Following the correct detection of S, the vector SA is known and therefore also the matrix Z. F is best derived from R by means of the point Gauss-Seidel iterative process, implemented as described in Section 5.4.11. As before, S is first detected from R, using a prior knowledge of V in the form of a stored estimate, and V is then estimated from R, using a prior knowledge of S. The estimated value of V is used to update the stored estimate.

6.3.2 Systems 2C and 2D

In these systems the transmitted signal at the input to the baseband channel in Figure 6.1 is a continuous sequence of impulses, at the time instants $\{iT\}$, there being no time gaps between adjacent groups of impulses. Each group of m parallel signal-elements is transmitted as the corresponding sequence of n impulses, the first impulse of one group of elements occurring T seconds after the last impulse of the preceding group, so that nT seconds are allocated to each group of m elements, as before. Consider again the transmission and detection of just one group of m signal-elements.

The coder and multiplexer in the transmitter of Figure 6.1 convert the m-component input signal-vector S (Equation 6.3.3) into the corresponding n-component output signal-vector SA, where A is now an $m \times n$ matrix of rank m, whose ith row is given by the n-component row vector

$$A_i = a_{i1} \ a_{i2} \dots a_{in} \tag{6.3.17}$$

SA is the sum of m signal-vectors $\{s_iA_i\}$, each having n components.

The resultant signal fed to the baseband channel in Figure 6.1 is the sequence of n regularly spaced impulses

$$\sum_{i=1}^{m} \sum_{j=1}^{n} s_i a_{ij} \delta(t-jT)$$

assuming for convenience that the first impulse occurs at time $t = T$. It can be seen that the received signal, obtained after the transmission of this signal over the baseband channel and sampling at the receiver input, is contained in a sequence of $n+g$ received samples $\{r_i\}$, starting with r_1 at time $t = T$ and ending with r_{n+g} at time $t = (n+g)T$. However, the sample values r_1 to r_g contain intersymbol-interference components from the preceding transmitted vector SA, and the sample values r_{n+1} to r_{n+g} contain intersymbol-interference components from the following transmitted vector SA, so that only the sample values r_{g+1} to r_n are free from intersymbol interference. Thus the buffer store in Figure 6.1 now holds the m-component row vector

$$R = r_{g+1} \; r_{g+2} \ldots r_n \qquad (6.3.18)$$

from which S is detected, without interference from the preceding or following transmitted vectors $\{SA\}$.

Suppose that the $m \times n$ matrix A is such that its ith row is

$$A_i = b_{i(m-g+1)} \; b_{i(m-g+2)} \ldots b_{im} \; b_{i1} \; b_{i2} \ldots b_{im} \qquad (6.3.19)$$

for $i = 1, 2, \ldots, m$, and let B be the $m \times m$ matrix formed by the last m columns of A. Thus the component in the ith row and jth column of B is b_{ij}.

The sampled impulse-response of the baseband channel is now taken to be the m-component row vector

$$V = y_0 \; y_1 \ldots y_{m-1} \qquad (6.3.20)$$

where $m > g$ and $y_i = 0$ for $i > g$. Also Y is taken to be the $m \times m$ circulant matrix whose ith row is

$$Y_i = \overbrace{y_{m-i+1} \; y_{m-i+2} \ldots y_{m-1}}^{i-1} \; y_0 \; y_1 \ldots y_{m-i} \qquad (6.3.21)$$

Clearly, $V = Y_1$.

It is shown in Section 2.4.6 that Y is nonsingular if and only if the z transform of the sampled impulse-response of the channel

$$Y(z) = y_0 + y_1 z^{-1} + \ldots + y_g z^{-g} \qquad (6.3.22)$$

has no roots on the unit circle in the z plane, at the points $\left\{ \exp\left(\sqrt{-1} \dfrac{2\pi i}{m} \right) \right\}$ where $i = 0, 1, \ldots, m-1$.

It can be seen that when the n-component signal vector SA is transmitted over the baseband channel in Figure 6.1, the central m components of the corresponding $(n+g)$-component received vector, that is, the m signal-components present in the m received samples r_{g+1} to r_n, are given by the m-component row vector SBY. This is a consequence of the fact that the components along each row of the $m \times n$ matrix A are repeated such that $a_{ij} = a_{i(j+m)}$, for $j = 1, 2, \ldots, g$. Thus the m-component vector R in Equation 6.3.18 is given by

$$R = SBY + W \tag{6.3.23}$$

where W is now the m-component row vector whose ith component is w_{g+i}.

In System 2C the matrix B is an identity matrix, and in System 2D the matrix B is a Hadamard matrix. In the latter case the components of the matrix B are multiplied by the appropriate constant, such that the lengths of the vectors formed by the rows and columns of B are all equal to unity, as is the case in System 2C.

The demultiplexer and decoder in the receiver of Figure 6.1 operate on the m-component vector R that is held in the buffer store, to obtain the maximum-likelihood estimate X of the signal-vector S. Thus, from Equations 3.3.56 and 6.3.23, the demultiplexer and decoder together form the linear network $(BY)^{-1}$, and the output signal is the m-component vector

$$X = R(BY)^{-1} = S + U \tag{6.3.24}$$

where

$$U = W(BY)^{-1} \tag{6.3.25}$$

It is assumed here that $Y(z)$ has no roots on the unit circle in the z plane, so that both Y and BY are nonsingular matrices, these being known at the receiver.

In the detector, each s_i is detected from the sign of the corresponding x_i, as before.

The linear transformation $(BY)^{-1}$ is best performed by means of the point Gauss-Seidel iterative process, as previously described. The arrangement is as shown in Figure 6.2, where the $m \times n$ matrix AY is now replaced by the $m \times m$ matrix BY.

When the receiver has prior knowledge of the $\{|s_i|\}$ and therefore of the 2^m possible vectors $\{SBY\}$, more effective detection processes can be applied, as previously explained. The detection process that minimizes the probability of error in the detection of S from R, selects the one of the 2^m possible values of S for which $|R - SBY|$ has the minimum value. This detection process replaces the demultiplexer, decoder and detector in Figure 6.1. It will, however, be assumed here that Systems 2C and 2D are implemented with the linear network $(BY)^{-1}$ for the demultiplexer and decoder.

6.3.3 System 2E

System 2E is basically the same as Systems 2C and 2D. It differs from the latter systems only in the matrix B and in some operational details that are a direct consequence of this difference in B. In System 2E the matrix B is a real orthogonal matrix which is such that, regardless of the channel response and hence of the value of the matrix Y, as given by Equation 6.3.21, the m individual signal-element vectors, at the input to the demultiplexer and decoder in the receiver, are always orthogonal.[19] These vectors are given by the $\{s_i B_i Y\}$, where B_i is the m-component row vector formed by the ith row of B. The required matrix B is derived as follows.

From Equation 6.3.24, the received m-component vector R is given by

$$R = XBY \qquad (6.3.26)$$

where X is the maximum-likelihood estimate of S. If now R is fed to the $m \times m$ linear network $(BY)^T$, the output m-component signal vector is

$$R(BY)^T = XBYY^TB^T \qquad (6.3.27)$$

Since Y is a real $m \times m$ circulant matrix, which may be singular or nonsingular, YY^T is a real $m \times m$ circulant which is in addition symmetric and nonnegative definite (that is, either positive definite or positive semi-definite). Let B be the real orthogonal matrix such that BYY^TB^T is a diagonal matrix. Clearly, $B^T = B^{-1}$. The components along the main diagonal of BYY^TB^T are now the eigenvalues of YY^T and are therefore real and nonnegative. Thus

$$BYY^TB^T = D \qquad (6.3.28)$$

where D is an $m \times m$ diagonal matrix, such that the ith component along the main diagonal is $d_i \geqslant 0$.

Furthermore, since YY^T is a circulant, B depends *only* on m (the order of YY^T) and not on the component values of YY^T. This is a most important result which means that a given matrix B can be used to reduce YY^T to a diagonal matrix according to Equation

6.3.28, regardless of the component values of Y and therefore for all values of the sampled impulse-response of the baseband channel within the limits assumed. Since YY^T is a real symmetric matrix and also a circulant, the m rows of B are formed by the different possible values of the two vectors

$$\text{Re}\big[e_i(1 \ \omega^i \ \omega^{2i} \ldots \omega^{(m-1)i})\big]$$

and $$\text{Re}\big[\sqrt{-1}f_i(1 \ \omega^i \ \omega^{2i} \ldots \omega^{(m-1)i})\big]$$

for $i = 1, 2, \ldots, m$. Re means the 'real part of'. e_i and f_i are real scalars which set the lengths of the corresponding resultant real vectors to unity, and

$$\omega = \exp\left(\sqrt{-1}\frac{2\pi}{m}\right) \tag{6.3.29}$$

so that ω^i is the ith of the m distinct mth roots of unity.

For example, when $m = 8$,

$$B = \frac{1}{\sqrt{8}}\begin{bmatrix} 1 & 1 & 1 & 1 & 1 & 1 & 1 & 1 \\ 1 & \sqrt{2} & 1 & 0 & -1 & -\sqrt{2} & -1 & 0 \\ 1 & 0 & -1 & -\sqrt{2} & -1 & 0 & 1 & \sqrt{2} \\ \sqrt{2} & 0 & -\sqrt{2} & 0 & \sqrt{2} & 0 & -\sqrt{2} & 0 \\ 0 & -\sqrt{2} & 0 & \sqrt{2} & 0 & -\sqrt{2} & 0 & \sqrt{2} \\ 1 & -\sqrt{2} & 1 & 0 & -1 & \sqrt{2} & -1 & 0 \\ 1 & 0 & -1 & \sqrt{2} & -1 & 0 & 1 & -\sqrt{2} \\ 1 & -1 & 1 & -1 & 1 & -1 & 1 & -1 \end{bmatrix} \tag{6.3.30}$$

Even values of m lead to simpler structures of the matrix B and are therefore to be preferred in practice.

The demultiplexer and decoder in the receiver now form the $m \times m$ linear network $(BY)^T$, so that when R is fed to the m input terminals, the m output terminals hold the vector

$$R(BY)^T = XD \tag{6.3.31}$$

as can be seen from Equations 6.3.27 and 6.3.28. Thus the signal at the ith output terminal of the network $(BY)^T$ is $d_i x_i$, where $d_i \geqslant 0$. s_i is now detected from the sign of $d_i x_i$. X is, of course, the linear estimate of S obtained at the m output terminals of the network $(BY)^{-1}$, when R is fed to the m input terminals.

It is clear from Equation 6.3.28 that, when the individual received signal-elements $\{s_i B_i Y\}$ are nonzero, they are always orthogonal, and they are nonzero so long as the matrix Y is nonsingular, that is, so long as the z transform of the baseband channel has no roots on the unit circle in the z plane, at the points $\left\{\exp\left(\sqrt{-1}\ \dfrac{2\pi i}{m}\right)\right\}$ where $i = 0, 1, \ldots, m-1$. When the z transform of the channel has one or more roots at these points on the unit circle, Y becomes singular and now one or more of the m $\{d_i\}$ are equal to zero. Under these conditions the corresponding $\{s_i\}$ cannot be detected from the $\{d_i x_i\}$. What has happened is that the corresponding received signal-elements $\{s_i B_i Y\}$ have gone to zero and have therefore been lost in transmission.

In System 2E the demultiplexer and decoder form the 'matched' network $(BY)^T$ in place of the 'inverse' network $(BY)^{-1}$ used in Systems 2C and 2D, and they can only do this because in System 2E the received signal-elements are orthogonal. The matched network in System 2E gives the output vector XD, whereas the corresponding inverse network in Systems 2C and 2D gives the output vector X. From Equations 6.3.31, 6.3.23 and 6.3.28,

$$XD = (SBY + W)(BY)^T$$
$$= SD + W(BY)^T \tag{6.3.32}$$

Thus, when R is fed to the input of the matched network $(BY)^T$, the signal $d_i x_i$ at the ith output terminal contains no intersymbol interference, being the sum of $d_i s_i$ and a Gaussian noise component. However, XD is a *biased* estimate of S.

The receiver is assumed to know $(BY)^T$, which is used to obtain XD. No iterative process is required, even with a time-varying channel, since the receiver does not now need to carry out the linear transformation $(BY)^{-1}$ on R. The required transformation $(BY)^T$ is determined *directly* from the received signal, using a prior knowledge of B, in a manner to be described presently. It follows that the detection of S from R can be carried out very rapidly.

Furthermore, since the individual received signal-elements $\{s_i B_i Y\}$ are orthogonal and the detection process of System 2E is one of matched-filter detection, this is the best available detection process for the given received signal. It achieves the same tolerance to additive white Gaussian noise as the detection process that selects the one of the 2^m possible values of S for which $|R - SBY|$ has the minimum value, using prior knowledge of the 2^m possible vectors $\{SBY\}$.

6.3.4   *Estimation of the sampled impulse-response of the channel*

The prior knowledge of BY that is used in any of the Systems 2C, 2D and 2E to obtain X (or XD) from R, must itself be obtained from the received signal by assuming that the detected vector S is correct. The receiver is, of course, assumed to know the matrix B.

Following the correct detection of S from R, the receiver knows the m-component row vector SB. It now needs to estimate the sampled impulse-response of the baseband channel from R, using its knowledge of SB, in order to determine Y and hence BY. But SBY is the 'circular' convolution of the m-component row vectors SB and V. Thus, if Z is the $m \times m$ circulant matrix whose ith row is

$$Z_i = \overbrace{z_{m-i+2} \; z_{m-i+3} \cdots z_m}^{i-1} \; z_1 \; z_2 \cdots z_{m-i+1} \qquad (6.3.33)$$

where z_i is the ith component of SB, and if Y is the $m \times m$ convolution matrix whose ith row is Y_i in Equation 6.3.21 and whose first row is V in Equation 6.3.20, then

$$SBY = VZ \qquad (6.3.34)$$

and the m-component received vector is

$$R = VZ + W \qquad (6.3.35)$$

The required linear estimate of V from R is the m-component row vector

$$F = RZ^{-1} \qquad (6.3.36)$$

F is derived from R by means of the point Gauss-Seidel iterative process and is used to update the stored estimate of V. Since Z is a circulant matrix, each row being formed from that above by a cyclic shift of one place to the right in its components, the iterative process can be implemented very simply, as before.

Unfortunately, in Systems 2C and 2D, where B is an identity and Hadamard matrix, respectively, the matrix Z in Equation 6.3.36 becomes singular for certain of the possible values of S. This means, of course, that Z now has no inverse, so that Equation 6.3.36 no longer holds and F cannot be obtained from R in the manner just described. More conventional cross-correlation techniques, of the type used with linear and nonlinear equalizers, must here be used to estimate V.

However, when B is the orthogonal matrix of System 2E, it appears[19] that $(1/k\sqrt{m})Z$ is an orthogonal matrix for all positive integer values of m. Clearly Z must now be nonsingular, and

$$k\sqrt{m}Z^{-1} = \frac{1}{k\sqrt{m}}Z^T \qquad (6.3.37)$$

so that from Equation 6.3.36, the linear estimate of V, obtained from R, is the m-component row vector

$$F = \frac{1}{k^2 m}RZ^T \qquad (6.3.38)$$

Thus V can be estimated from R by feeding the latter to the 'matched' network $(1/k^2m)Z^T$ which is obtained directly from the detected vector S and the matrix B, without the need for an iterative process. This is a most important result which shows that V can here be estimated simply and very rapidly from R, using the same technique as that involved in the estimation of S. In each case, the estimation process is one of matched-filter estimation and so maximizes the output signal/noise ratio. Furthermore, the matched network used for the estimation process is determined directly from the received signal, without the need for the inversion of a matrix, so that no iterative process is required.

6.3.5 System 2F

It is shown in Section 2.4.6 that whenever the z transform of the baseband channel in Figure 6.1 has any roots on the unit circle in the z plane, at the points $\left\{ \exp\left(\sqrt{-1}\dfrac{2\pi i}{m} \right) \right\}$ where $i = 0, 1, \ldots, m-1$, the $m \times m$ matrix Y in Equation 6.3.23 becomes singular. It is otherwise nonsingular. However, when Y becomes singular it nearly always has rank $m-1$ or $m-2$. Furthermore, in System 2E, when Y has rank $m-l$, for $l = 1$ or 2, it always happens that l of the individual received signal-elements $\{s_i B_i Y\}$ become zero. The detector now recognizes which of the m elements have been lost, since it knows BY, and it ignores these elements, treating them as undetectable. It detects the remaining $m-l$ elements in exactly the same way as previously described for the m received elements.

If m is even and the rows of the matrix B of System 2E are arranged in the order of increasing frequency, as in Equation 6.3.30, then when two of the received signal-elements $\{s_i B_i Y\}$ are lost or highly attenuated, one of these is an odd-numbered element (i odd) and the other an even-numbered element (i even). Indeed, it appears that of the two most severely attenuated elements, regardless of how severely these are attenuated, one is an odd-numbered element and the other an even-numbered element. The reason for this can be seen from the fact that the rows of B are generated by the regularly spaced sample values of the appropriate sine waves[20] and except for the first and last rows, that correspond to zero frequency (d.c.) and the highest frequency, the rows occur in pairs having the same frequency of the sampled sine wave but a relative phase shift of 90°. Clearly if the individual received signal-element $s_i B_i Y$, where i is even and less than m, is severely attenuated, then so also is $s_{i+1} B_{i+1} Y$, since Y is a circulant matrix and therefore operates in a similar manner on both $s_i B_i$ and $s_{i+1} B_{i+1}$.

If now two of the m elements are parity-check elements, and if one of these is an odd-numbered element whose binary value s_i is determined by the remaining odd-numbered elements, and the other is an even-numbered element whose value s_j is determined by the remaining even-numbered elements, then, when either one or two of the received elements are severely attenuated, their binary values can be determined from the detected binary values of the *remaining* elements. Thus correct operation is maintained. No significant equipment complexity is involved and there is only a small reduction in the effective transmission rate, so long as $m \gg 1$. This technique can with advantage be extended to operate so that the detector ignores the received signal-element having the lowest level of the different elements involved in a parity check, *whatever* its attenuation.[19] As before, the detector determines the binary value of this element from the detected binary values of the remaining elements involved in the parity check. This development of System 2E is known as System 2F.

In the case of System 2D, where B is a Hadamard matrix, the technique just described becomes less effective, since a given matrix Y not only affects the relative levels $\{|s_i B_i Y|\}$ of the received signal-elements but also affects the angles between the element vectors $\{s_i B_i Y\}$, since they are not now necessarily orthogonal. Thus the conditions for the correct operation of the arrangement do not apply quite so often.

In the case of System 2C, where B is an identity matrix, correct operation can be maintained when Y has rank $m-l$, for $l = 1$ or 2, simply by omitting l of the m signal-elements in each group. Only $m-l$ elements are now transmitted in a group, but these are detected from the m-component received vector R, as before.

It is clear that when Y has rank $m-l$, only $m-l$ signal-elements can in effect be transmitted over every interval of nT seconds, whatever the orthogonal matrix used for B.

6.3.6   System 3

In the systems previously considered, the operation of the coder and multiplexer on the signal element values $\{s_i\}$, at the input of the transmitter, is fixed and independent of the transmission path. In each case the decoder and demultiplexer at the receiver are appropriately adjusted to allow for the signal distortion introduced in transmission. Thus the various systems can all be used over slowly time-varying channels.

It is now of interest to consider the case where the sampled impulse-response of the baseband channel is fixed and known. Each group of m element-values $\{s_i\}$ at the transmitter are processed *linearly* to give, at the input to the baseband channel in Figure 6.1, the corresponding sequence of $n = m+g$ impulses, regularly spaced at intervals of T seconds. These are such that after transmission over the baseband channel and sampling at the receiver input, each group of m signal-elements give a sequence of m sample values equal to the corresponding transmitted $\{s_i\}$, neglecting for the moment the effects of noise.

As before, there is no intersymbol interference in the detection of any one received group of m signal-elements, from any other received group, so that it is sufficient to consider the transmission and detection of just one group of m signal-elements. Suppose therefore that the buffer store at the transmitter holds the m element-values $\{s_i\}$ given by the m-component vector

$$S = s_1\ s_2 \ldots s_m \qquad (6.3.39)$$

The coder and multiplexer perform a linear transformation on S to give at the output the n-component row vector

$$C = SA \qquad (6.3.40)$$

where A is an $m \times n$ matrix. No restrictions are here imposed on A, which no longer necessarily satisfies Equation 6.3.19. The vector C, whose ith component is c_i, is sampled at the time instants $\{iT\}$, for $i = 1, 2, \ldots, n$, to give the corresponding sequence of impulses $\{c_i \delta(t - iT)\}$. These are fed to the baseband channel whose sampled impulse-response is given by the $(g + 1)$-component vector

$$V = y_0 \ y_1 \ldots y_g \qquad (6.3.41)$$

In general, the signal just described, after transmission over the baseband channel and sampling at the receiver input, affects a sequence of $n + g$ received samples $\{r_i\}$, starting with r_1 at time $t = T$ and ending with r_{n+g} at time $t = (n+g)T$. The central m of these received samples, r_{g+1} to r_n, are free from intersymbol interference from the preceding and following transmitted vectors $\{C\}$, and are therefore used for the detection of S. Let

$$R = r_{g+1} \ r_{g+2} \ldots r_n \qquad (6.3.42)$$

The matrix A is now selected so that

$$R = S + W \qquad (6.3.43)$$

where W is the m-component noise vector whose ith component is w_{g+i}.

Finally, for each received signal-element, s_i is detected as k or $-k$, depending upon whether $r_{g+i} > 0$ or $r_{g+i} < 0$, respectively.

It can be seen from Equations 6.3.41 and 6.3.43 that

$$S = CG^T \qquad (6.3.44)$$

where G is the $m \times n$ matrix whose ith row is

$$G_i = \overbrace{0 \ldots 0}^{i-1} \ y_g \ y_{g-1} \ldots y_0 \ 0 \ldots 0 \qquad (6.3.45)$$

In order to optimize the tolerance of the system to noise, it is necessary to minimize the average transmitted energy per signal element, for a given noise level and a given error probability in the detection of S. Since the noise level and error probability are fixed, as can be seen from Equation 6.3.43, it is necessary to select the vector C to minimize the total transmitted energy of the m signal-elements at the input to the transmission path. But the transmitted energy of the m elements is $|C|^2$, where $|C|$ is the length of the vector C. Thus it is necessary to minimize $|C|$, for the given vector S in Equation 6.3.44. From this equation,

$$CG_i^T = s_i \tag{6.3.46}$$

for $i = 1, 2, \ldots, m$. Now the length of the vector G_i is

$$|G_i| = |V| = d \tag{6.3.47}$$

for $i = 1, 2, \ldots, m$. Thus in the n-dimensional Euclidean vector space containing the m vectors $\{G_i\}$, the point corresponding to each vector is at a distance d from the origin. This distance is of course fixed by the channel.

CG_i^T is the inner product of the vectors C and G_i, so that it is d times the value of the orthogonal projection of C on to the one-dimensional subspace spanned by the vector G_i in the n-dimensional vector space. From Equations 6.3.46 and 6.3.47, this orthogonal projection is the vector or point $(s_i/d^2)G_i$, and C lies on the hyperplane ($(n-1)$-dimensional plane) which contains the point $(s_i/d^2)G_i$ and which is orthogonal to the vector given by this point. Clearly the hyperplane is orthogonal to the vector G_i. The relationship between the vectors C and G_i is now as shown in Figure 6.3, where O is the origin of the n-dimensional vector space. Since $s_i = \pm k$ and $|G_i| = d$, the point $(s_i/d^2)G_i$ is at a distance k/d from the origin.

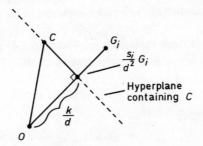

Fig. 6.3   Relationship between the vectors C and G_i

The vector C must lie on *each* of the m hyperplanes given by Equation 6.3.46 and shown in Figure 6.3, so that it must lie on the *intersection* of the m hyperplanes. But each of the m vectors $\{G_i\}$ is orthogonal to the corresponding hyperplane, as in Figure 6.3, and from Equation 6.3.45, the m vectors $\{G_i\}$ are linearly independent. Thus no one of the m hyperplanes is coincident with or parallel to any other, so that the intersection of the m hyperplanes is an $(n-m)$-dimensional plane in the n-dimensional vector space. The required vector C is the point on this plane at the *minimum distance* from the origin and, by the projection theorem, it is therefore the *orthogonal projection* of the origin on to the plane. Since each of the m vectors $\{G_i\}$ is orthogonal to the corresponding hyperplane, the orthogonal projection of the origin on to the $(n-m)$-dimensional plane must itself lie in the m-dimensional subspace spanned by the m vectors $\{G_i\}$. Thus the required vector C lies in the *intersection* of the m-dimensional subspace spanned by the m $\{G_i\}$ and the $(n-m)$-dimensional plane formed by the intersection of the m hyperplanes.

The fact that C lies in the m-dimensional subspace spanned by the m $\{G_i\}$ means that C can be expressed as a linear combination of the m $\{G_i\}$, so that

$$C = \sum_{i=1}^{m} f_i G_i = FG \qquad (6.3.48)$$

where

$$F = f_1\, f_2 \ldots f_m \qquad (6.3.49)$$

The fact that C lies in the $(n-m)$-dimensional plane formed by the intersection of the m hyperplanes, means that C satisfies Equation 6.3.44, as can be seen from Equation 6.3.46 and Figure 6.3.

Thus, from Equations 6.3.48 and 6.3.44,

$$S = FGG^T \qquad (6.3.50)$$

where GG^T is an $m \times m$ nonsingular matrix, so that

$$F = S(GG^T)^{-1} \qquad (6.3.51)$$

Now, from Equation 6.3.48,

$$C = S(GG^T)^{-1}G \qquad (6.3.52)$$

The $m \times n$ matrix $(GG^T)^{-1}G$ has rank m, and so there is a unique one-to-one relationship between S and C. The two conditions satisfied by C, that have been used in the derivation of Equation 6.3.52, therefore uniquely determine the required vector C having the minimum length. This implies, of course, that the m-dimensional subspace spanned by the m $\{G_i\}$ and the $(n-m)$-dimensional plane, formed by the intersection of the m hyperplanes, intersect at a *point* in the n-dimensional vector space.

Equation 6.3.52 means that the coder and multiplexer at the transmitter perform the linear transformation A on the input vector S, where A is the $m \times n$ matrix of rank m, given by

$$A = (GG^T)^{-1}G \qquad (6.3.53)$$

It can be seen that if the n-component vector C is transmitted over the baseband channel with sampled impulse-response V, the corresponding m-component received vector R is given by

$$\begin{aligned} R &= CG^T + W \\ &= S(GG^T)^{-1}GG^T + W \\ &= S + W \end{aligned} \qquad (6.3.54)$$

which agrees with Equation 6.3.43.

The arrangement just described is known as System 3. It is of interest because all the signal processing is now carried out at the transmitter, which could be useful in some applications. Furthermore, it is the arrangement of linear signal processing at the transmitter that maximizes the tolerance to additive white Gaussian noise, for a given transmitted signal level and given that no signal processing, other than threshold-level detection, is carried out at the receiver.

6.4   COMPARISON OF THE DIFFERENT CDM SYSTEMS

The tolerances to additive white Gaussian noise of the different systems studied here, have been tested by computer simulation, for various channels and with $m = 8$, $g = 4$ and $n = 12$. In every case the data-transmission system is as shown in Figure 6.1 and described in Section 6.2. The average transmitted energy per signal element at the input to the transmission path is set to k^2, by multiplying each component of the matrix A by the appropriate constant, and the level of the additive white Gaussian noise is adjusted for an average error rate of 4 in 10^3 in the detection of the $\{s_i\}$. The sampled impulse-response of each channel is given by a vector of unit length, so that no signal attenuation (or gain) is introduced by any channel, the different channels being the same as those in Table 5.3.

The results of the tests are given in Table 6.1.[19] This quotes the number of dB reduction in tolerance to additive white Gaussian noise, at an element error rate of 4 in 10^3, when the given channel

Table 6.1 Reduction in tolerance to additive white Gaussian noise, measured in dB, when the given channel replaces one introducing no distortion or attenuation and the given system replaces System 1C

Channel	Sampled impulse-response of channel	System 1C	System 1D	System 2C	System 2D	System 2E	System 2F	System 3
1	$2^{-\frac{1}{2}}(1,1,0,0,0)$	6.0	9.0	7.4*	5.0*	5.5*	4.6	5.2
2	$1.5^{-\frac{1}{2}}(1,0.5,0.5,0,0)$	3.3	3.7	4.9	5.5	6.5	4.6	2.6
3	$1.5^{-\frac{1}{2}}(0.5,1,0.5,0,0)$	13.7	17.2	12.8*	12.5*	12.8*	11.9	12.8
4	$1.5^{-\frac{1}{2}}(0.5,1,-0.5,0,0)$	0.3	0.2	2.0	2.1	2.4	1.8	0.0
5	$2^{-\frac{1}{2}}(0.235,0.667,1,0.667,0.235)$	17.6	19.4	24.0	27.3	29.1	19.4	16.8
6	$2^{-\frac{1}{2}}(-0.235,0.667,1,0.667,-0.235)$	4.9	6.7	23.6	27.3	29.3	4.8	4.3

replaces one that introduces no distortion or attenuation and the given system replaces System 1C. In the presence of no distortion or attenuation, Systems 2C–2F have a tolerance to additive white Gaussian noise that is 1.8 dB below that of the Systems 1C, 1D and 3, because in the Systems 2C–2F an uninterrupted signal is now transmitted, with no time gaps between adjacent groups of elements, but in the detection of each group of elements only eight of the twelve received samples are used. The 95% confidence limits for the results quoted in Table 6.1 are about ± 0.4 dB. Where an asterisk is placed against an entry in the table, it means that the corresponding $m \times m$ matrix Y has rank $m-1$ and only seven signal-element values $\{s_i\}$ are effectively transmitted for each group of elements. The eighth element value in each group is neglected here. In System 2F only six element values are effectively transmitted for each group of elements.

The best overall performance is achieved by System 3, closely followed by Systems 1C and 2F. However, the information rate in System 2F is only $\frac{3}{4}$ of that in Systems 3 and 1C, on account of its two parity-check elements. Thus the best overall performance is in fact achieved by Systems 3 and 1C. These two systems are duals of each other. In System 3 a process of linear equalization of each group of m signal-elements is carried out at the transmitter, whereas in System 1C the corresponding linear equalization process is carried out at the receiver. Furthermore, the linear equalizing networks of the two systems are very closely related. The slight advantage of

System 3 over System 1C is due to the fact that the error rate in System 1C is effectively determined by the *largest* noise variance in the *m* output signals from the linear equalizing network at the receiver, the probability of error in the detection of any one of the *m* signal-elements of a group not being necessarily the same as that in the detection of any other. In System 3 the *average* transmitted energy per signal element, taken over all *m* signal-elements of each group, is set to k^2 at the input to the transmission path in Figure 6.1, as before, there being now the same probability of error in the detection of each signal-element at the receiver. This means that, in System 1C, the tolerance to noise is determined by the *largest* value of the sum of the squares of the components along a row of the $m \times n$ matrix $(YY^T)^{-1}Y$, where Y is given by Equation 6.3.5 and the transpose of $(YY^T)^{-1}Y$ is, of course, $Y^T(YY^T)^{-1}$. On the other hand, in System 3, the tolerance to noise is determined by the *average* value of the sum of the squares of the components along a row of the $m \times n$ matrix $(GG^T)^{-1}G$, where G is given by Equation 6.3.45. If the transmitted signal level in System 3 were reduced to set the *largest* of the *m* transmitted energies of the *m* individual signal-elements of a group to k^2, then Systems 3 and 1C would become exactly equivalent.

Systems 1D and 2D are on balance slightly inferior to Systems 1C and 2C, respectively, so that the replacement of the identity matrix by the Hadamard matrix, within the matrix *A*, has in each case the effect of slightly reducing the performance of the system. System 2E has a slightly inferior performance to System 2D, so that the replacement of the Hadamard matrix by the orthogonal matrix in Equation 6.3.30 further reduces the performance of the system. The transmission of two appropriate parity-check elements in each group of eight elements of System 2E, to give the System 2F, improves the tolerance to noise to that of System 1C, but at the penalty of a reduction in the effective transmission rate.

The important property of Systems 2E and 2F is that no iterative process is here required at the receiver in the estimation of either *S* or *V*. The systems can therefore be used at much higher transmission rates than any of the other systems. Furthermore, in each case the estimation process is achieved by means of a linear network that is matched to the received signal and therefore maximizes the output signal/noise ratio in the presence of additive white Gaussian noise. System 2E is in fact the CDM equivalent of a synchronous parallel FDM system that makes an efficient use of bandwidth and achieves a good tolerance to distortion.[23] Both in System 2E and in the parallel FDM system, successive groups of parallel elements are transmitted consecutively, with no time gaps, but successive detection processes are separated by the appropriate time gap. Each received group of elements is in fact detected only over the central portion, over which there is no intersymbol interference from the neighbouring received groups and over which the individual elements remain orthogonal for any signal distortion within the design limits of the system. Individual signal-elements may, however, be severely attenuated, an effect that occurs in the presence of frequency-selective fading.

Clearly Systems 2E and 2F have all the important properties of parallel FDM systems, without the considerable equipment complexity involved in these systems, since of course only a single process of modulation and demodulation is required by a CDM system in a one-way data link. However, at transmission rates of up to 10 000 bits/s, System 1C, which is a serial system, is much less complex than Systems 2E and 2F and has at least as good a performance. System 1C is, of course, the same as the System 2B in Chapter 5. Furthermore, its tolerance to additive noise can be greatly improved by suitable modifications to the detection process, as can be seen from Chapter 5, whereas no improvements in tolerance to additive white Gaussian noise can be obtained by any modifications to the detection processes of Systems 2E and 2F. Therefore, through the appropriate modifications to the System 1C, its overall performance can be made considerably better than that of either System 2E or 2F.

The Systems 1D, 2C and 2D are more complex than System 1C and have inferior performances. Just as the tolerance to noise of System 1C can be improved by suitable modifications to the detection process, so also can the tolerances to noise of Systems 1D, 2C and 2D. But extensive tests, carried out by Tang, have shown that for any useful modification and in particular for the detection process that minimizes the probability of error in the detection of S, System 1C in general increases its advantage in tolerance to noise over the other systems.

Ghani has shown that the tolerance of System 1C to additive white Gaussian noise, when operating over a baseband channel that introduces amplitude distortion, can be improved through the appropriate *sharing* of the linear signal processing between the transmitter and receiver. The improvement in tolerance to noise is not great, being similar to that achieved in a conventional serial system using a linear equalizer, when the equalization of the channel is appropriately shared between the transmitter and receiver (Section 4.6).

It seems that if a parallel CDM system is in fact able to achieve a better tolerance to additive white Gaussian noise than System 1B (Section 5.4.2), three conditions must be satisfied. Firstly, the channel must introduce amplitude distortion. Secondly, the signal processing at the transmitter must be *nonlinear*, and thirdly, this signal processing must be dependent on (a function of) the channel impulse-response. Unfortunately, such a system is likely to be complex and is only suitable for use over a time-invariant channel.

In conclusion, it appears that at transmission rates of up to 10 000 bits/s, a serial system is more cost-effective than the corresponding parallel CDM system, which is, in turn, more cost-effective than the corresponding parallel FDM system.

REFERENCES

1. Zadeh, L. A. and Miller, K. S. 'Fundamental aspects of linear multiplexing', *Proc. IRE*, **40**, 1091–1097 (1952)
2. Shnidman, D. A. 'A generalised Nyquist criterion and an optimum linear receiver for a pulse modulation system', *Bell Syst. Tech. J.*, **46**, 2163–2177 (1967)
3. Smith, J. W. 'A unified view of synchronous data transmission system design', *Bell Syst. Tech. J.*, **47**, 273–300 (1968)
4. Franks, L. E. *Signal Theory*, 1–65, Prentice-Hall, New Jersey (1969)
5. Kaye, R. A. and George, D. A. 'Transmission of multiplexed PAM signals over multiple channel and diversity systems', *IEEE Trans. Commun. Technol.*, **COM-18**, 520–526 (1970)
6. Dijon, B. and Hansler, E. 'Optimal multiplexing of sampled signals on noisy channels', *IEEE Trans. Inform. Theory*, **IT-17**, 257–262 (1971)
7. Timor, U. 'Equivalence of time-multiplexed and frequency-multiplexed signals in digital communications', *IEEE Trans. Commun.*, **COM-20**, 435–438 (1972)
8. Harmuth, H. F. 'On the transmission of information by orthogonal time functions', *AIEE Trans. Commun. Electron.*, **79**, 248–255 (1960)
9. Judge, W. J. 'Multiplexing using quasiorthogonal binary functions', *AIEE Trans. Commun. Electron.*, **81**, 81–83 (1962)
10. Glenn, A. B. 'Code division multiplex system', *IEEE Int. Conv. Rec.*, **12**, Pt. 6, 53–61 (1964)
11. Golomb, S. W., Baumert, L. D., Easterling, M. F., Stiffler, J. J. and Viterbi, A. J. *Digital Communications*, Prentice-Hall, New Jersey (1964)
12. Pierce, W. H. 'Linear real coding', *IEEE Int. Conv. Rec.*, Pt. 7, 44–53 (1966)

13. Chang, R. W. 'Precoding for multiple speed data transmission', *Bell Syst. Tech. J.*, **46**, 1633–1649 (1967)
14. Harmuth, H. F. *Transmission of Information by Orthogonal Functions*, Springer-Verlag, Berlin (1969)
15. Brayer, K. 'Euclidean orthogonal data transmission', *Proc. IEEE*, **59**, 79–80 (1971)
16. Lackey, R. B. and Meltzer, D. 'A simplified definition of Walsh functions', *IEEE Trans. Comput.*, **C-20**, 211–213 (1971)
17. Weinstein, S. B. and Ebert, P. M. 'Data transmission by frequency-division multiplexing using the discrete Fourier transform', *IEEE Trans. Commun. Technol.*, **COM-19**, 628–634 (1971)
18. Davies, A. C. 'On the definition and generation of Walsh functions', *IEEE Trans. Comput.*, **C-21**, 187–189 (1972)
19. Clark, A. P. and Mukherjee, A. K. 'Parallel data transmission systems using code-division multiplexing and adaptive detection', *IERE Conf. Proc.*, No. 23, 391–408 (1972)
20. Mukherjee, A. K. 'A new set of orthogonal periodic sequences', *Proc. IEEE*, **61**, 483–484 (1973)
21. Yuen, C. K. 'Interchannel interference in Walsh multiplex systems', *IEEE Trans. Commun.*, **COM-22**, 255–261 (1974)
22. Scolaro, R. J. and Hullett, J. L. 'Intersymbol interference with orthogonal signalling', *IEEE Trans. Commun.*, **COM-23**, 472–474 (1975)
23. Clark, A. P. *Principles of Digital Data Transmission*, Pentech Press, London (1976)

Index